G. Kaupp

Atomic Force Microscopy, Scanning Nearfield Optical Microscopy and Nanoscratching

ISBN 978-3-540-28405-5

Erratum

We regret that, due to a mishap in the production process, this book does not contain the author's final corrections. The text on the pages listed below should read as follows:

1. p. IV, Author's affiliation: Diekweg 15, D-26188 Edewecht, Germany, gerd.kaupp@uni-oldenburg.de
2. p. IV Cover background image: "Deep craters going down in concentric molecular-sized steps as modelled by a chemical reaction starting from a monolayered organic crystal surface". Concept: G. Kaupp, Department of Organic Chemistry, University of Oldenburg, image: G. Kaupp and J. Boy, University of Oldenburg, Germany.
3. p. 3, line 12: which shows the topology of silica on polyethene terephthalate (PET)
4. p. 8, line 12: on a 25 nm pentacene film
5. p. 15, line 14: Also single gratings with well defined pace as in Fig. 1.12
6. p. 24, line 1: leads to islands of considerable height
7. p. 39, line 1 of caption (Fig. 1.42): on (010) of the (R, R)-cyclohexylspirotaddol
8. p. 62, line 16: as suggested in [42]
9. p. 97, line 3: between 10 and 120 μm s^{-1} (200 μm s^{-1} at larger scans) can be achieved
10. p. 97, Sect. 2.4, line 6: with a pixel size of 20 nm.
11. p. 99, line 1 of caption (Fig. 2.7): (633 nm)
12. p. 125, line 2: crystal axis [010]
13. p. 140, line 2 of caption (Figs. 2.52 and 2.53): latices
14. p. 159, line 5: this highly superior technique. Also confocal femto- and picosecond two-laser scanning STED (stimulated emission depletion using focal and doughnut-focal beam shapes) microscopy with a claimed axial resolution of 33–46 nm of fluorescence spots [83a] is inferior. It is limited by orientation

requirements not allowing for considerable roughness and by photobleaching. A scanned bacteria membrane was stained with a vertically oriented fluorescent dye. It appears highly preferable to use the shear-force apertureless SNOM for the distinction of the constituents in biological cell materials rather than staining, the selectivity of which might be questionable.

15. p. 175, new reference:
    83a. M. Dyba, S.W. Hell: Focal spots of size lambda/23 open up far-field fluorescence microscopy at 33 nm axial resolution. Phys. Rev. Lett. **88**, 163901-1 – 163901-4 (2002)
16. p. 205, line 16: Very large average pressures calculate for high values of the elastic modulus, but the mean contact pressure is actually limited to the hardness value $H = p_{\mathrm{av}}$ if this is smaller than calculated by (3.19) or (3.20). As the situation changes upon increasing phase transition around the indenter tip these are useful estimates, though.
17. p. 218, caption of Table 3.6: Comparison of cube corner (R = 200 nm) mechanical parameters of nanoindentations on glassy polymers
18. p. 221, Sect. 3.12, line 2: The new $F_{\mathrm{N}} \sim \mathrm{h}^{3/2}$ relation
19. p. 222, line 12: the same $F_{\mathrm{N}} \sim \mathrm{h}^{3/2}$ relationship
20. p. 233, line 12: It was tried
21. p. 239, line 26: [16]
22. p. 240, line 8: $F_{\mathrm{L}}\text{–}\mathrm{F}_{\mathrm{N}}^{3/2}$
23. p. 243, lines 21-22: is larger when compared
24. p. 243, lines 22-23: But the still high energetic effort
25. p. 265, line 3 of caption (Fig. 4.41): cleavage planes under (010) and (102)
26. p. 271, scale of Fig. 4.49(b): 100,000 200,000 300,000
27. p. 272, line 6: linear polymers at 310 μN load
28. p. 290, missing entry: STED microscopy, 159

NanoScience and Technology

NanoScience and Technology

The series NanoScience and Technology is focused on the fascinating nano-world, mesoscopic physics, analysis with atomic resolution, nano and quantum-effect devices, nanomechanics and atomic-scale processes. All the basic aspects and technology-oriented developments in this emerging discipline are covered by comprehensive and timely books. The series constitutes a survey of the relevant special topics, which are presented by leading experts in the field. These books will appeal to researchers, engineers, and advanced students.

Nano-Optoelectonics
Concepts, Physics and Devices
Editor: M. Grundmann

Noncontact Atomic Force Microscopy
Editors: S. Morita, R. Wiesendanger, E. Meyer

Nanoelectrodynamics
Electrons and Electromagnetic Fields in Nanometer-Scale Structures
Editor: H. Nejo

Single Organic Nanoparticles
Editors: H. Masuhara, H. Nakanishi, K. Sasaki

Epitaxy of Nanostructures
By V.A. Shchukin, N.N. Ledentsov and D. Bimberg

Applied Scanning Probe Methods I
Editors: B. Bhushan, H. Fuchs, S. Hosaka

Nanostructures
Theory and Modeling
By C. Delerue and M. Lannoo

Nanoscale Characterisation of Ferroelectric Materials
Scanning Probe Microscopy Approach
Editors: M. Alexe and A. Gruverman

Magnetic Microscopy of Nanostructures
Editors: H. Hopster and H.P. Oepen

Silicon Quantum Integrated Circuits
Silicon-Germanium Heterostructure Devices: Basics and Realisations
By E. Kasper, D.J. Paul

The Physics of Nanotubes
Fundamentals of Theory, Optics and Transport Devices
Editors: S.V. Rotkin and S. Subramoney

Single Molecule Chemistry and Physics
An Introduction
By C. Wang, C. Bai

Atomic Force Microscopy, Scanning Nearfield Optical Microscopy and Nanoscratching
Application to Rough and Natural Surfaces
By G. Kaupp

Applied Scanning Probe Methods II
Scanning Probe Microscopy Techniques
Editors: B. Bhushan, H. Fuchs

Applied Scanning Probe Methods III
Characterization
Editors: B. Bhushan, H. Fuchs

Applied Scanning Probe Methods IV
Industrial Application
Editors: B. Bhushan, H. Fuchs

Nanocatalysis
Editors: U. Heiz, U. Landman

G. Kaupp

Atomic Force Microscopy, Scanning Nearfield Optical Microscopy and Nanoscratching

Application to Rough and Natural Surfaces

With 239 Figures and 7 Tables

Springer

Professor Dr. Gerd Kaupp
University of Oldenburg, Department of Chemistry
26129 Oldenburg, Carl von Ossietzky-Str. 12–14, Germany
E-mail: kaupp@kaupp.chemie.uni-oldenburg.de

ISSN 1434-4904
ISBN-10 3-540-28405-2 Springer Berlin Heidelberg New York
ISBN-13 978-3-540-28405-5 Springer Berlin Heidelberg New York

Library of Congress Control Number: 2006923691

Cover background image: "Single-wall nanotube METFET inverter: organic molecules wrap around a metallic tube channel and modify it to a semiconductor while the second tube is used for gating". Concept: S.V. Rotkin, Physics Department, Lehigh University, image: B. Grosser, Imaging Technology Group, Beckman Institute, UIUC.

Typesetting: SPi, Pondicherry, India
Cover design: *design& production*, Heidelberg

Printed on acid-free paper SPIN: 11323839 57/3100/SPi - 5 4 3 2 1 0

Preface

The introduction of the scanning tunneling microscope (STM) for electrically conducting surfaces by Binnig and Rohrer in 1981 was followed by the invention of atomic force microscopy in 1986 by Binnig, Quate, and Gerber also for nonconductive samples. The initial interest was in atomic and molecular resolution, and it rightfully continues also with building molecular machines by atomic-scale manipulation and industrial applications in the field of high-density data storage on ultraflat surfaces. The latter topic also uses the various techniques of scanning near-field optical microscopy (SNOM) for surfaces without roughness. It was not until 1990 that the capabilities of AFM for the third dimension started to become recognized, even though commercial instruments were available since 1989, but funding of projects aiming for submicroscopic characterization of topology was still not achievable then and early publications were impeded by referees and editors. Many authors in the first half of the 1990s did not even give a z-scale to their plots but stressed only on the lateral resolution on top of 3D objects. It was not appreciated then that most industrial and biological applications would be on real-world samples at ambient conditions rather than with artificial test samples under high vacuum, or with other artificial treatment. Similar experiences were encountered with SNOM when a new instrument became available in 1995 with the capability of reliable scanning on rough real-world surfaces with height differences of more than a micrometer, and again similarly in 1999 when two-dimensional transducers were added to nanoindentation equipment enabling nanoscratching for novel nanomechanical testing. The early pioneers who immediately recognized the unprecedented capabilities and necessities for real-world applications were again not granted the funding for the initial basic research but had to privately purchase all of these instruments and were left alone in establishing the foundations for scientific and industrial use. Funding was only granted for applied research after the basics had been pioneered. However the initially disliked topology is by far the most important topic in practical and industrial applications of AFM and SNOM, and the new quantitative treatment that has been empirically developed in nanoscratching for all kinds of solid materials is revolutionizing the field. The latter also provided the entry for quantitative treatments of nanoindentation, which was unduly regarded as the "smaller-scale microindentation" and therefore relied on the multiparameter iterative data treatment of Oliver and Pharr from 1992 that was adopted in the ISO 14577 standard.

This book is devoted to the practical and industrial applications of the nanotechniques AFM, SNOM, nanoindentation, and nanoscratching for real-world samples in ambient environment. It does so in four independent chapters. It also deals with theory as far as it is necessary for the practitioner and as far as new physical effects, laws, and relations have been empirically found, deduced, or derived and that are not covered in previous reviewing books or monographs on the subjects if available. Further scanning probe microscopy techniques (SPM) that have not yet been developed for application to real-world surfaces are not covered. They are treated in the cited books. The present book does not treat atomic and molecular resolution, molecular Langmuir–Blodgett films (notwithstanding molecular steps on rough crystal surfaces), atomic or molecular machines, and further nanotechnology devices that exclude (significant) topology. Exhaustive treatises on these subjects are available and there is no need for repeating them. None of the books with particular emphasis on theory on the basis of model systems and simulations treat the special requirements and additional expertise for reliable scanning on (high) topology with AFM and SNOM or for quantitative measurements in nanoindentation, and no such treatment existed for nanoscratching with its wealth of information on mechanical properties of industrial interest.

Practical applications from almost all sciences, industries, and daily life are collected, described, and cited with particular emphasis on the topology of the real-world surfaces and on the quantitative study of nanomechanical properties on the basis of new standards for all kinds of materials. Unfortunately, many authors who are scanning now AFM on rough surfaces, after AFM is directly connected to much less resolving (confocal) microscopy, do not still stress the topology properties in the abstracts of their papers. Therefore, some highly relevant reports must have been inevitably lost in the literature retrieval of the author. But the author tried his best to approach as many different applications as possible and hopes that he did not miss any important fields of practical application.

The book intends to close the gap between theory and practice of the four major tools for nanotechnology and to further advance the interest of industry in SNOM on rough surfaces, nanoscratching, and nanoindentation. AFM and microindentation are already widely accepted, although the latter on the basis of a multiparameter iterative ISO standard, while quantitative treatments with highly improved information and reliability on the basis of simple apparently universal laws would be obtained from nanoindentation. The book should therefore be of interest to students and scientists in academia and industry in all fields of surface investigations not only in natural and life sciences or industrial engineering, but also in apparently unrelated fields such as astrophysics, historical sciences, arts, forensics, jurisdiction, etc. The purpose of the book is to give a systematic physical and technical background on the basis of well-documented experimental results on all kinds of materials surfaces using uncomplicated robust equipment in ambient atmosphere. It does so also with the probably most exhaustive collection of varied applications from the literature beyond the atomic and molecular resolution issues, while previous treatises restrict their application parts largely to DNA imaging or sectioning and industrial data storage aspects. AFM, SNOM on rough surfaces, nanoindentation, and nanoscratching will certainly enjoy a rapid further development with applications to real-world problems.

Oldenburg,
February 2006

Gerd Kaupp

Contents

1

Atomic Force Microscopy

1.1 Introduction

Atomic force microscopy (AFM) is also known as scanning force microscopy (SFM). AFM is a basic technique and inevitable for all nanoscopic research. For example all chapters of this book rely on AFM measurements. However the scientific community became interested in the many uses of AFM for (very) rough surfaces at large scan ranges only in 1990 [1–4], and it took two years before these first results were accepted and published [5–7]. Actually, it was not possible in 1989 to convince funding agencies to support the idea of using the three-dimension capabilities of AFM and the author had to purchase all of his nanoscopic instruments privately.

Measurements had been made on large biological objects such as whole cells, bacteria, etc. [8, 9]. However, the authors of these studies were only interested in the smallest features (<50 nm height) on the surfaces that could be laterally resolved, notwithstanding the two-dimensional projection of the whole object without a z-scale, and the migration of living cells on glass slides in buffer solution. Early high three-dimensional AFM topologies with assessment of the z-scale were made for a dental implant [10], wear marks on single-crystal silicon [11], and silver bromide cubic crystals [12].

Like all other scanning probe microscopies (SPM), AFM works by scanning with a tip (or more general a probe) very close to the sample surface. It operates by measuring attractive or repulsive forces between the tip and the sample in constant height or constant force mode. Most spectacular are atomic resolution and manipulation, but most practical applications deal with the (sub)micrometer x/y- and nano z-range.

The first atomic force microscope (AFM) was made in 1986. It was of the dynamic type with vertical tip vibration [13]. However, the first commercial instruments (since 1989) were static-force or contact AFMs. Shear-force AFM microscopes with horizontal tip vibration for distance control were developed in 1992 [14, 15]. The dynamic-mode AFM found a new application with the commercialization of noncontact and tapping-mode AFM [16, 17]. Other surface properties such as friction forces, sample elasticity, adhesion, or chemical differences by lateral and torsional force sensing, force modulation, frequency modulation, and phase imaging are available. Magnetic nanostructures are scanned by magnetic force microscopy using cantilever tips that are coated with a ferromagnetic film of a few nm thickness, with most applications of MFM being in the storage media industry. Virtually all solid surfaces from all branches of science, industry, medicine, daily life are accessible to nanoscopic investigation with unprecedented information. The resolution here is much higher than with a microscope and the three-dimensional information is another most important distinguishing feature of AFM. It supplements electron microscopy and increases the possibilities, as it does not require surface treatment and is able to measure at ambient conditions. Unlike scanning tunneling microscopy (STM) the samples need not be electrically conductive.

Dynamic mode AFM lacks a quantitative theory. It is described in great detail in [18, 19] and will not be repeated here. Further SPM techniques are treated in [20, 21]. Neither do we deal with atomic or molecular resolution, which is amply described in [22, 23], or self-assembled molecular layers and Langmuir–Blodgett films. We concentrate here on contact AFM and shear-force AFM, the merits of which are not covered in [19–21]. However, these techniques are most easily used and interpretable as they directly yield the interaction force itself between tip and sample or generate a constant shear-force damping for the topologic measurement and the topology may be (very) large. For these reasons contact and shear-force AFM are most suitable for rough and very rough surfaces as they usually occur in real-world samples. Practical applications are legion and thousands of user-friendly instruments are in active use. A new development is piezoless (that means electromagnetically controlled) AFM. This versatile new technique is briefly addressed in Sect. 1.12.

1.2 Basics of Contact AFM

Contact AFM is run at a prefixed force for tip–sample interaction, usually in an attractive range (repulsive range for atomic resolution). The major forces derive from van der Waals, electrostatic, hydrophobic, hydrophilic, and capillary interactions with the moving tip, which is usually multiatom at its apex. We do not enter into a discussion of single atom contact in atomic resolution and the interpretation of such images, but deal here only in the nanometer range (not subnanometer). While the separation of the various forces may be useful, a qualitative empirical approach might be sufficient for practical applications (e.g., the influence of moisture, etc.). Most measurements of contact AFM are at ambient conditions. Inert gas or reactive gas atmosphere might be applied as well. The vertical resolution is routinely in the range of molecular monolayers even on rough surfaces. Measurements under solvent will mostly not reveal the original surface due to nanoscopic dissolving or

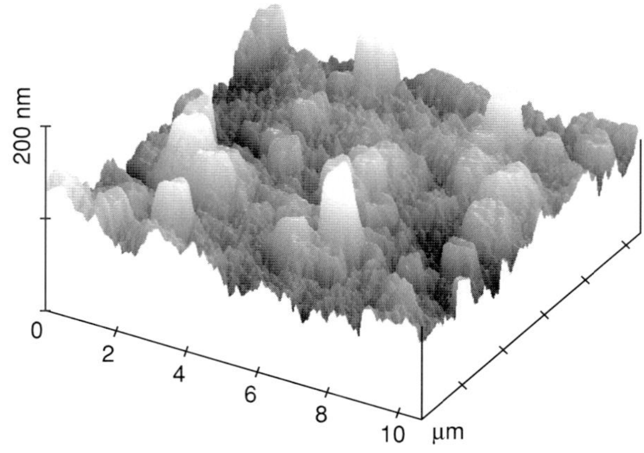

Fig. 1.1. Contact AFM surface of PET with a protecting PVD silica cover, showing high roughness (z-range: 128 nm, R_{ms} 17.3 nm) and considerable submicroscopic detail in a 10 µm scan

addition of material and swelling. Such measurements in liquid cells should be restricted to studies of dissolution, crystallization, (electrochemical) corrosion, and the like. Furthermore, samples that are in constant contact with the liquid, such as for example living cells, are favorably measured in biological environment when adhered to a support. For practical use, the aim is not an ideal surface, but the undisturbed natural surface with all of its deficiencies resolved such as oxide impurities and water layers, where applicable. But it can also be important to trace the effects of staining or other purposeful surface modification with unmatched AFM precision. Also artificial surfaces such as plastic polymers may become very rough by moisturizing, environmental pollutants, irradiation, or PVD coating for surface protection. An example is given in Fig. 1.1, which shows the topology of silica on polyethylene terephthalate (PET) that is widely used in beverage containers, for example. The surface is very stable and does not change upon 30 scans or more, even with an asymmetric and therefore scratchy (Sect. 1.6) tip. This underlines the necessity of contact AFM prior to applying all of the other surface characterization techniques.

An exhaustive report on contact AFM measurements on treated technical polymers such as PC, PMMA, CR39, PU 161 can be found in [24].

1.3 Instrumental

The atomic force microscope is a relatively simple and versatile instrument. Figure 1.2 shows Nanoscope-III (DI-VEECO) with a rubber band vibration protection mount. It is also available in multimode version for various other scanning techniques. This is easy to use and numerous other brands are commercially available at reasonable price. The general block diagram of a laser deflection-type instrument as in Fig. 1.2 is sketched in Fig. 1.3.

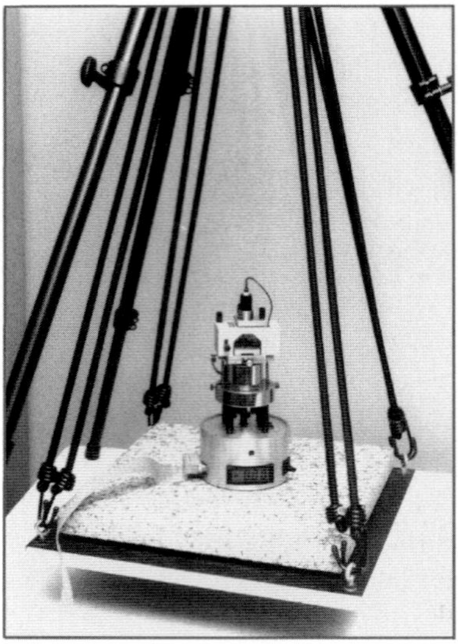

Fig. 1.2. Ready-to-use vibration protected typical AFM (Nanoscope III) with stage, including the XYZ-piezo unit, sample mount, cantilever tip, laser focused to the backside of the reflecting cantilever and guided to a split photodiode for the feedback loop that controls the constant force and provides the topologic information

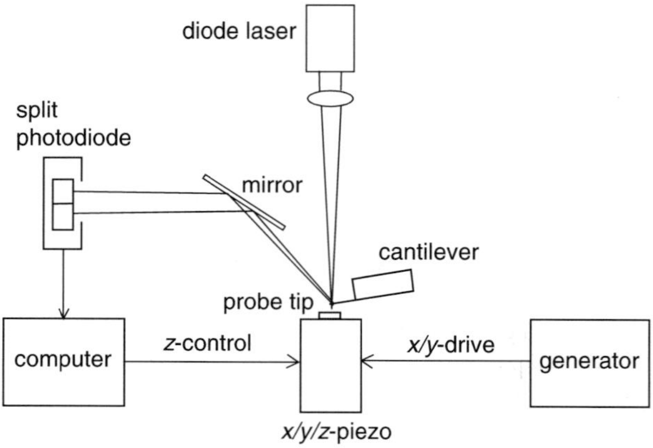

Fig. 1.3. Block diagram of a laser deflection contact AFM

The necessary parts are the x/y and z piezo that are separately actuated by x/y drive and z-control with extreme precision, so that atomic distances can be measured. The sample is mounted on the xyz piezo [25, 26], close to a sharp tip under the inclined cantilever with its mount. The diode laser light is focused at the end of the cantilever, reflected via mirror to a split diode that provides the feedback signal (topologic information) for maintaining the force by z-piezo response. If the tip is scanned instead of the sample the instrument is called a *standalone microscope*. Data sampling is made at discrete steps by means of an analog-to-digital converter. A computer reconstructs the three-dimensional topological image or projections from the data matrix. Imaging software adds color, height contrast, and illumination from variable directions. Commercial instruments are supplied with all piezo calibration values, and various supply for different SPM techniques as well as instruction manuals and software for imaging. Different piezos are used for different xy scan ranges ($1\,\mu m \times 1\,\mu m$ up to $200\,\mu m \times 200\,\mu m$) and z ranges (up to $15\,\mu m$). A number of hardware and software linearization facilities are provided for correction of piezo hysteresis that creates stretching at the start of the scan in comparison to the end of the scan, line by line always starting from the same side. Furthermore, piezoelectric materials with low hysteresis are available [27]. Calibration should be regularly checked with standard gratings for width and height. Nonlinearities of the piezo are corrected. They can be run in constant height mode (usually predetermined average flat surface and keeping the scanner height at constant distance to the average surface at every point of measurement and detecting light deflection variations), but more useful for rough surfaces is constant force mode where the movement of the cantilever is used as the input to the feedback loop that moves the scanner up and down for constant light deflection to the phototube.

More recently developed cantilevers combine tip arrays with self-detecting and active capabilities. For example, 2×1 microfabricated arrays have been successfully used in parallel imaging (with integrated actuators for compensation of the height differences between the two tips) and sequential imaging (by pulling back the inactive tip by a thermal bimorph actuator) for test grids in constant height, constant force (an integrated actuator for z-direction feedback motion maintains the interaction force between the tip and the sample to be constant during the scanning), and tapping mode. The z-ranges were 180 nm, $1.3\,\mu m$, and 30 nm, respectively. The integration of stress-sensing metal-oxide semiconductor transistors, thermal bimorph actuators, and complementary metal-oxide semiconductor amplifiers has been demonstrated [28]. Using the second cantilever as a reference during AFM operation can compensate the crosscoupling signal from the actuator and sensor. Arrays of tips are primarily important for wafer-derived samples with uniform basic surface level in industrial environments.

Planefit filtering of AFM images is a necessity unless the tilt of the surface is to be determined. Such a tilt should be smaller than $10°$ for reliable topologic measurements, but isolated steep features may be successfully measured at purposely high tilt of the average surface. Spikes of statistical noise can be removed by sparing the use of the low-pass filter, while flattening should only be used with flat uncorrugated surfaces. The feedback system is high enough for rapid scanning with typical times of two to four minutes per image frame, and considerable roughness can be safely measured provided the feature steepness does not reach or exceed the slope of the tip. Optimal scan angles are adjusted in dual-trace mode by putting the forward and backward traces as closely coincident as possible and simultaneously by putting the propensity for swinging as low as possible. Also the scan rates and force have to be optimized on rough irregular surfaces within narrow ranges in the stable regime but close to swinging at high feedback gains while avoiding input filters. For example, 10 μm scans can be routinely recorded at scan rates of 3.5–5 Hz without change of the image. Lower contrast is the result if such optimization is neglected.

The AFM tips are usually microfabricated silicon cones (sometimes edged) or silicon nitride four-sided pyramids, the latter with high durability, that are commercially available mounted to cantilevers with spring constants of 0.05–1.5 $\mathrm{N\,m^{-1}}$ for contact AFM and 40–80 $\mathrm{N\,m^{-1}}$ for tapping-mode AFM.

A broad range of different AFM cantilever tips are commercially available. Most frequently used is the particularly long-life silicon nitride pyramid that is microfabricated by vapor deposition into the {1 1 1} silicon planes in the wafers that serve as molds during cantilever production [29]. The pyramidal molds are produced in various steps. A silicon wafer on (100) is coated with a suitable masking material (SiO_2 may be used). Then a small square opening (4 μm × 4 μm) down to the silicon is etched away, followed by anisotropic etching with KOH solution. The faces of the pyramidal mold are the {1 1 1} planes of the exposed silicon, since the (100) face is etched more than 100 times faster than the four {1 1 1} planes. This guarantees a very sharp apex. Finally, the pit is used as a mold for the Si_3N_4 film by vapor deposition techniques also for the cantilever by deposition at the surface, and the cantilever tip obtained by etching away of the mold material. These tips are not solid but hollow, which improves their performance. The pyramidal faces are very flat. Also in frequent use are etched silicon tips, but they vary in shape even within single batches, due to difficulty in controlling the manufacture process [30]. Very sharp tips are obtained by electron beam- or ion-beam deposition [31,32]. More recent approaches attach single-walled carbon nanotubes to conventional AFM cantilever tips [33]. These have to be homemade, but they are stiff and gentle, do not easily break, give high resolution, and can be used for imaging in deep structures. Carbon nanotubes can be sharpened by radiation-induced necking [34].

Figure 1.4 depicts SEM images of a silicon nitride tip (a) with cantilever mount (b) and an electron beam-deposited diamond-like tip with two enlargements (c,d).

Silicon nitride tips have a well-defined geometry with very smooth sidewalls with slope angles of 54.7°. The angle between opposing edges is 90° and between opposing faces 70.5°. That means the steepness of topologic features must not exceed about 55° or 45° if the tip is mounted with a side in front (Fig. 1.4a) or with an edge in front, respectively. The cantilever inclination angle adds to upward slopes and subtracts to downward slopes. Therefore, separate measurements in forward and backward reaction are useful if steep slopes occur. For steeper topology either the sharper tips or shear-force AFM

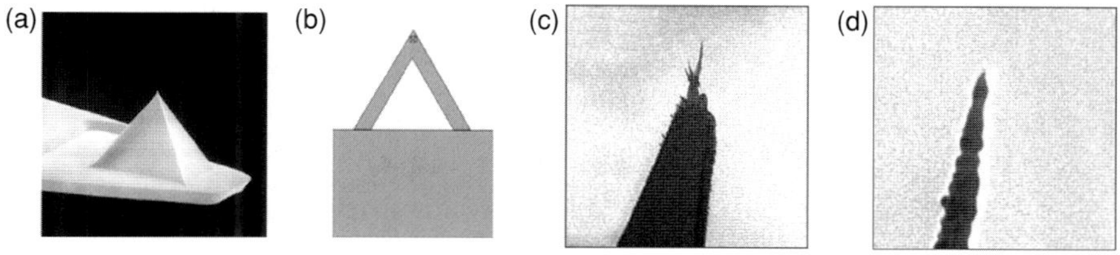

Fig. 1.4. SEM images of AFM tips and mount, (**a**) Si_3N_4 pyramid, (**b**) cantilever mount, (**c**) extremely sharp diamond-like extra tip at the apex (600 nm scale), (**d**) enlarged tip end (80 nm scale)

(Sect. 1.10) have to be used. The base of the silicon nitride tip is typically 4–5 μm long and the heights are 0.707 ($1/\sqrt{2}$) times of that length. The cantilevers may be one-leg, triangular (Fig. 1.4b), or square. They vary in their spring constants also by the thickness and width of the legs as is required for various application modes.

Etched silicon tips are conical or more angular and do not have well-defined geometries. There is a risk in having double tips and they are less stable. This detracts from their advantage of having higher aspect ratio and lower tip radius.

Electron-beam deposited or nanotube tips have extremely high aspect ratio and very small tip radius, which is immediately measured by electron microscopy after fabrication. However, they are much more involved than the highly economic pulled tips in shear-force mode AFM for the sake of minimizing tip–sample convolution at steep features. Tip characterization techniques are described in [20] (pp. 147–168).

Special tips may be focused by ion beam techniques. Cone angles below 12° and radii of curvature of 4 nm have been achieved [32]. ZnO whiskers with cone angles of 2°–4° with tip radii <10 nm [35] may have flexing problems. This improves with ZnO nanowires grown by metallorganic chemical vapor deposition as studied in [36]. These are structurally compatible with typical AFM cantilevers and maintain mechanical stability during operation. They are promising candidates for high aspect ratio probes. A totally different approach is tetrahedral tips with two vertical sidewalls that can be used for the imaging of nearly vertical surface steps [29, 37].

Stressed metal probes are of particular interest for extremely high roughness [38]. A fabricated metal cantilever out of MoCr bends away from a transparent probe chip and its tapered end is used as the probe tip. Light is focused through the transparent chip to the bent cantilever for reflection from a selected section to the split photodetector, which can be adjusted in a wide range from tens to hundreds of micrometers. Several of these can be mounted to the horizontal transparent chip. The cantilevers are typically 100–400 μm long, 40 nm wide, and 1 μm thick. This corresponds to tip heights of 10–150 μm at a radius of cantilever curvature of 520 μm. But the latter may be decreased

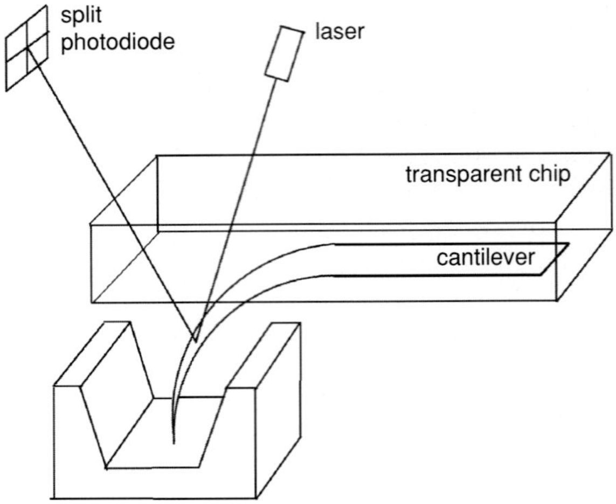

Fig. 1.5. Sketch of the AFM setup with stressed probes for very deep structures

to 280 µm so that the 400 µm cantilever exhibits a tip height of 300 µm. The spring constant of this 400 µm cantilever is 0.06 N m^{-1} and the resonance frequency 7.2 kHz. Figure 1.5 sketches the setup of the bent cantilever with the tip at the end. Its uncomplicated manufacture also in arrays of tips with various materials and numerous application proposals are claimed in some patents [39, 40]. A particular advantage for tip arrays is the parallel alignment of chip and sample surface. Most obvious is the contact AFM mode. However, a stiffer stressed cantilever out of NiZr with larger spring constants and clamp holder has been fabricated for noncontact operations [41].

These are AFM tips for very deep and narrow structures and measurement at their bottom. But the resolving power is still very good. Structures as small as 20 nm have been imaged by an 80 µm high tip on a 25 µm high pentacene film due to the micro-roughness of the MoCr material. It is to be expected that this technique, which can be used with common AFM equipment (e.g., Nanoscope III), will find widespread use in metrology and AFM on very rough surfaces. But shear-force AFM (Sect. 1.10) is even simpler in the preparation of sharp tips and may serve for the same purposes. Both techniques are far from being fully acknowledged.

A further application of stressed cantilever tips is imaginable for application in the 'Milliscope' [42, 43]. This instrument is cantilever based and uses four audio loudspeakers (ALS) as electrodynamic positioners instead of piezos. The scanning range covers enormous $1.1 \times 1.1 \times 0.14\,\mathrm{mm}^3$ [maximum range of other cantilever SFM (commercially: DualScope^{TM} Scanner DS 95–200) is $200 \times 200 \times 15\,\mu\mathrm{m}^3$], and the linearity is good up to an image size of 600 µm × 600 µm and acceptable over the whole scan area. The lateral resolution is only limited by the acquisition grid (pixel size) and is of the order of 10 nm for smaller scale scans. This has been verified with a test grid, gold-coated polystyrene spheres (270 nm in diameter), and human sperm (maximum height 1.9 µm). Furthermore, a thin section of a tomato bud effectively imaged the sharpened tip at the stiff and steep cell walls. However, the enormous height capabilities of 140 µm can only be used for very shallow extended features due to cantilever sliding at feature heights greater than 3 µm and growing at a slope greater than 15°–20°. The complicated three-dimensional scanner did not find widespread application even if some problems with low-scale scans could be solved. It has to compare with stressed cantilever (Fig. 1.5) and shear-force AFM (Sect. 1.10).

1.4 Validity Checks

AFM measurements on (very) rough surfaces must be represented with three-dimensional surface plots (Sect. 1.9) rather than top-view projections. This is necessary for the assessment of quality and detection of possible artifacts in some regions of the image by tip–sample convolution or tip imaging. AFM imaging of large and steep features in wide scans is only possible in constant force mode. The validity checks are easily obtained with perspective imaging, while two-dimensional projections may hide numerous deficiencies. In particular, tip-derived asymmetries of large features should be absent. In good measurements there is practically no asymmetry of the features except for the tilt of the cantilever, which makes the tip decline. More pronounced asymmetries point to nonoptimal conditions during the scan. A sensitive validity test is provided by the comparison of perspective images at 0° and 90° rotation (in addition to the pitch angle) and further orientations. Two examples for perfect imaging of large and frequent cones on hydrated hydroxylammonium chloride (Fig. 1.6) and floes on photolyzed 9-chloroanthracene single crystals (Fig. 1.7) are shown [44]. The cones in Fig. 1.6a,b are almost symmetric, stand vertically, and look similar from all sides. However, the image cannot be free from artifacts in the depths, which means that they are actually larger and taller in the deeper regions than when imaged. Calculating the image inversion can easily test this. The inverted Fig. 1.6c reveals sharp edges due to some tip imaging in the depths and it not reaching the ground. Such information is not obtainable from the top view (Fig. 1.6d), which more or less hides all of these important facts. It is, however, useful for counting features and assessing the lateral arrangements. Clearly, there is no high symmetry in the arrangement of the columns.

The cones in (a) or (b) with heights of up to 800 nm at a frequency of one per square micron do not show scan-derived asymmetries, scan tracks, or other tip-derived deficiencies. Nevertheless only the tops of the cones are free of tip–sample convolution. The structures at the bottom of the perspective image trace the negatives of the tip. This could be more clearly shown by the inverted images. Clearly, the geometric limits of contact AFM with pyramidal

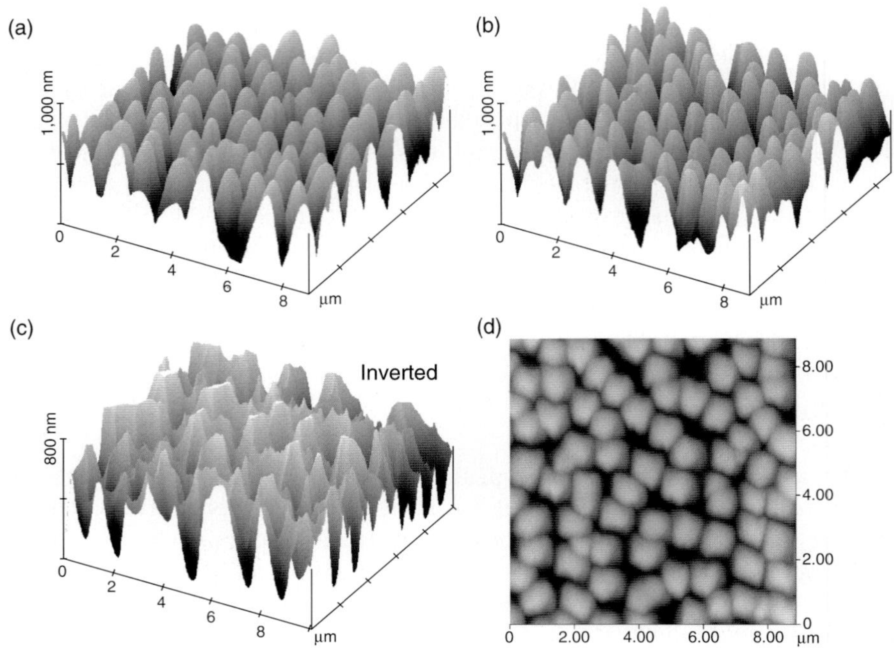

Fig. 1.6. Contact AFM surface plots of large cones on hydrated hydroxylammonium chloride (crystallized from methanol) with Si_3N_4 tip at a scanning rate of 3.55 Hz; the views are (**a**) at 30° rotation, (**b**) at 120° rotation, (**c**) inverted at 30° rotation and (**d**) a top view of (**a**)

tips have been exceeded and the apparent slope angles of 55° must be steeper in reality. Thus, image inversion is another valuable means of image check. The stability of the image is the result of proper adjustments of all parameters in the scan. More realistic imaging in the lower part would require sharpened tips or the high aspect tips in shear-force AFM (Sect. 1.10).

Also the (prismatic) floes in Fig. 1.7a and Fig. 1.7b are undistorted from scan or tip despite their sharpness and have slopes of 41°–44° and 39°–42° on opposing sides. That is they are unusually steep but do not reach the gradient of the tip. Importantly, the image is stable for 10 consecutive scans and next day. The height of the floes is valid and there can be no doubt concerning the validity of the shapes of these features [44] with the exception close to their base, where the floes should be thinner.

The check is most sensible if the directions differ by 90° as in Fig. 1.6a,b and Fig. 1.7a,b, as tip-derived asymmetries and scan deficiencies would be most evident. The best scan direction must be found by trace/retrace technique. Extended steep features should not be measured nearly parallel to the scan direction in order to minimize scan errors. If that is not possible due to

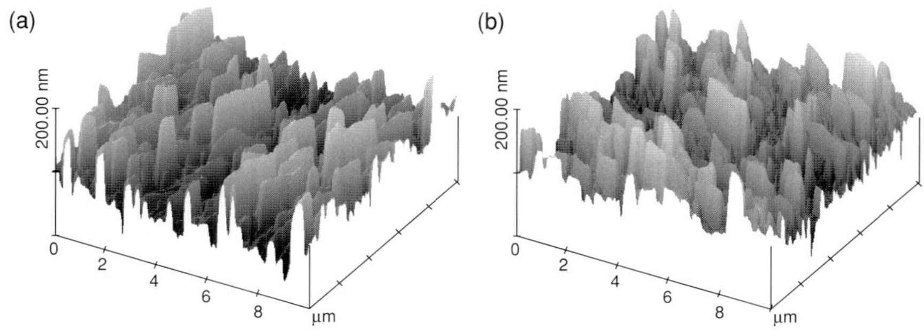

Fig. 1.7. Contact AFM surface plots of steep prismatic floes on photolyzed 9-chloroanthracene with Si_3N_4 tip at a scanning rate of 3.55 Hz; the views are at 30° rotation in (**a**) and 120° rotation in (**b**)

diversity of the topology, various parts should be measured at their optimal scan angle in the smaller range and handled as inserted image.

1.5 Artifacts

There are unavoidable topologic artifacts in particular situations and artifacts that can be avoided by suitable scanning conditions. All artifacts that derive from geometric effects have been termed *tip–sample convolution.*

As long as the slopes on rough surfaces are below about 50° or 40° (depending on the tip orientation) there is no risk of severe tip–sample convolution in measurements that do not aim at atomic or molecular resolution. Many natural surfaces do not reach such steepness even if the height differences might be large, up into the multimicrometer range. If the surface slopes are steeper, the side of the tip will touch an upper point of the surface and there will be imaging of the tip at that point. In extreme cases it might even happen that the cantilever legs slide on the surface. The tip–sample convolution gives artificial images that can be mathematically corrected in favorable cases [45, 46]. The problems are different in the highly resolved range below the radius of curvature. No reconstruction is possible when the tip contacts the sample at more than one point simultaneously. Such calculations allow deduction of tip shape for example from AFM on photoresist gratings. Tip imaging was developed with fabricated indium phosphide columns that were constructed using aerosol and dry-etching techniques. Scanning with a silicon nitride tip gave an image showing a perfect image of the tip at the positions where the etched columns were present, except at the top of the pyramids (these regions were flat due to the width of the columns). This is reverse AFM imaging: the vertical column becomes the tip and the silicon nitrite pyramid becomes the sample [47]. A very characteristic example for tip imaging is given in Fig. 1.75a. All tips are more or less truncated and this can be seen if sharper

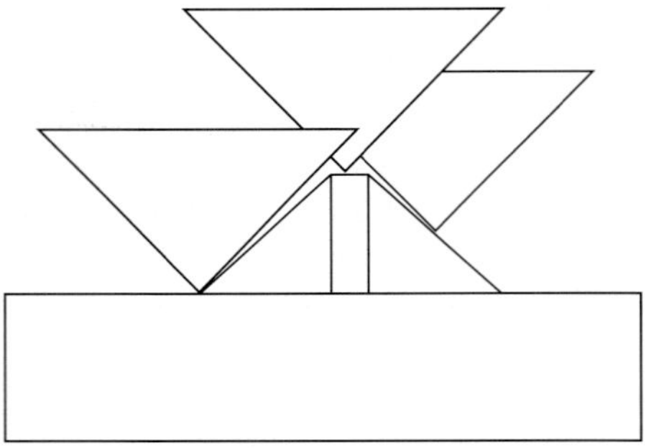

Fig. 1.8. Schematic representation of the tip–sample convolution at a vertical column; *solid line* represents the obtained image of the Si_3N_4 pyramid and the flat part

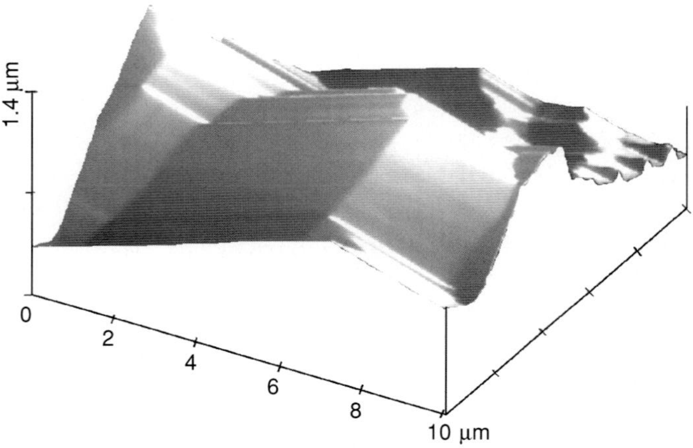

Fig. 1.9. Artificial contact AFM on an irradiated *trans*-3-benzylidene-butyrolactone (010) face exhibiting slip of the cantilever legs over the rims of deep clefts

features than the tip apex show up as tip-imaging features that have only one orientation in a scan. The sketch in Fig. 1.8 gives the explanation. It indicates that even in tip imaging the height of the sharper feature is correctly obtained, unless further close-by features hinder the tip to touch ground before and after the convolution. Si_3N_4 tips with their well-characterized shape are the best choice for the identification of topologic artifacts. Truncations of the tip from manufacture or by breakage upon approach or abrasion upon scanning have to be controlled separately.

Different problems occur between very high floe structures and deep clefts, and inside cylinders or funnels that are deeper than the tip height. The tip loses the ground and hangs at the rims, or the cantilever slides on the surface. Such sliding of the cantilever leg along extended rims outside the scanning area is exemplified in Fig. 1.9. Irradiation of *trans*-3-benzylidene-butyrolactone on its (010) face with a 2.8 μm long Si_3N_4 tip produced very high features with sharp edges so that the tip touched the surface only briefly. Most of the time the cantilever slid over closely distant rims of clefts, while feedback went off scale and it did so repeatedly at the next rim after intermittent coming back of the tip to the surface when the feedback briefly resumed and tip contact was lost again due to the next rim. Only the occurrence of very high edge floe features between deep clefts can be deduced from the artificial image in Fig. 1.9 [48]. Longer or sharper tips (higher aspect ratio) such as the ones used in shear-force AFM (Sect. 1.10) should be used in order to obtain real topology in that situation. But even then there will be artifacts if the walls are (near) vertical and the ground cannot be reached in extreme cases.

Another artifact with a powder grain larger than the tip length has been described [12]. The grain imaged the structure of the whole cantilever (leg edges and Si_3N_4 tip) in a 20 μm scan with z-scale of 4 μm.

Further difficulties occur upon measurement of the pores of membrane filters even if the slopes are not extreme. Two examples are given in Fig. 1.10 [44].

The images of Fig. 1.10 contain small shallow funnels with width/depth ratios from 200-230/10-30 to 340/120 nm and also larger funnels, mostly in the shape of craters. The latter fit to the nominal pore size. Some of them do not reach the tip gradient at their slopes, whereas others exhibit slopes approaching 55°. These are tip–sample convolutes representing open voids in their lower parts. Apparently, the pyramidal tip fills the whole cross-section of the void at a certain level and cannot descend further. The flat regions in Figs. 1.10a,b indicate sliding of the cantilever at heights outside of the scan

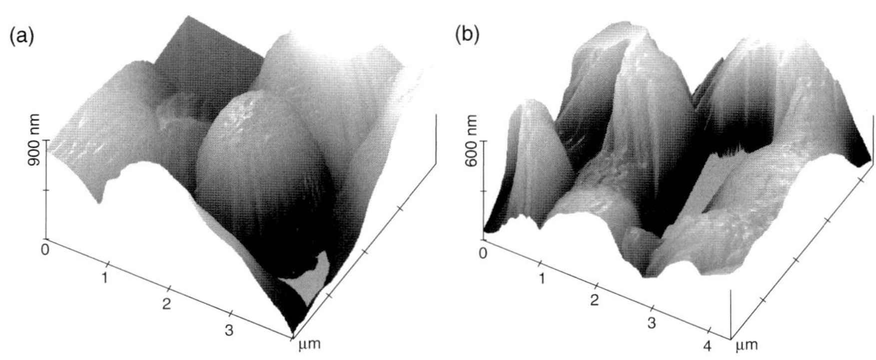

Fig. 1.10. Two different surface plots (**a**) and (**b**) of contact AFM images of a dry microporous membrane filter out of regenerated cellulose (Schleicher & Schüll, RC 59, 0.6 μm pore size) showing very deep funnels between hills as well as artifacts produced by a cantilever sliding over features outside the scan area; (**a**) scan direction from left to right, (**b**) scan direction from front to back; scan speed was 16 μm s^{-1}; z-range of the piezo was 2.7 μm

area on the very rough membrane (see also the descent artifact in Fig. 1.27b). Clearly, feedback control does not work during sliding of the cantilever as enforced by the x–y drive, but regular feedback returns as soon as the tip touches the ground again. Three sliding regions are recognized as wedge-like artifacts in Fig. 1.10b. Apparent gradients around the lake-like depression measure up to 80°–84°, values that cannot be obtained regularly with pyramidal silicon nitride tips. The coexistence of valid topology and artifacts within the same scan proves the high dynamics and fast response of the laser deflection setup (Fig. 1.3). The artificial bottom in Fig. 1.10a is 1.15 µm and that in Fig. 1.10b is 0.84 µm below the highest image point, respectively, and these are by far the deepest funnels ever seen by SXM with regular Si_3N_4 tips of 2.8 µm height (further references in [44]). Sharpened tips are not usually applied for measurements in that range of roughness. Other AFM images of membranes were restricted to z-scales of 200 nm [49, 50] or 400 nm [51].

Further improvement of the measurements in the depths of membranes can be expected by shear-force AFM (Sect. 1.10) or by contact AFM using stressed metal probes of suitable length and bending curvature [38]. The former has been applied to porous silicon with very deep descents (Figs. 1.76a,b). The performance of the latter for deep and narrow samples has been demonstrated for deep reactive ion etching (DRIE) and microelectromechanical systems, in which tip heights of 100 µm or more are required for nondestructive measurements at the bottom of deep and narrow trenches [38]. For example, a 50 µm wide and 100-µm deep trench was successfully imaged at the bottom with a 150 µm high tip. 10 µm stressed tip height would be suitable for the expected depths in Fig. 1.10. The stressed metal probe is an invaluable depth gauge in the technically interesting range using common AFM equipment. However, an easier alternative is available by shear-force AFM (Sect. 1.10) for such extreme topology.

Scanning techniques cannot image at verticals and overhangs (even if pyramidal tips with two vertical faces are applied). While the artifacts at vertical slopes in single gratings or overhangs lead to (partial) tip imaging at the upper edge and give at best a scan line with the tip gradient, the practical problems are more involved with double gratings and worse with irregular impossible topologies. Unfortunately, no general technique is available for corrections of tip–sample convolution at irregular or very high topologies [12, 44, 46]. However, there is also application of deep vertical features in test samples: They have been used for tip orientation adjustment in contact AFM and tapping-mode AFM via experimentally determined cross-sections for mathematical fitting curves [52]. Without knowing or being able to control the actual tip orientation the AFM image results may be misleading at the bottom of deep features. The actual tip orientation is particularly important for the measuring of semiconductor devices exhibiting deep features, and the scanner head needs such calibration for that purpose. Clearly, this technique also detects the actual tip angles.

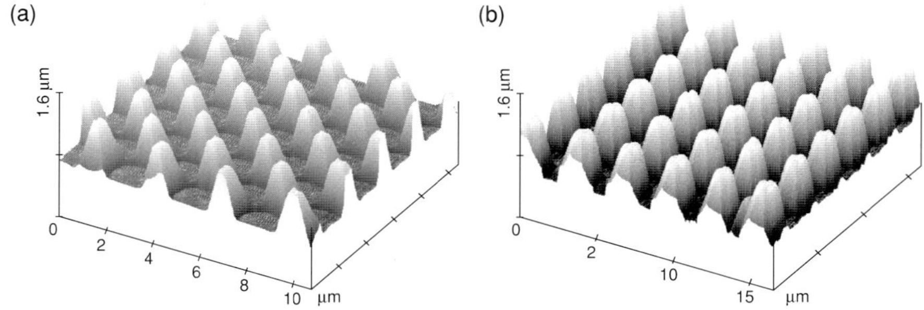

Fig. 1.11. Contact AFM topology of a commercial 3 µm × 3 µm × 0.5 um double calibration grating with an Si_3N_4 cantilever tip; (**a**) excellent scan indicating that adjustment of y-derating was necessary, (**b**) after the y adjustment but suboptimal scan though sufficient for the calibration purposes

Focused ion beam milling may pattern verticals. An example is given in [53]. The bending at the edges, which may be as high as 9 nm with respect to the flat inner area, has been stressed. Preferred calibrating double gratings for AFM scanners have similar shape. Their measurement cannot avoid tip–sample convolution due to the tip slope that is imaged as soon as the tip loses ground and touches at the edge with its sidewalls. This is clearly seen in the excellently resolved double grating in Fig. 1.11a, where the slopes of the tip pyramid are clearly recognized (±50°, also after correction for the y-direction). Fig. 1.11b shows the same grating (after the necessary adjust of y-derating) with a less optimized scan (scan direction less favorable than in the excellent image). This scan exhibits further artifacts from the corner edge. However, it is nevertheless valid for the calibration in x, y, and z, because the top of the blocks is artifact free.

Also single gratings with well-defined space as in Fig. 1.12 are well suited for x and y calibration (two orientations), notwithstanding an actually different shape at slopes that equal or exceed the tip slope.

A typical problem of unavoidable artifacts in scanning techniques concerns with spheres that occur frequently in test samples. This includes verticals and overhangs. Figure 1.13 depicts contact-mode AFM images of a hexagonal close-packed latex sphere monolayer on mica at equal x, y, and z scales scanned with an Si_3N_4 tip. Clearly, 220 nm diameter latex spheres do not show up as balls or calottes but as flattened calottes, because tip–sample convolution increases in steeper parts up to tip imaging. The tip imaging can be nicely seen in the upside-down inverted perspective image (Fig. 1.13b), which exhibits sharp rims (actually trenches) and peaks (actually depths) that are images of the tip at these sites. The tip cannot reach the ground between the spheres. The medium depth at the triangles between three spheres is only 83 nm, and the medium depth between two spheres is only 61 nm in Fig. 1.13a.

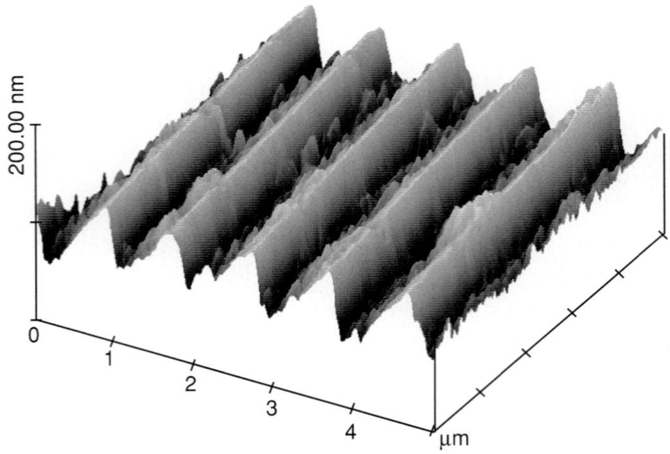

Fig. 1.12. Contact AFM calibrations scan on a 1 µm single grating with an Si_3N_4 cantilever tip

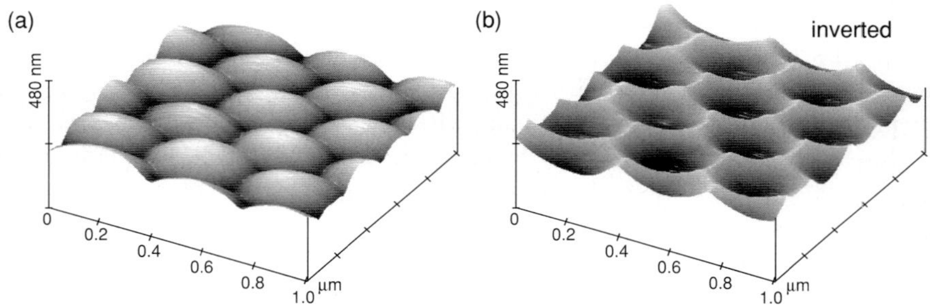

Fig. 1.13. Contact AFM images of 220 nm latex beads with standard Si_3N_4 probe tip at equal x, y, z scaling; (**a**) surface plot, (**b**) inverted surface plot showing tip imaging in the depths of the distorted topology

Such artifact cannot be seen in top-view projections. Imaging can be improved with sharper tips but artifacts will remain (Fig. 1.74).

The pronounced artifacts at cubes' sides and at cubes' edges (where the imaged pyramidal sides of the tip change), as well as flexing or bending of extremely sharpened tips, are described in great detail for blunt tips also in [12]. Importantly, flexing of tips often causes exaggerated slope gradients and fraying. Tip shapes cannot be inferred directly from the steep slope gradients observed at verticals when measured with long thin tips since the tip may have flexed. Fortunately, the occurrence of extended verticals is not very common in natural surfaces, but slopes are in most cases low and the surfaces rather smooth. If verticals occur, shear-force AFM with tapered tips is the proper choice for less convolution (Sect. 1.10). Another important point is the optimization of scan direction if extended rims/trenches occur. It should be avoided to parallel their directions or cross them at right angles, if possible, in order to minimize tip–sample convolution. The 45° crossing is best. If necessary, different partial images at changed scan direction are required.

The usual publication of SFM data as two-dimensional projection is unsuitable for rough surfaces as it is impossible to analyze potential artifacts and to identify the regions of valid measurement. Three-dimensional imaging is much superior, but quality and reliability tests and further analyses are still not possible with published pixel graphics or printed images, but require access to the primary XYZ-data [54]. The first way to reach that goal for everybody was the disclosure of AFM data in the format of interactive VRML data. Their access to all major analyzing software has been realized and is now also commercially available.

1.6 Surface Scratching and Plowing, Liquids on the Surface

Composites of hard and soft materials may produce elasticity artifacts in the topology that can be traced due to the extraordinarily high resolution in the vertical direction. This can be judged by repeated measurements with varying preset loads. It excludes damage of the surface and it is also a means to detect chemical variation on the surface. Actually, the elasticity feature is used for force modulation microscopy (FMM) as an extension of AFM imaging by detection of different amounts of deflection at different elasticity. Adhesion is obtained from the force–distance measurements. An oscillating voltage signal from a separate piezo is applied to the z-control signal. However, if the same piezo is used for the modulation the deflection tends to be unstable. With the force modulation technique, the sample assembly is scanned with a small vertical oscillation, which is significantly faster than the scan rate [55]. The average force on the sample is equivalent to that in contact mode, and both a topologic image and a force modulation image are simultaneously obtained. Stiffness differences can be located. Another approach is the so-called interfacial force microscopy (IFM). It uses capacitor sensors for that purpose [56]. The pull-off force is set equal to the peak attractive force for the work of adhesion calculation.

This technique clearly enters the nanoindentation field, where diamond tips and μN up to mN load ranges are used (Chap. 3), if forces between different materials are increased to achieve plastic deformation. A similar extension of contact AFM is lateral force or friction force microscopy (LFM or FFM). In that mode the tip or sample is moved perpendicular to the long axis of the cantilever and the amount of bending measured by the light deflection in a four sections phototube. Various procedures exist and have been theoretically analyzed [57, 58].

The standard force settings range for contact AFM is 1–30 nN. However, there were numerous complaints of surface modification upon scanning with

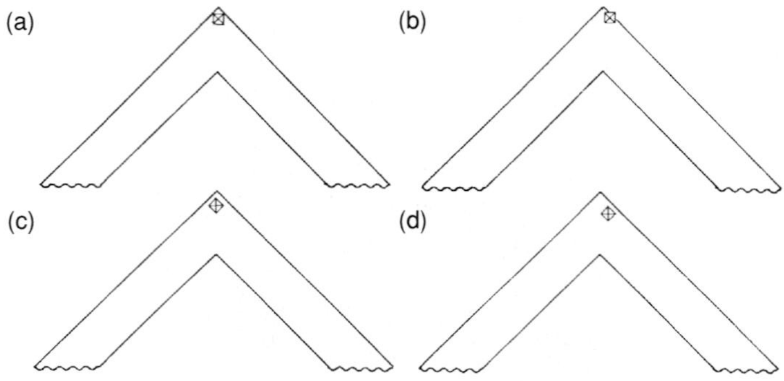

Fig. 1.14. Symmetrical (**a**,**c**) and asymmetric cantilever tips (**b**,**d**)

forces up to 10 nN (e.g., in [22]). Such erratic scratching, scraping, plowing has a definite reason. Regular cantilever tips of types (a) and (c) in Fig. 1.14 do not scratch, scrape, and plow on solids up to 20–30 nN in constant force mode. Even polymer surfaces are sufficiently elastic. On the other hand, tips of types (b) and (d) do scratch, scrape, and plow at and below 10 nN load in contact AFM [44, 59–61].

It was clearly not acknowledged that noncentric cantilever tips or unequal cantilever legs must be avoided. For example, Fig. 3.12 in [22] or Figs. 1a,b and 2c in [12] depict asymmetrically placed tips without comment. There is ample experimental proof that such asymmetric tips scrape even below 10 nN and sometimes very efficiently (this depends on the crystal packing) [44, 59–61]. For example, the quite hard and elastic crystal surfaces of *trans*-benzylidene-cyclopentanone (indentation hardness $H = 110$ MPa, reduced modulus 4.9 GPa) can be easily scraped out upon repeated scanning in 2 μm × 2 μm regions of a 10 μm × 10 μm area. The nominal force was 10 nN with an asymmetric tip in the face in front orientation. Intermediate recording was omitted to avoid fraying out of the rims by the scratchy cantilever tip.

The flat square holes after five scans are 50 nm deep with a slope angle of 10°–14°. The big hole after ten scans is 120 nm deep with slope angles of 22°–27° (Fig. 1.15). Further 10 μm scans would both show and increase fraying. On the other hand, symmetrical tips up to 20 nN load and more did not alter the surface of the same material upon repeated scanning [44]. The scraping efficiency depends strongly on the crystal packing [59]. Clearly, scraping can be purposely obtained with asymmetric cantilever tips, cantilevers that have a drop of liquid, or a dust grain on one of the cantilever legs [44, 60, 61]. Conversely, selecting symmetric tips under a microscope with magnification larger than 400-fold avoids scraping. Nevertheless, checks of scraping by several scans in a zoomed-in region are advised due to the risk of cantilever contamination that may occur upon instrument setup. Seven papers reporting surface scraping are listed in Ref. 18 of [44] and numerous reports in [22] claim that "their force" was too high. The corresponding authors did not

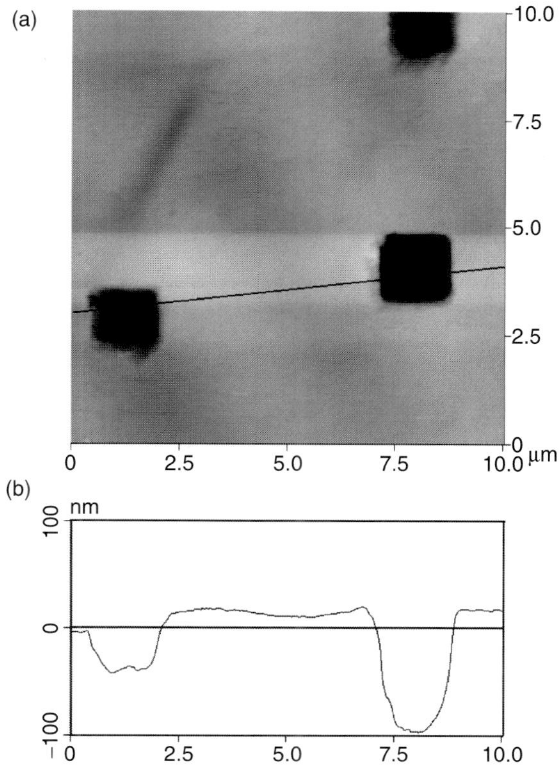

Fig. 1.15. AFM top-view image (**a**) and cross-section (**b**) of scraping experiments on the main surface of *trans*-benzylidene-cyclopentanone with a cantilever tip of type (**b**) of Fig. 1.14 at 10 nN force, two experiments at five, and one at ten 2 μm × 2 μm repeated scans

acknowledge the tip-symmetry feature and this is also not acknowledged in [62]. These authors should have carefully selected their tips. A more recent report on muscovite mica scratching at loads above 70 and 80 nN was related to water content and pH, while others did not find scratching wear up to 400 nN under vacuum [63]. Unfortunately, the symmetry of the cantilever tip was again not addressed, which makes the quantitative assessments doubtful at least. Conversely, purposeful plowing on polycarbonate (PC), polymethylmethacrylate (PMMA), or epoxy resin (ER) with the AFM required loads of 500 nN with a diamond tip. The radius of curvature was given to be 100 nm [64]. But the real shape of the tip apex (curved or flat truncation) will play an important role in scratching as well (Chap. 3, Sect. 3.11).

AFM on rough surfaces is very sensibly reproducible if symmetrical tips are applied. This may be illustrated by a photostability test at the nanoscopic level with a rough and diverse surface of tetraphenylethene, an organic hydrocarbon crystal. There was 10 min irradiation between Fig. 1.16a and b with

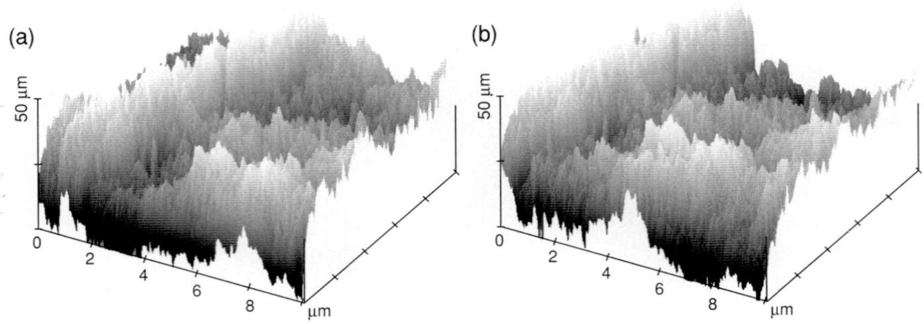

Fig. 1.16. Contact AFM reproducibility test with symmetric Si_3N_4 cantilever tip; (**a**) fresh crystal of tetraphenylethene, (**b**) the same sample after 10 min photoirradiation in order to demonstrate photostability at the nanoscopic level

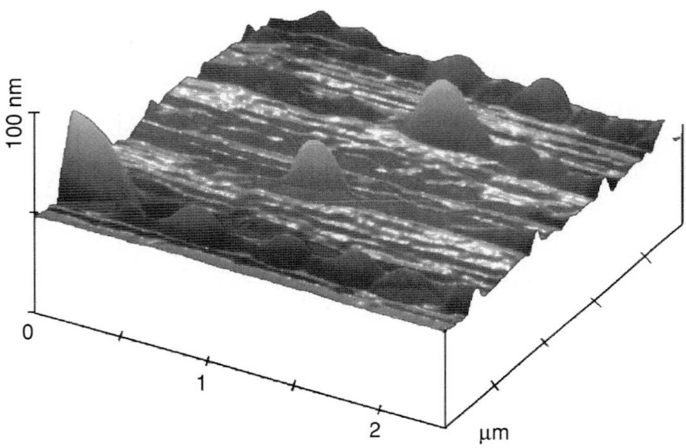

Fig. 1.17. Contact AFM on the viscous surface of an *N*-vinylcarbazole crystal that had been exposed to methylamine gas, showing spreading of material by the tip

a water-cooled Hg high-pressure lamp (Hanau 150 W) through Solidex from 5-cm distance [65]. The original site was found again with high precision and all minor details are unchanged, which is very remarkable, indeed.

A particular advantage of contact AFM is the most direct information about the surface qualities. It may be viscid and the tip will stick to it and eventually snap free (no measurement is possible under such circumstances). If a highly viscous liquid is present or formed during the scan, the tip spreads it along the scan direction (as shown in Fig. 1.17) [7], and it will stay there for some time.

If a liquid of low viscosity is formed on the surface, there will be gradual remodeling of the surface upon continuous AFM scanning also with symmetric tips: the liquid dissolves some of the crystalline material and this solution

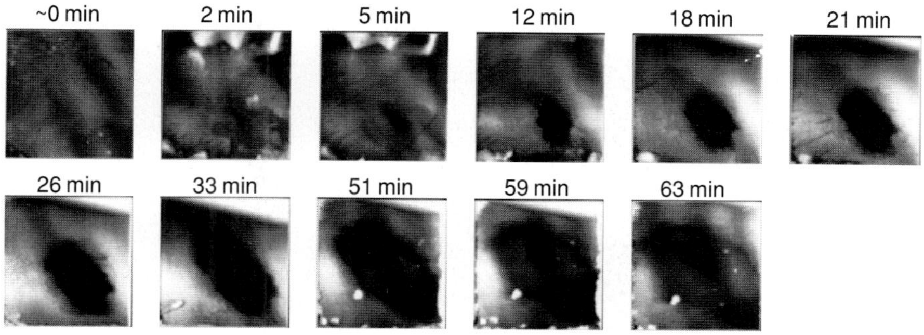

Fig. 1.18. AFM traces (10 μm, z-scale: 200 nm) on an α-cinnamic acid crystal surface after exposure to chlorine gas with a nanoliquid formed on it at the times given, showing development of an increasing hole, a secondary hole, and desolving and recrystallization of materials due to transport of the liquid by the AFM tip

is transported back and forth with all signs of nano-recrystallization [59, 66]. This is shown in Fig. 1.18. The nanoliquid was generated by brief reaction of α-cinnamic acid on the (010) face with chlorine, blowing away of residual chorine with air and continuous AFM measurement at the times given. Measurements started 10 min after the reaction. This is an important general method for the detection of minor liquids on surfaces that may not be visible under strong light microscopes. These results enforce repetition of AFM measurements on rough surfaces for exclusion of nanoliquids that are different from adsorbed water layers.

Another case of surface liquid formation is encountered if the very strong organic base benzimidazole-2-guanidine is reacted with gaseous carbon dioxide. A freshly cleaved face (Fig. 1.19a) is preferably used, as naturally grown faces tend to be very irregularly rough [67]. Fig. 1.19b shows the interaction with carbon dioxide (probably formation of the carbonate salt by additional reaction of a water layer) that provides flat layers with steps (for higher resolution of the molecular steps see Fig. 1.24). If however the carbon dioxide was blown away with moist air a viscous liquid started to form. It starts to spread in scan direction (left to right) as is already seen in (c). 36 min later at a rotation angle of 35° the viscous liquid had largely increased. The irregular features are not solid, because they change their shape upon scanning. The "drops" are not regular and do not fully unite. Clearly, there are some problems with the definition of the terms viscid, viscous, liquid, or fluid in the nanoscopic range. Importantly, solid features do not change upon five to ten consecutive contact AFM scans with symmetric cantilever tips in the 10–30 nN load range, and that is a clear phenomenological definition of a solid.

A special case is water layers on hydrophilic surfaces in ambient atmosphere. If they do not show similar features as in Figs. 1.17–1.19d, they do not distort the contact AFM image but belong to the natural surface. The tip simply penetrates through humidity-dependent water multilayers. The water

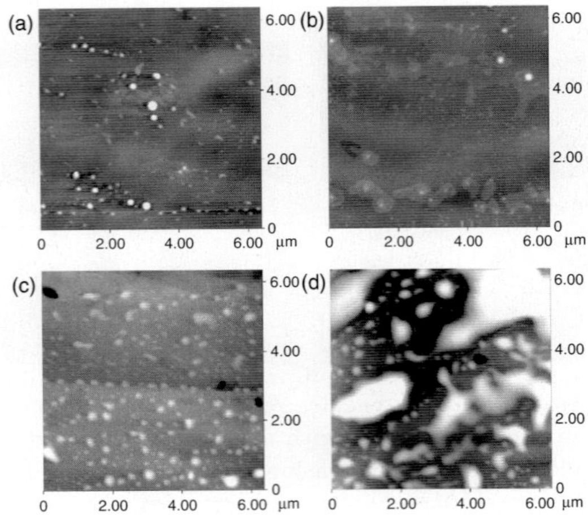

Fig. 1.19. Contact AFM traces on the cleavage plane of benzimidazole-2-guanidine; (**a**) fresh; (**b**) after application of gaseous carbon dioxide; (**c**) after blowing away the carbon dioxide with moist air; (**d**) 36 min later; the z-scales are 50 nm in (**a**) and (**b**) and 100 nm in (**c**) and (**d**); a viscous liquid is clearly formed in (**d**)

layer may be adsorbed as inorganic (hydr)oxides (these are also as layers on numerous metals), or solid surface hydrates may be present on polar organic molecular crystals, or hydrolyzed surface layers may be hydrated. Also surface hydration to give solid surface hydrates may be found if fresh molecular crystals are exposed to moist air for the first time during AFM measurements.

1.7 Molecular Steps

The z-piezo response in contact AFM is so sensitive that molecular steps of organic molecules (5–10 Å range) are routinely vertically resolved. It is remarkable to obtain reproducible images with Si_3N_4 cantilever tips at 10–20 nN force under ambient conditions even on rather corrugated surfaces at the correct height when compared with crystallographic data in large scans. Such molecular steps are frequent with layered crystal structures on faces that are parallel to the layers. A typical example is 3,5-diphenyltetrazine ($P2_1/c$) with four nitrogen atoms per molecule. It exhibits its skew molecular monolayers at an effective height of 5.4 Å (100), but natural organic single crystals are not perfectly flat [65]. Two images (a) and (b) of different crystal specimens that were routinely taken at a time distance of three years are shown in Fig. 1.20. The measured step heights are consistently 10, 15, 20, 25, and 30 Å, which corresponds to steps of 2–6 molecular layers (Fig. 1.20).

The crystal packing of 3,5-diphenyltetrazine is visualized in Fig. 1.21. It shows the skew (54°) molecular layers with alternating orientation and some interlocking (about 2 Å). The correct distance tells that there is no surface

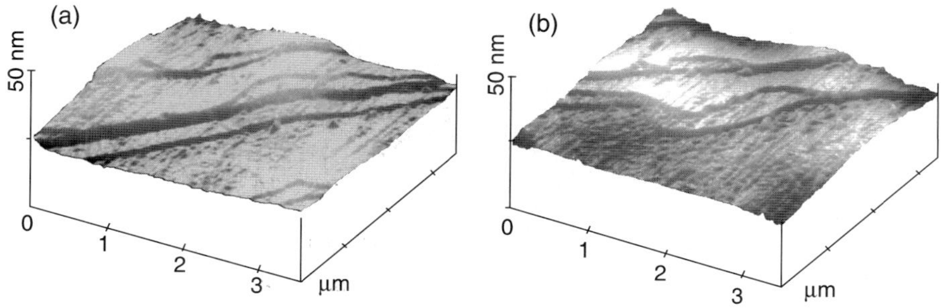

Fig. 1.20. Contact AFM surface of 3,5-diphenyltetrazine on (100) scanned with an Si_3N_4 square pyramid, showing molecular steps of 10, 15, 20, 25, and 30 Å heights

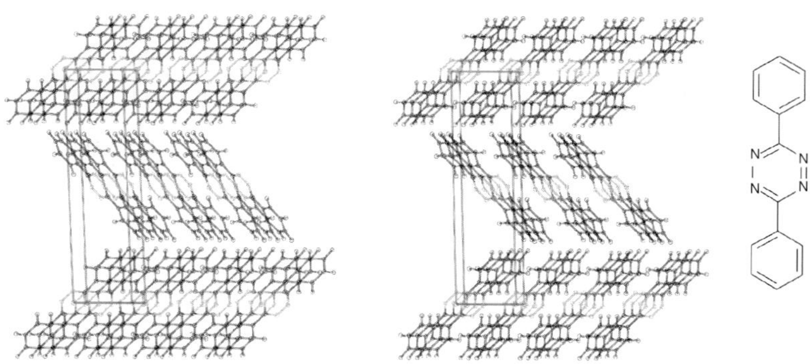

Fig. 1.21. Stereoscopic packing diagram of 3,5-diphenyltetrazine with (001) on top along [100] but rotated around y by 10° for a better view. The layered structure type is clearly seen as well as the outer hydrocarbon arrangement, but some access of the uppermost nitrogen, e.g., for protonation appears still possible

reconstruction. The uppermost layer clearly indicates that the external phenyl groups shield the four nitrogens per molecule. It appears not impossible to reach the uppermost nitrogens by the protons of water, but the hydrophobic phenyl groups should prevent the formation of a closed water layer.

Salt formation experiments under AFM control serve the clarification of this point. Protonation of 3,5-diphenyltetrazine has been investigated with gaseous hydrogen chloride under AFM control on a surface of equal quality as in Fig. 1.20. Clearly, the protonation to give the 3,5-diphenyltetrazine hydrochloride would be easiest at the molecular steps. However, no change is seen at the steps that are unaffected at this large scan (Fig. 1.22) probably because the chemical reaction does not continue there after protonation of the front molecules, because there is no need to leave the layer: the protonated cation molecule is more attracted by its inner neighbor and the chloride counterion is well accommodated at the step. However, the harder way, which is protonation from the surface, can start at nucleation centers and

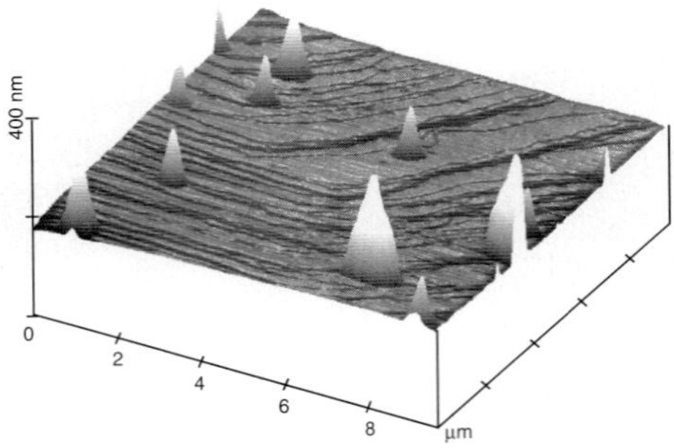

Fig. 1.22. AFM topology of 3,5-diphenyltetrazine after two additions of diluted hydrogen chloride gas (1.2 and 3.2 ml) for salt formation via proton transfer; the salt islands grew after the second application of the acid; the scanning (after blowing away of the HCl) did not influence the result; no continuing protonation occurred at the molecular steps

lead to islands of considerable height around such points of initial attack. After application of 1.2 ml of 10% HCl gas (diluted with dry air) from a syringe in 5-cm distance, no islands were observed. The 12 islands that are seen in Fig. 1.22 are the result of a second application of 3.2 ml of the diluted acid. The AFM scanning did not influence the result. The acid was blown away before the scanning. These results seem to exclude a water layer on the surface of 3,5-diphenyltetrazine. It is assumed that the islands in Fig. 1.22 consist of the monoprotonated salts, but the possibility for diprotonated salt formation cannot be excluded presently [68].

No risk of a water layer is encountered on anthracene single crystals (grown from ethanol solution), which also exhibit a layered structure. Again these layer surfaces are corrugated. The crystal packing under the (001) face is depicted in Fig. 4.37 (similar to Fig. 1.21, but all molecules inclined to the right side). The molecules in the monolayers are inclined by 67° and all point in the same direction. The topologic image has been taken with contact AFM using regular Si_3N_4 cantilever tips under ambient conditions [66], and a similar image was obtained three years later [69]. The terrace steps in Fig. 1.23a are 10, 20, 30, and 40 Å high. This corresponds precisely to 1–4 monolayers, all within one scan and it excludes surface reconstruction.

Molecular steps with molecules that stand skew and free to one side are sites for release of internal pressure from inside the crystal where the molecules can freely exit. Such internal pressure can be produced by light-induced [4 + 4]-photodimerization of anthracene. The photodimer is no longer a planar molecule and therefore considerably distorts the lattice. The applied light penetrates down along a gradient through about 200 molecular layers, so it is a crystal bulk effect. The bulky dimer molecules migrate and shift also molecules of anthracene in front of them. Fig. 1.23b shows that indeed

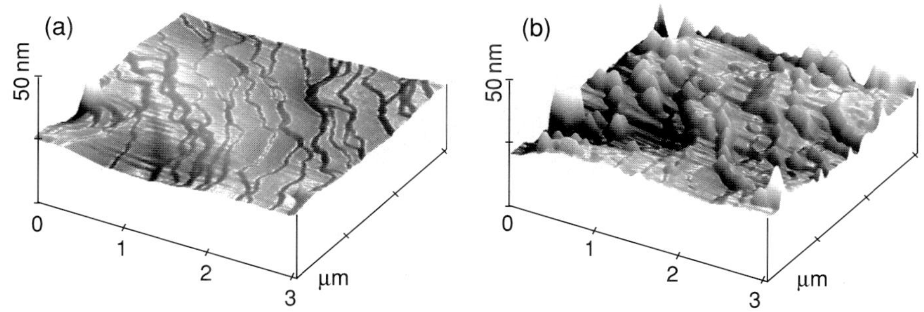

Fig. 1.23. Contact AFM image on (001) of an anthracene crystal, (**a**) showing molecular steps, (**b**) exit of molecules at the molecular steps after irradiation, which produces internal pressure due to [4 + 4]-dimerization reaction

molecules exit along the molecular steps by forming walls that increase at later stages of the photolysis [61, 65, 66].

Alternatively, it might be asked if the features at the steps are the actual reaction sites. However, the question of the chemical composition at the exit regions has been definitely answered by SNOM on the rough surface (Chap. 2) [69]. The whole surface of Fig. 1.23b and in later stages of the photolysis is chemically uniform (equal chemical composition), as no chemical contrast occurs over all features. Therefore, the mixture of anthracene and its photodimer changes constantly all over the surface, at the exit walls and at the plain surface that experiences vertical transport in the later stages. This result disproves earlier defect-site theories claiming energy migration to such defects. The molecular steps are certainly crystallographic defect sites, but the product is not specifically formed here (at the sites of exit) but at the sites of light absorption. Such conclusion is only possible if AFM evidence is combined with SNOM evidence.

Molecular steps may also be formed upon chemical reaction if flat layers ensue. Thus, the reaction of the strong base benzimidazole-2-guanidine (including water layer) with carbon dioxide in dry air (Fig. 1.24) produces layers that exhibit terrace-like steps that are stable and more highly resolved than those in Fig. 1.19b (in moist air). The step height is typically 2 nm, some steps measuring 1.5 and 2.5 nm. Clearly, this is in the molecular range of the resulting benzimidazole-2-guanidinium carbonate or hydrogencarbonate [67].

Crystallization on the cleavage plane of calcite ($CaCO_3$) was studied by contact AFM. Growth and dissolution occurred in a layer-by-layer fashion in monomolecular steps that were 3-Å high [70]. Terraces with heights of 4 Å have been obtained by layer-by-layer growth of $SrTiO_3$ and $BaTiO_3$ films using laser molecular beam epitaxy (MBE) and measured by AFM [71]. Also the molecular layer-by-layer growth of Al_2O_3, and BaO thin films occurred in terraces [72]. Two unit-cell thick (6.6 Å) layers of MBE-grown $SrCuO_2$ and $CaCuO_2$ were recorded on $SrTiO_3$ (100) [73].

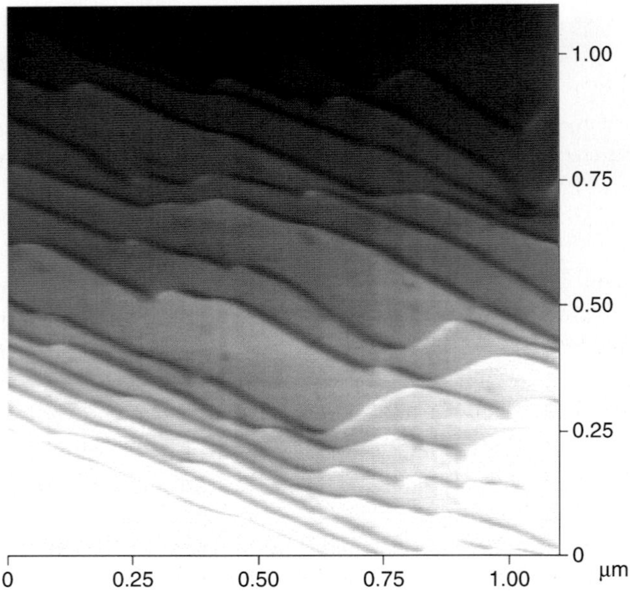

Fig. 1.24. Contact AFM topology of a step structure in the reaction of benzimidazole-2-guanidine with carbon dioxide with molecular step heights of 2 nm; the z-scale is 25 nm

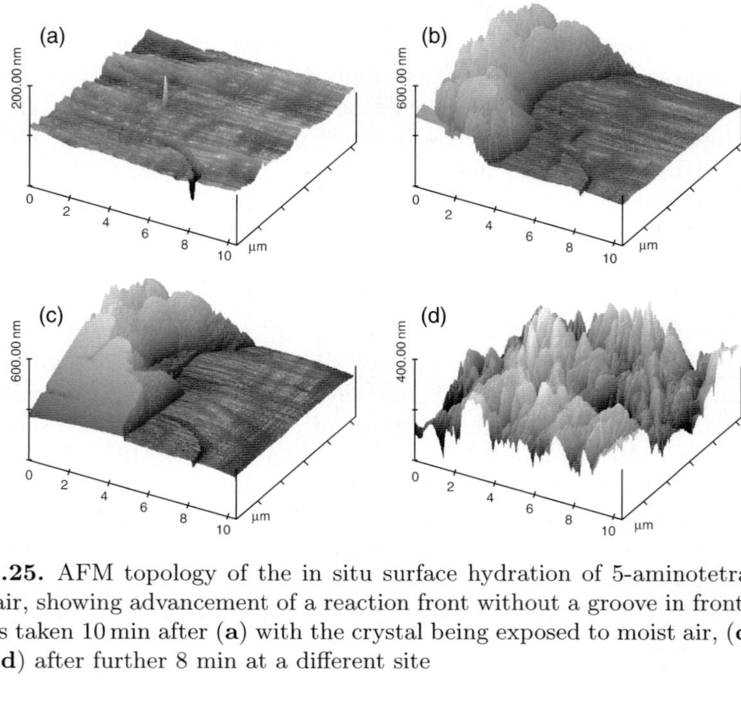

Fig. 1.25. AFM topology of the in situ surface hydration of 5-aminotetrazole in moist air, showing advancement of a reaction front without a groove in front; image (**b**) was taken 10 min after (**a**) with the crystal being exposed to moist air, (**c**) 2 min later, (**d**) after further 8 min at a different site

1.8 Ambient Surface Modifications

Ambient conditions mean moist air (mostly undefined relative humidity) at room temperature and 1 bar atmospheric pressure. Carefully crystallized samples with numerous polar substituents may efficiently catch water from the atmosphere and form solid surface hydrates. This is followed by AFM in Fig. 1.25 with surface hydration of 5-aminotetrazole (CH_3N_5, crystals from dry methanol) [7]. During contact AFM scanning a broad reaction front arrived as is seen in Fig. 1.25b that still contains the landmark fissure. There were 10 min between the measurements (a) and (b), 2 min between (b) and (c), and 8 min between (c) and (d). This is a very efficient surface hydration reaction from the gas phase forming a crystalline hydrate. Unfortunately some sliding of the cantilever at a remote obstacle occurred that could not be taken care of in the very rapid scanning routine. Two further images between (c) and (d) continued the migration of the broad front as judged from zoomed-in images leaving out the artificial area (not shown), because the cantilever sliding grew into the image more badly. Image (d) at a different site shows that the final size of the hydrate structures is reached already while it runs and a rough hydrate coating covers the crystal at the end. No groove is formed before the broad front and very rapid molecular migrations above the surface occur. This is unusual. Either the crystal packing under the most prominent face of 5-aminotetrazole impedes hydration or the feature formation is impeded in the absence of crystal seeds that are available wherever the reaction front arrives. In that case it would be recrystallization of an unstable preformed solid phase. The scanning does not induce it, and there is no nanoliquid on the crystal.

A reacting front upon surface hydration with the result of a crystalline surface hydrate is also observed if hydroxylammonium sulfate crystals (crystallized from methanol) are exposed to moist air [74]. In that example a groove ahead of the slowly proceeding wedge-shaped front is recognized (Fig. 1.26). It takes several hours to pass the $10\,\mu m \times 10\,\mu m$ surface, and Fig. 1.26d does not change after one more day in moist air. These features point to a different mechanism or different crystallographic influences than in Fig. 1.25. Again, the very rough surface layer assumes its final height while it proceeds. The surface hydrate has different chemical properties from the unhydrated surface, but a bulk crystalline hydrate of hydroxylammonium sulfate is not known. It crystallizes unhydrated also from water [74]. Surface hydration may protect the crystal or it may increase its reactivity. AFM shows that there is no liquid on the surfaces, but this does not exclude additional firmly adhered water layers. Importantly, surface changes in ambient air at the submicron level are very important for all surface properties and should therefore routinely be tested by contact AFM before application of any perhaps more well-known surface techniques such as ellipsometry, interferometry, UV/VIS-, or IR- spectroscopy, etc.

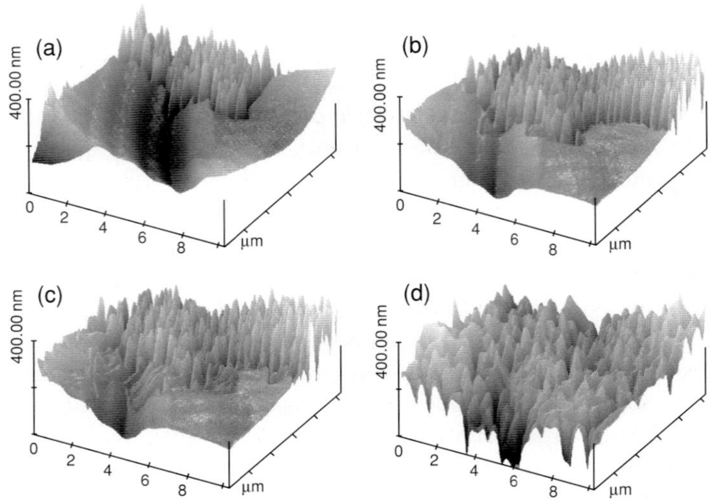

Fig. 1.26. Contact AFM topology of the *in situ* surface hydration of hydroxylammonium sulfate $[(NH_2OH)_2 \cdot H_2SO_4]$ in moist air, showing advancement of a reaction front with a groove in front; image (**a**) was taken 40 min after mounting and exposure of crystal to the moist air, (**b**) 23 min later, (**c**) after further 28 min (four scans in between), and (**d**) after 20 h

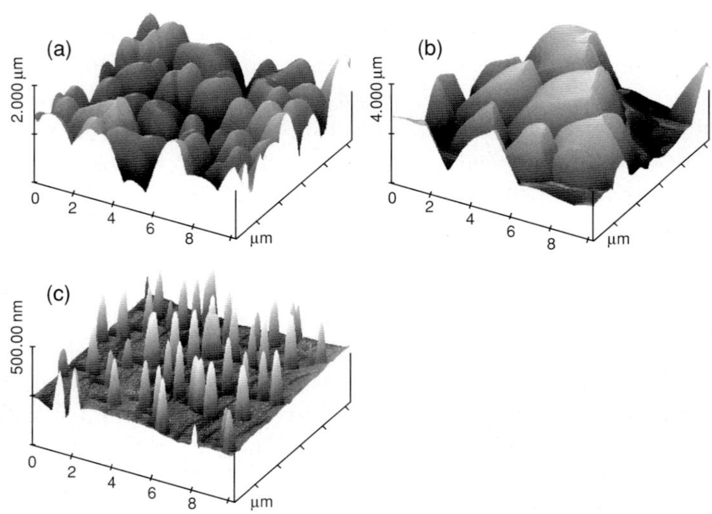

Fig. 1.27. Contact AFM surfaces (Si_3N_4 tip) of the dominant face of hydroxylamine hydrochloride, (**a**) after final hydration in moist air for 24 h at 18°C; (**b**) after reaction of the surface with diluted acetone vapor for 2 min with descent artifact to the right as the limits of the piezo was exceeded (z-scale is 4 µm!); (**c**): pristine (not hydrated) surface that was exposed to acetone for 5 min showing much slower reaction and island formation

Moving reaction zones are not very frequent. But various types of surface changes occur on polar crystal surfaces. Surface hydration in moist air may occur at random and so rapidly that it is hard to obtain an area without previous hydration such as in Fig. 1.6, which shows the surface hydration of hydroxylamine hydrochloride (dense volcano-type features). In that case an enormous increase in reactivity toward the reaction with acetone gas to give acetone oxime hydrochloride has been found [74]. Figure 1.27 demonstrates the enormous reactivity of the final and stable hydrated surface of hydroxylamine hydrochloride (a) after the cones of Fig. 1.6 have grown together. These solid hydrate features change to the huge volcano-like features in Fig. 1.27b after application of diluted acetone vapor from a close-by filter paper that was soaked with some drops of acetone for 2 min and blowing away of the gas mixture. The height of these also microscopically detectable hills exceeds the capacity of the applied piezo and this is the reason for the artifact in the descent of the features (scan from left to right). But the size of the presumed symmetric hills is nevertheless roughly discernible.

The materials transport is enormous and the product structure has been secured by isolation and chemical characterization in large-scale runs that go to completion with 100% yield. Conversely, the reaction of acetone gas with unhydrated hydroxylamine hydrochloride is much slower as shown in Fig. 1.27c starting from a pristine surface after 5-min exposure. Importantly, the cones in (c) are not at random, but stay in line, though in various directions. The fact that gaseous water (also water that is stoichiometrically liberated by the reaction) catalyzes the solid-state reaction had to be taken into account for large-scale runs in flow systems for removal of highly dilute acetone from spoiled atmosphere [74]. Numerous related reactions have been tested there.

Random groups of volcanoes with different sizes are described for the hydration of 4-aminobenzoic acid in Fig. 1.32. Furthermore, random islands with different sizes (for example 1.5–6 and 10–15 μm wide, and up to about 1 μm high) in 100 μm AFM scans have been observed after aging of α, ω-dihydroxy

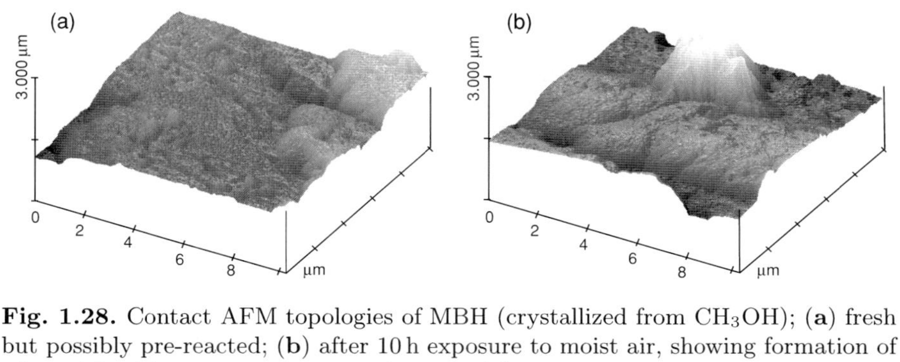

Fig. 1.28. Contact AFM topologies of MBH (crystallized from CH_3OH); (**a**) fresh but possibly pre-reacted; (**b**) after 10 h exposure to moist air, showing formation of a huge island but not at the same site as in (**a**)

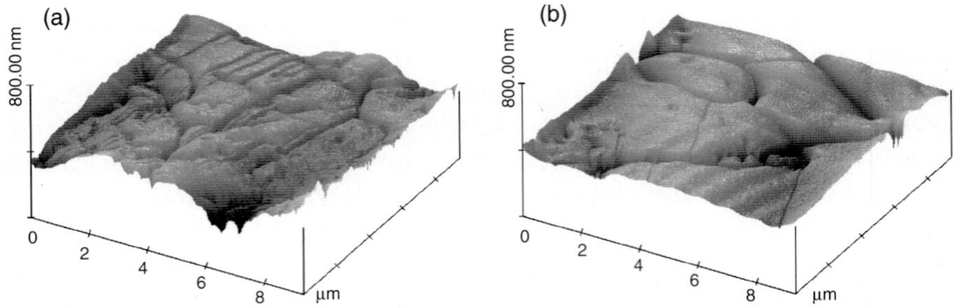

Fig. 1.29. Contact AFM topologies of DABCO after exposure to moist air; (**a**) was only measurable after considerable loss of material by sublimation; the time difference between the measurements (**a**) and (**b**) is 1 h; in between were four additional scans

functionalized poly(dimethylsiloxane) (PDMS) and the ethoxysiloxane mixture for six months under ambient conditions [75]. Obviously this "unexpected" feature formation must be the result of chemical reactions with water vapor. Not at random and very slow is the surface hydration of *N*-methylbenzothiazolone-2-hydrazone (MBH) with ambient moist air. Huge isolated islands (not the same site in Fig. 1.28b) are formed within 10 h [67]. Apparently, this is an impeded surface hydration while forming a solid hydrate. There is again no liquid on the surface. It is assumed that the crystal packing hides the functional groups under the surface and that hydration proceeds only around rare nucleation centers.

Island formation has also been observed by autoxidation on (001) of 2-mercaptobenzothiazole crystals (Fig. 1.41) or on (110) of anthracene crystals (Fig. 2.29) with oxygen. Much smaller islands have been found with shear-force AFM in the hydration of phthalimide, but only after hydrolytic ring opening and not in the initial surface hydration. This was clarified by SNOM and is described in Chap. 2 (Fig. 2.34).

It is also possible that surface hydration protects the crystal by nanoscopic covers. Thus, hydration may smooth the surface by forming flat layers. This is shown by the hydration of 1,4-diazabicyclooctane (DABCO) crystals (sublimed) in Fig. 1.29. These crystals are highly volatile so that only large crystals (>3 mm) can be studied under ambient conditions. The first image that can be taken in air shows a smooth flat covered with many separated domains apparently from a hydrate (Fig. 1.29a). At that time the vaporization losses come to a stop, and after 1 h the smooth surface covers have increased (Fig. 1.29b). After that, the crystal is protected from further evaporative losses. It stays unchanged for days in air and does not smell intensely anymore. Again, no nanoliquid is on the surface, but this does not exclude a strongly adhered water layer. Clearly DABCO crystals protect themselves from losing weight in moist air by forming a nanoscopic protective hydrate cover [7]. This finding may be of practical importance for its handling.

Another example for flat cover formation at in-situ reactions with atmospheric constituents [7] is described in Fig. 1.24.

Importantly, surface layers and particularly the flat ones often (but not always) protect the crystal from chemical reaction with reactive gases. Striking examples have been described in the reaction of penicillamine or cysteine with acetone vapor. AFM clearly indicated initial reaction followed by formation of flat layers. These prevented further reaction by surface passivation. It was also shown that the application of penicillamine hydrochloride or cysteine hydrochloride and acetone vapor did not produce surface passivation and the corresponding thiazolidine salts could be isolated with 100% yield [74]. Hence, these reactions can be engineered, because the free bases can easily be liberated from the salts.

The examples of this section show that contact AFM control of crystal surfaces in ambient environment or upon aging (e.g., Figs. 1.52 and 1.64) is indispensable prior to performing other surface investigations. The submicroscopic features or layers that may form can decisively influence physical properties and chemical reactivity also on the macroscopic scale.

1.9 Chemical Surface Modeling and Typical Shapes of Topologic Features

A great number of different surface features occur on natural surfaces, but these are normally too complicated for a classification of their shapes. However, anisotropic surface modeling can isolate the basic shape components. Most successful are chemical reactions on surfaces of (single) crystals without involving liquids or solvents. These proceed anisotropically and are a good means for creating well-defined shapes or a limited number of shapes of submicroscopic features. Solid-state reactions are strictly controlled by the crystal structure and they are face selective. Apart from creating well-defined surface nanostructures the pronounced anisotropy is the background for chemical application in waste-free quantitative syntheses, which are really benign and sustainable. The classification of naturally grown surface features according to their geometric shape is of scientific and technical importance. The nanostructures are formed by self-assembling far above the molecular level by lattice-directed transport phenomena. The solid-state reactions can be initiated by heating, photoirradiation, gas–solid reaction, or solid–solid reaction. All of these processes may become profitable surface modeling procedures.

1.9.1 Volcano or Cone Type

The volcano type describes regular cones with sharp apex such as in Fig. 1.30 and vertical elevations with more or less pronounced summit curvature in dense packing. The term is also reminiscent of their formation by "eruption" of molecules over the initial surface. They occur also as one type of islands

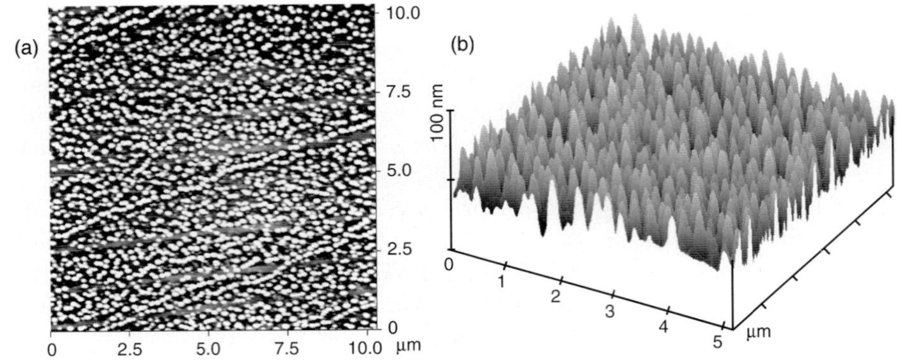

Fig. 1.30. Contact AFM topology of α-cinnamic acid (crystallized from benzene) on (010) after 30 min irradiation with 365-nm light (6 mW cm^{-2}) exhibiting densely packed volcano structures; (**a**) top view with 1,520 stable volcanoes, some of them aligned along the cleavage plane direction; (**b**) surface plot of a section of image (**a**)

where the volcanoes are well separated from each other (Sect. 1.9.2). The same slope angles all around are a quality mark of proper scanning (Fig. 1.6). The volcanoes may be tall, of similar size (Fig. 1.30), and so densely packed that the bottom of the surface is tip imaging by necessity [76]. The features of Fig. 1.30 are created by bulk photolysis (light penetration down by 1,300 layers) of planar hydrogen-bound cinnamic acid molecules to give three-dimensional four-ring dimer molecules. The internal pressure produced is released by upward transport of molecular material through the monolayers of the skew hydrogen-bound planar molecules parallel to the previously flat surface with terrace structure.

Considerably taller, in random large groups and with downhill fissures are the volcanoes of Fig. 1.31. They were achieved by reaction of a layered planar aromatic hydrocarbon (anthracene) in a gas–solid reaction. Again the molecular transport was across the horizontal layers of skew molecules [67].

Volcanoes of very different sizes are obtained on (101) of a highly polar organic crystal of 4-aminobenzoic acid that was previously exposed to moist air (Fig. 1.32).

Interestingly, volcanoes can also be generated by gas–solid reactions with plastic polymers. For example injection-molded PMMA and gaseous chlorine give the rather blunt volcanoes in Fig. 1.33 by chlorination of the surface region [24].

1.9.2 Islands

Island formation occurs if the crystal surface shields the functional groups of organic molecules. The crystal protects itself from being chemically attacked and damaged at the molecular level. Reaction can only start at nucleation

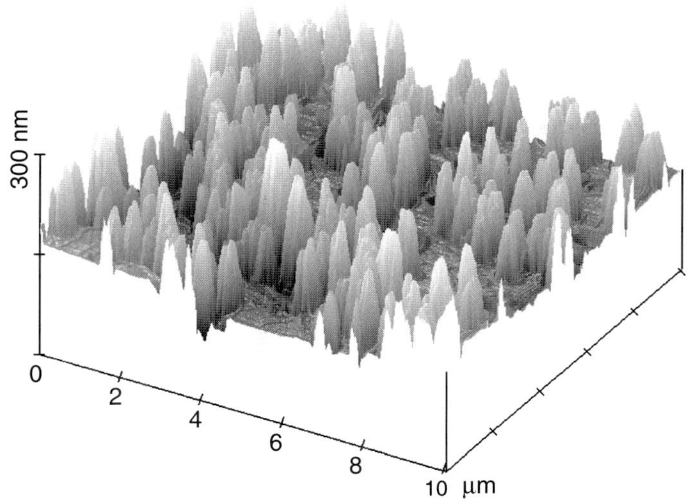

Fig. 1.31. AFM topology on (001) of anthracene after exposure to 0.05 ml chlorine in 3 ml air with a syringe from 3 cm distance and blowing away of the reactive gas after 2 min

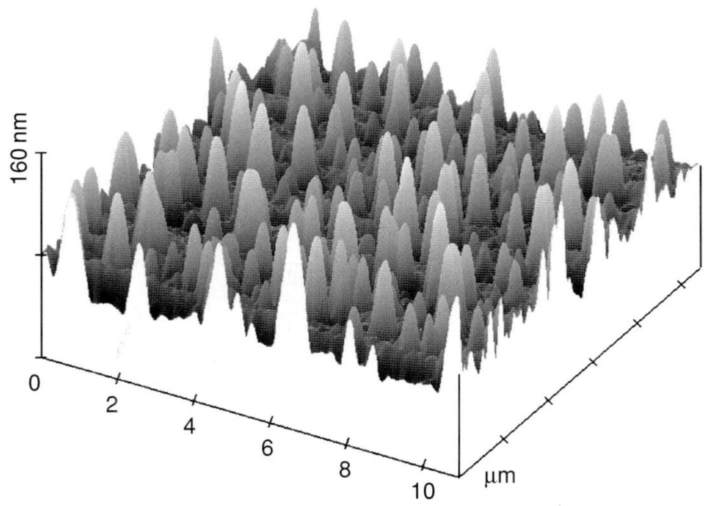

Fig. 1.32. AFM topology on a (101) surface of 4-aminobenzoic acid ($P2_1/n$, crystallized from ethanol) that was previously exposed to moist air indicating surface hydration by the formation of volcanoes of different heights

centers and it spreads around them. The crystal packing of benzimidazole is a good example for the hiding of the functional N and NH groups that would be ready for reaction with cyanogen chloride as a very potent agent. The benzene

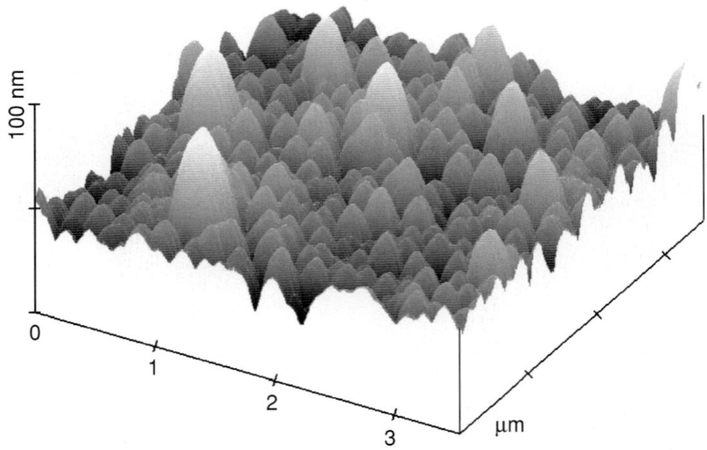

Fig. 1.33. AFM topology on PMMA after treatment with chlorine gas for 1 min

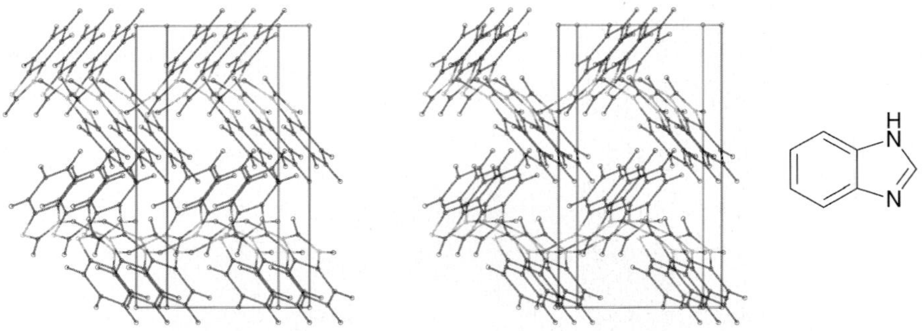

Fig. 1.34. Stereoscopic packing diagram of benzimidazole with the (100) face on top, showing the shielding of the functional groups by the benzene rings

rings of benzimidazole point outward on its overwhelming natural (100) face (Fig. 1.34). That is enough to effectively impede the reaction with cyanogen chloride or other reagents. However, it can be seen that the functional groups are available at the side faces.

These side faces are present at molecular steps on the (100) face as searched and found by contact AFM in Fig. 1.35a. Clearly, the huge islands build up at the molecular $(6+2)$ nm steps and grow from there, but not at the undisturbed parts of the surface. These islands ensue aligned along the molecular steps [77]. Related island formations at molecular steps of thiohydantoins and gaseous amines have been described in [78].

Molecular steps such as in Fig. 1.35 are prominent defect sites for reaction at self-protected surfaces, but there are further nucleation sites at the molecular level that may not always be detectable with AFM on rough surfaces away from lateral molecular resolution. Some examples are defects by missing

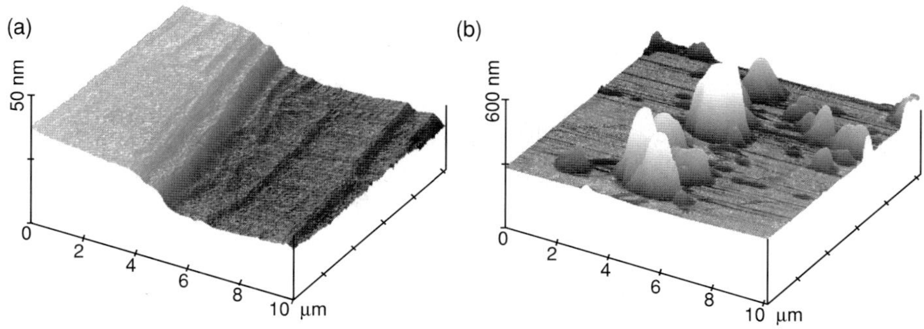

Fig. 1.35. AFM topologies of benzimidazole on the (100) face at a step site: (**a**) fresh surface crystallized from methanol; (**b**) same site after application of 3 ml 20% ClCN vapor in air, showing the development of huge islands at the 6 + 2 nm step site of (**a**)

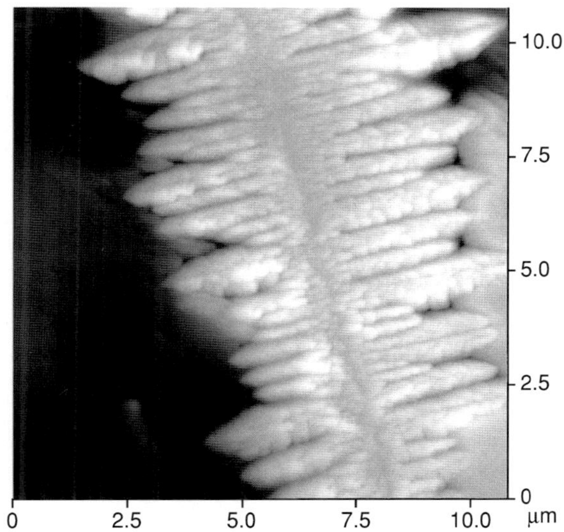

Fig. 1.36. AFM topology of an island structure that was formed on a benzimidazole crystal on (100) along a crack upon reaction with cyanogen chloride; z-scale is 800 nm

molecules, impurity molecules, pits, and cracks. Typically, such defect sites occur in low concentrations on crystal surfaces and prevent uniform reaction on it. An example of an island that spreads along a crack of benzimidazole in the same reaction with cyanogen chloride vapor is depicted in Fig. 1.36. Clearly, the reaction started along the crack of the self-protected single crystal and extended to both sides from there [79].

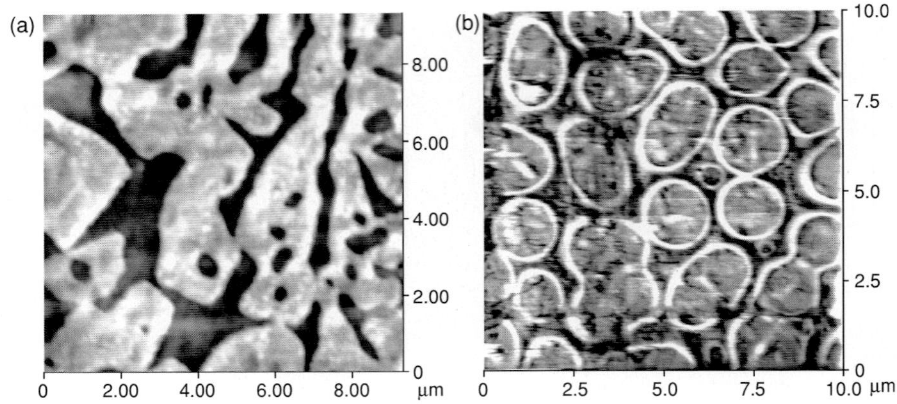

Fig. 1.37. Contact AFM topologies; (**a**) flat islands on sulfanilic acid monohydrate after cautious application of very diluted NO_2 (z-scale: 50 nm); (**b**) flat diked islands of 1,3,5-triphenylverdazyl and triphenylverdazylium nitrate complex by a solid-solid reaction (z-scale: 200 nm)

Low-height flat islands with enclosed "lakes" are obtained upon very minor reaction of sulfanilic acid monohydrate with dilute nitrogen dioxide as shown in Fig. 1.37a [80]. More intense application leads to huge volcano islands. Also diked flat islands can be produced and imaged by putting a small crystal of triphenylverdazylium nitrate on a single crystal of 1,3,5-triphenylverdazyl stable free-radical. Scanning is performed after 20 h at 0.5–1 mm distance from the edge of the crystal contact (Fig. 1.37b) [81]. The solid components quantitatively form a charge transfer complex upon comilling. This complex exhibits interesting magnetic properties [82].

Volcano islands may be created at random as in Fig. 1.38. They are obtained by surface modeling of solid aniline derivatives with nitrogen dioxide or nitrosylchloride gas [83]. An example is depicted in Fig. 1.38. The islands are not symmetrical in this case, probably because hydrogen-bonded ribbons in the crystal packing impede the exit of molecules over the surface.

Volcano islands can also occur aligned in cleavage directions of crystals. Huge regular and blunt islands in Fig. 1.39 are the result of a solid–solid azo coupling exemplifying this. These volcanoes are definitely aligned but not at random [67].

Sharp aligned fence-like islands consist of numerous volcanoes that touch every direct neighbor. The direction of the fences corresponds to the cleavage plane direction of α-cinnamic acid that was exposed to filtered daylight for half a year in a secured environment in order to achieve extremely slow partial photodimerization (Fig. 1.40) [76]. This should be compared with Fig. 1.30 at much more rapid photolysis. The anisotropic molecular migrations over far distances are very evident.

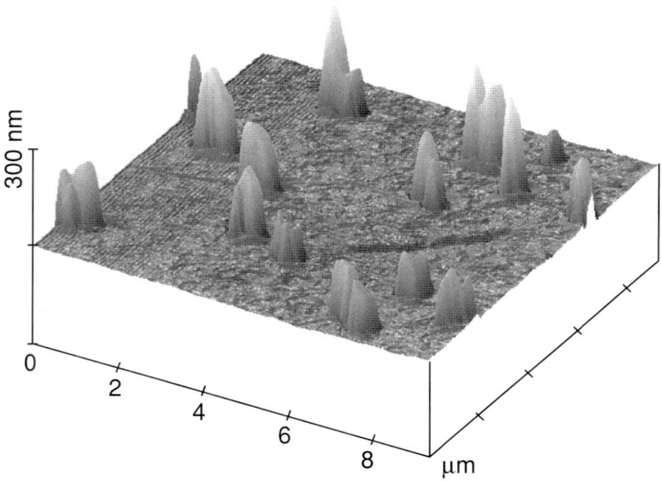

Fig. 1.38. AFM topology on (001) of 4-aminobenzoic acid after exposure to NOCl vapor giving somewhat irregular volcano islands as a consequence of the diazonium chloride formation

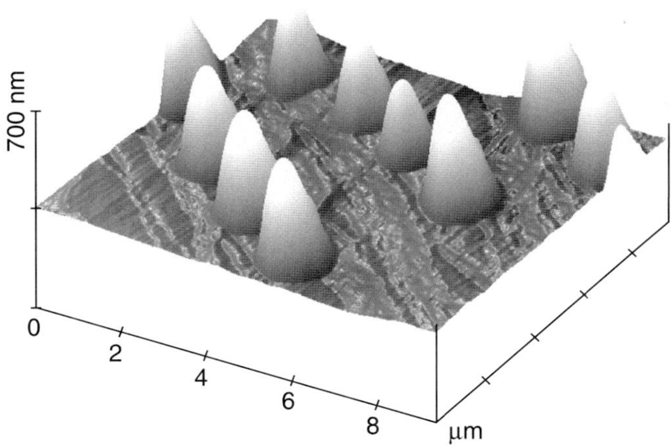

Fig. 1.39. Contact AFM topology of blunt aligned volcano islands that occurred on (001) of 2-naphthol about 1 mm apart from a solid grain of diazotized 2-aminobenzoic acid after 3 h; the cones are oriented along a cleavage plane direction of 2-naphthol

Still another island structure has been obtained in the form of concentric rings upon autoxidation of mercaptobenzothiazole (Fig. 1.41) [67]. Presumably, autoxidation on the self-protected surface started around a hole as the nucleation site. The circles are seen both in the top view (a) and in the surface plot (b), but only the latter indicates the volcano type of the densely packed segments.

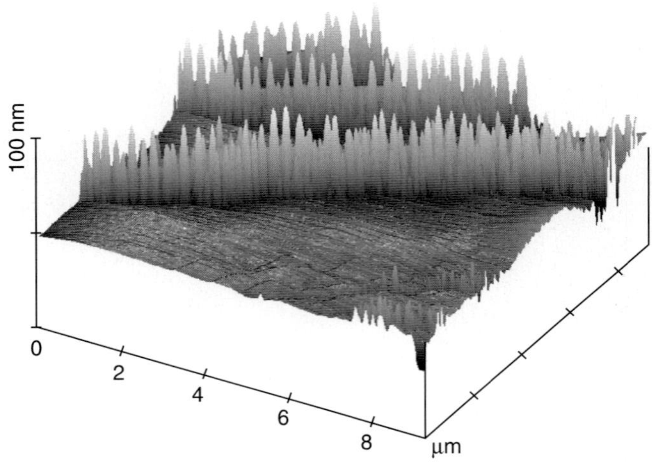

Fig. 1.40. AFM topology of α-cinnamic acid (crystallized from benzene) on (010) after 6 months' exposure to Pyrex-filtered daylight under argon exhibiting fence-like structures aligned along the cleavage plane direction

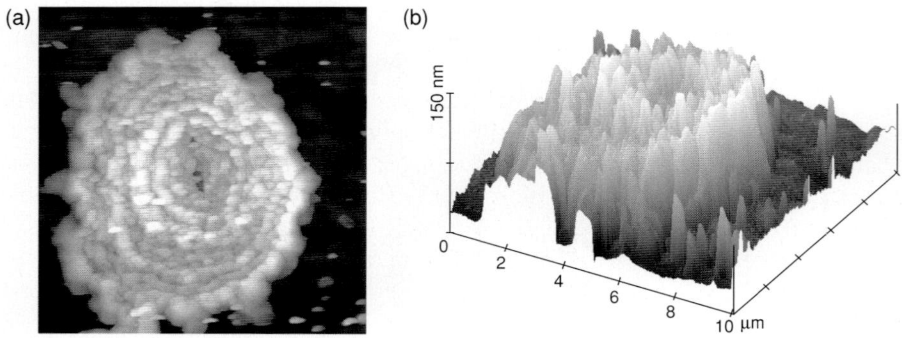

Fig. 1.41. AFM topology on (001) of 2-mercaptobenzothiazole after partial autoxidation in oxygen atmosphere; (**a**) top view (z-scale: 250 nm); (**b**) surface plot, showing circular island formation

1.9.3 Craters

Numerous shapes of what should be called craters exist. Small submicroscopic funnels are easily imaged by contact AFM if their slopes are not too high and if their depths do not reach the length of the tip. Figure 1.42a gives an example for the funnel-type craters at slopes $<20^\circ$ and depths of typically 40 nm. They were created by a solid-solid reaction with AFM scanning close to the contact edge of the reacting crystals in air at the exclusion of solvents. The shapes of these stable craters can only be judged from inverted images, which translate the craters into the inverted volcano shape (Fig. 1.42b) [84]. As there

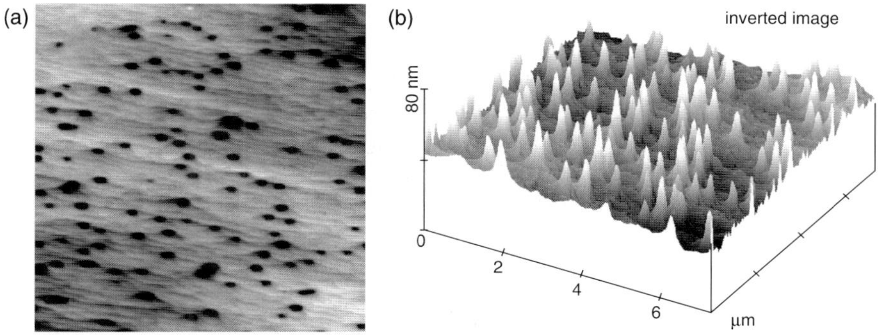

Fig. 1.42. AFM topologies on (010) the (R, R)-cyclohexylspirotaddol after 2 h treatment with solid (S)-pantolactone creating steep random craters; (**a**) top view, (**b**) inverted surface plot for the visualization of the shape of the craters; the slopes are <20°

is local mass and volume increase in solid-solid reactions that usually cannot be compensated by an increase of density the formation of craters must be accompanied by upward transport. This is evident from an increased surface corrugation with relation to the original surface in Fig. 1.42 having the form of shallow hills.

Such funnel-type craters may also assume shallower shapes and very different sizes as in a typical gas–solid reaction without solvent, which produces a linear dimer molecule with shrinking of the volume upon proton catalysis from an alkene. The craters exhibit depths from <20 up to 200 nm and slope angles <10° (up to 15° only at the front left funnel) (Fig. 1.43) [85]. There is no mass increase in the catalytic reaction and apparent density increase, but some compensating upward transport is seen by the hills between the craters.

Crevice-type craters with sharp corners are distinctly different (Fig. 1.44a). The actual shape is only revealed with the inverted image (Fig. 1.44b), which clearly indicates the sharp edges in various directions. The slopes are below 40° with the exception of a few crevice craters that have short parts in north-south orientation at 55°–60°. These few sites are artificially imaged. The craters were formed upon a photolytic rearrangement in the crystal bulk that severely changes the molecular shape. The corresponding upward transport is unspecific at that degree of conversion. Volcanoes and craters are present after 5 min irradiation [65].

In photochemical conversions without mass change crater formation must be accompanied by upward transport if the density change is low. This is more evident when there are larger craters. Quite often, craters and volcanoes of similar size are formed at the same time. Figure 1.45 gives an example from a solid-state *cis–trans* photoisomerization, where the crater rims unite to volcano-like structures [86].

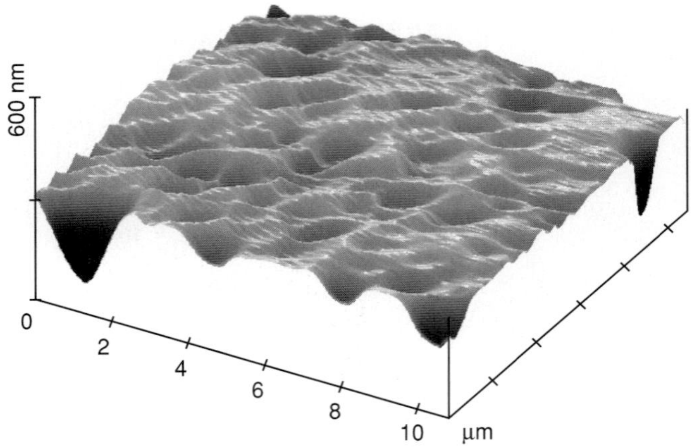

Fig. 1.43. AFM topology of shallow funnel-type craters on the main face of 1,1-dianisylethene after two applications of 0.5 ml gaseous HCl in 5 ml air from a syringe in 3 cm distance

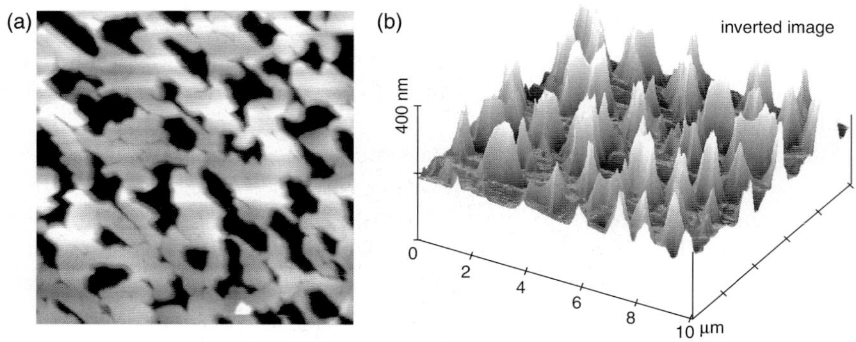

Fig. 1.44. Contact AFM topology on (001) or (00-1) of dicarbethoxy-dibenzobarrelene ($P2_12_12_1$) after 15 min irradiation with a high-pressure Hg lamp through a Pyrex filter producing an enantioselective di-π-methane rearrangement

Even better visible is the combination of large craters and large volcanoes (like egg trays) in the isomerization of ß-*trans*-cinnamic acid (Fig. 1.46) [65]. Further examples for this type have been published [48, 65, 66].

Three different types of large craters with flat bottom can be created. Figure 1.47 shows a rim-type flat crater. All upward transport goes to the rim upon a gas–solid nitration in the absence of solvent [87]. This is a multistep solid-state cascade reaction that also produces water and nitrogen monoxide. Therefore, the relation of the features to the original lattice may be not so direct as in the case of the one-step chemical reactions. However, the molecular packing appears suitable for the circular response.

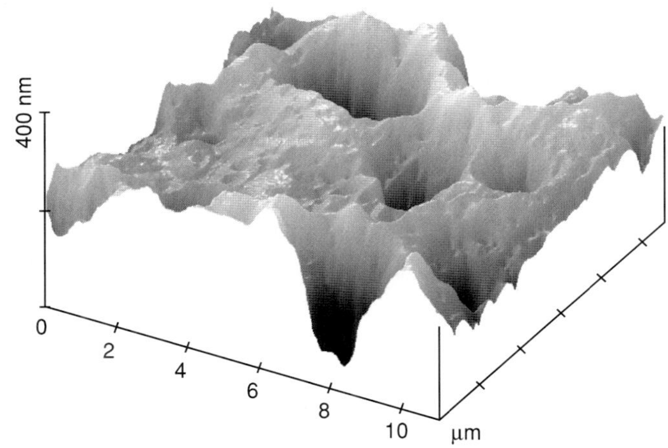

Fig. 1.45. AFM topology on (010) of *cis*-benzylidenebutyrolactone after irradiation at 365 nm (bandpass 6.4 nm; 6 mW cm^{-2}) for 120 min

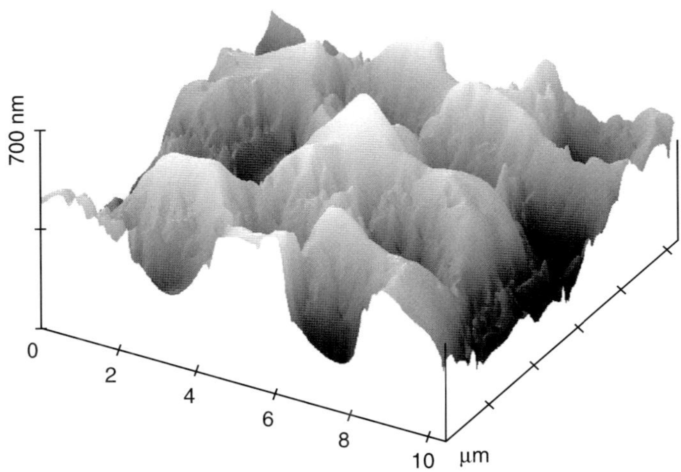

Fig. 1.46. AFM topology on (100) of ß-*trans*-cinnamic acid after 1 min irradiation ($\lambda > 300$ nm) and 20 h rest

The second type of a flat-bottomed crater goes down in concentric steps, the heights of which vary from 1.5 to 11 nm (Fig. 1.48) [81]. These are in the range of a few molecular layers probably of the product molecules from a gas–solid cyanation reaction, as the crater is built down gradually by increasing the number of steps. The initial crystal [88] consists of monolayers without hydrogen bonding. The layers are parallel to the (001) face, and that results in the spectacular shape of the craters.

It is also possible to obtain large craters with flat rims by gas–solid catalysis of linear alkene dimerization, which is the third type of large flat-bottomed

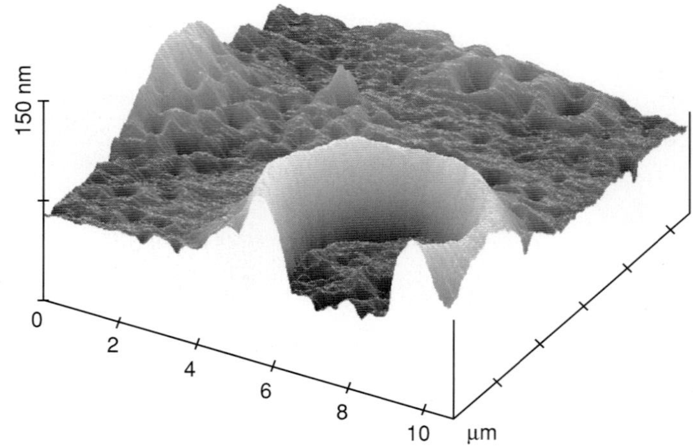

Fig. 1.47. AFM topology on (10-1) of tetraphenylethene after exposure to 0.2 bar of NO_2 for 10 min to give tetra-*p*-nitrophenylethene

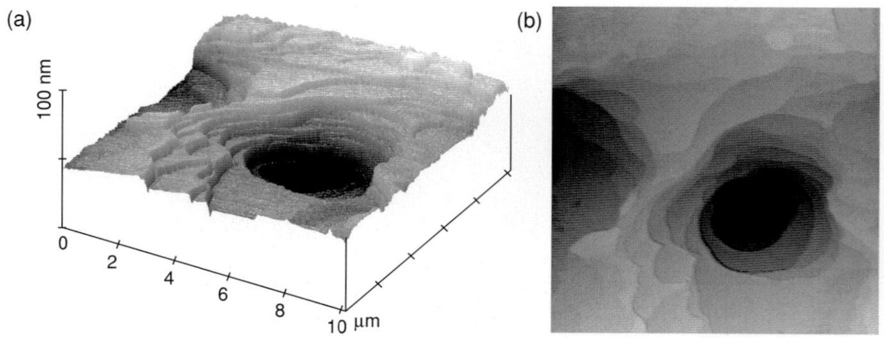

Fig. 1.48. AFM topology on (001) of diphenylamine (*P*-1) after five additions of 0.5 ml diluted ClCN gas in air to give *N*-cyanodiphenylamine and HCl gas; (**a**) surface plot; (**b**) top view (*z*-scale: 200 nm)

craters (Fig. 1.49) [85]. The starting surface was very flat with molecular terrace steps on it. No significant upward transport is recognizable, which is surprising, as it is hard to imagine that all of it can be accounted for by a density increase due to the overall shrinking. The crater width is 8 µm at a depth of 400 nm though. The E/Z-ratio is 15:85 in the 1,1,4,4-tetraarylbuten-1 products, if the same reaction is performed on a preparative scale to completion. The reason can be found in the crystal lattice of the starting material [89]. Its crystallographic density is rather low (1.17 g cm^{-3}). A packing diagram on (001) (Fig. 1.50) reveals numerous small channels orthogonal to it. Even though the layered alkene molecules are interlocked it appears comprehensible that, after the shrinking linear dimerization, the extended product molecule migrates along the channels and fills them, while compression to neighboring

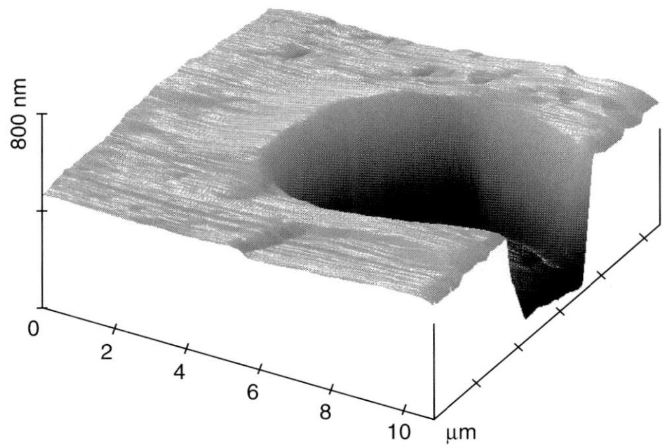

Fig. 1.49. AFM topology on (100) of 1-(4-anisyl)-1-phenylethene (*Cc*) after three applications of 1 ml HCl gas each to catalyze the stereoselective linear dimerization

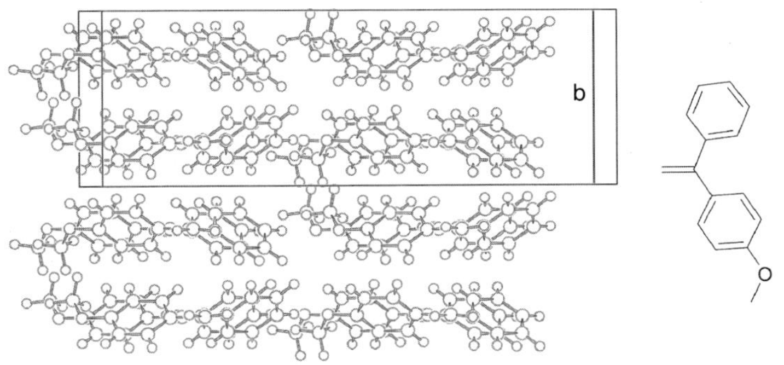

Fig. 1.50. Packing diagram of 1-anisyl-1-phenylethene (*Cc*) on (001) showing channels orthogonal to that face

channels will also occur in the distorted lattice. Unfortunately, we cannot assess the original level of the surface in Fig. 1.49. Some upward and probably downward transport must have occurred, but not concentrated at the rim and edge in the depth of the crater.

1.9.4 Pool Basin Type

If depressions have square, rectangular, or rhomboidal cross sections and flat bottoms these resemble "pool basins." Examples for the latter two are depicted as top views in Fig. 1.51. They were created by gas–solid reaction with the same reactive gas (cyanogen bromide vapor) but two different crystal substrates and are the result of advanced reaction. In the case of Fig. 1.51a the pool basins developed in a sudden event from numerous small flat molecular

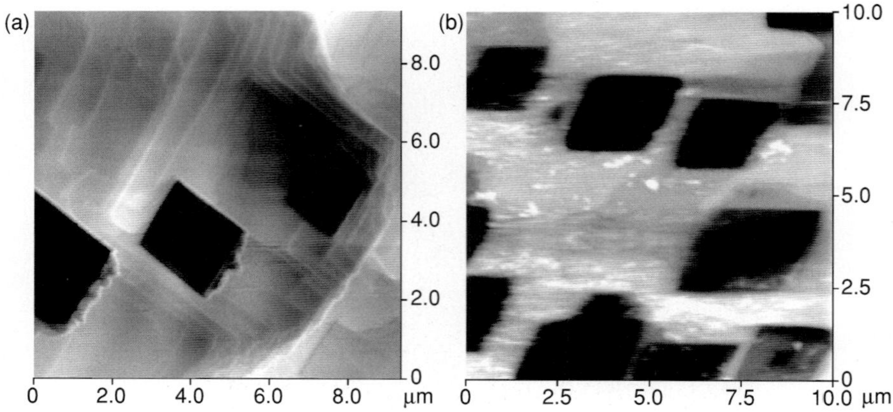

Fig. 1.51. AFM imaging of pool basin-type features; (**a**) generated by reaction on (001) of 2-hydroxyaniline (*Pbca*) with six portions of cyanogen bromide gas, depth <120 nm and (**b**) by reaction on (100) of benzoic hydrazide ($P2_1/c$) with three portions of cyanogen bromide gas, depth 60–100 nm

scale depressions [77]. The packing of the starting crystal for (a) has two rows parallel to the a and b axes. In the case of Fig. 1.51b the reaction started gradually with tall volcanoes up to more than 50 nm, before a sudden event upon continuing reactive gas exposure formed the rhombs [77]. The angle of the rhombs is about 70°. Importantly, the crystal packing of the starting crystal for (b) has two directions of preference crossing at an angle of 72°. The three-step behavior of gas–solid reactions, which is exemplified in both cases, is the common basis of the molecular solid-state reaction mechanisms. These are subdivided into steps of phase rebuilding, followed by phase transformation, and followed by crystal disintegration in the reaction layer of about 100 nm for generation of new surface for the next cycle, etc. [59, 90, 91]. These and many other AFM measurements replaced previous theories that would forbid and deny anisotropic molecular migrations within crystals as a prerequisite of solid-state reactions [92].

Less spectacular almost square pool basins with depths down to 350 nm are obtained with a totally different gas–solid reaction if a previously very flat single crystal of 4-nitroaniline is aged in air for some weeks (Fig. 1.52) [67].

1.9.5 Prismatic Floes

Vertical elongated prismatic features either in one, two, or three different directions have the appearance of floes even if the opposing surfaces are not parallel but inclined, and we actually have more or less straight three-sided prisms. However, in common imaging they usually appear more like floes. Such surface modeling by chemical interaction requires cleavage planes that

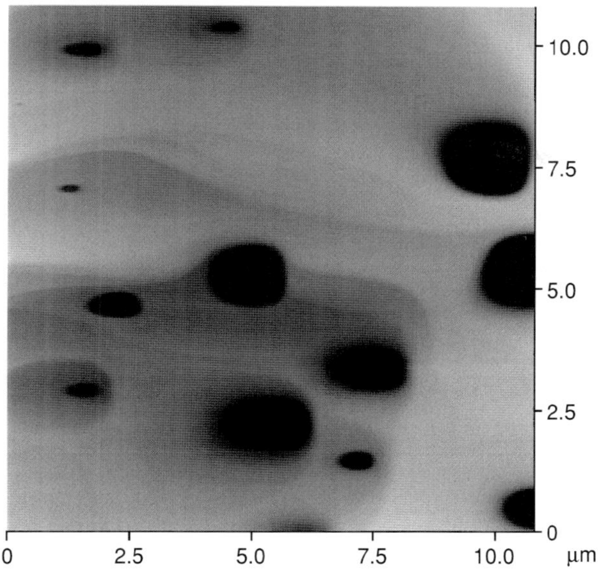

Fig. 1.52. AFM topology on (100) of an aged single crystal of 4-nitroaniline (crystallized from hot water by slow cooling; $P2_1/n$), showing pool-like depressions

end on the surface in question. The first example has been depicted at two different view directions on monoclinic 9-chloroanthracene [93] (unfortunately no atomic positions are made available from the X-ray analysis) in Fig. 1.7. These photochemically generated features align only in one direction, which is the cleavage plane direction. Similar prismatic floes can be generated by a gas–solid reaction of *rac*-camphene (cubic highly disordered plastic crystals) [94] and hydrogen chloride to give the elusive *rac*-camphene hydrochloride with *exo*-position of the chlorine substituent [95]. The AFM result is shown in Fig. 1.53. The slope angles are up to 25° here and the floes have different sizes.

Very steep floes are obtained by the solid–solid condensation of 4-hydroxybenzaldehyde with 4-toluidine. A small crystal of the amine is placed on a single crystal of the aldehyde, and AFM measurement is performed at 0.2 mm distance to the contact edge. The floes in Fig. 1.54 are created by crystal-guided removal of aldehyde molecules for reaction within the amine crystal to form the colored condensation product and water there. The remaining non-consumed aldehyde material is imaged by AFM. Two different directions are active for the aldehyde migrations that relate to the orientation of the layered molecules [96]. The floes in Fig. 1.54 reach heights of more than 1 μm and the slopes of the large ones are found steeper than the angle of the pyramidal tip (> 55°). Two aniline derivatives (4-toluidine and anisidine [96]) are equally active. What is seen in Fig. 1.54 are the remains of an initially smooth only slightly corrugated surface of 4-hydroxybenzaldehyde after its modeling down.

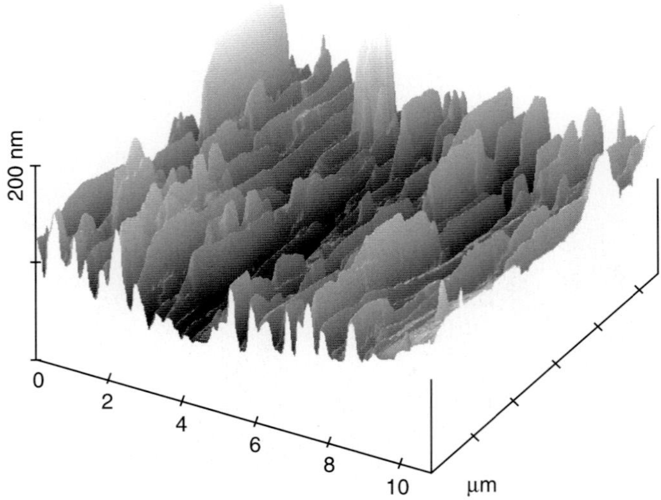

Fig. 1.53. AFM topology of a *rac*-camphene surface after exposure to 0.5 ml of gaseous HCl; scan direction was from left to right

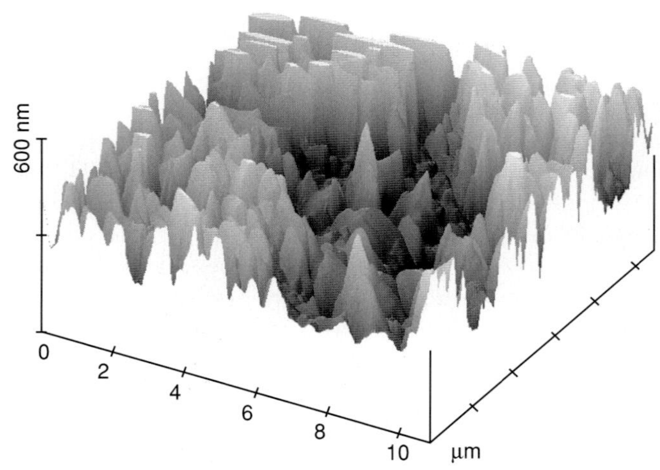

Fig. 1.54. AFM topology on (010) of 4-hydroxybenzaldehyde after placing a crystal of 4-toluidine (m.p. 45°C) with the edge at 0.2 mm from the scan site 10 min before measurement; chemical reaction occurred only in the 4-toluidine crystal, which is out of the scan range

The achieved steepness (>55°) appears to be not reachable by building up of pyramidal floes.

The prism structure is very evident in the solid–solid reaction of the same aldehyde with 4-aminophenol (Fig. 1.55) [79]. The mechanism is different here.

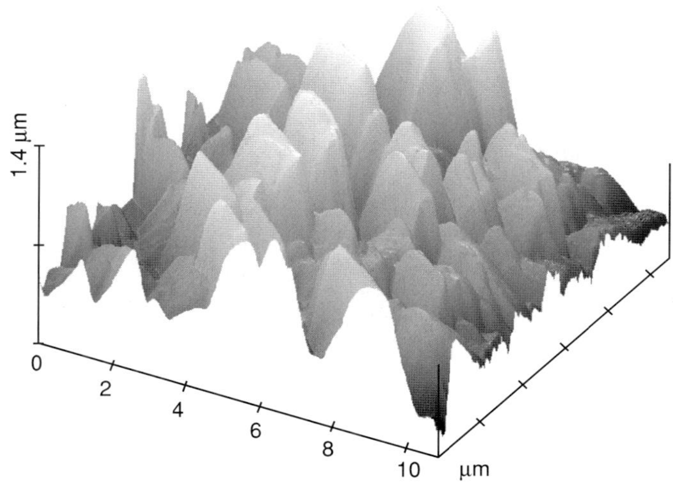

Fig. 1.55. AFM topology on (010) of 4-hydroxybenzaldehyde after placing a crystal of 4-aminophenol at 0.2 mm from the scan site

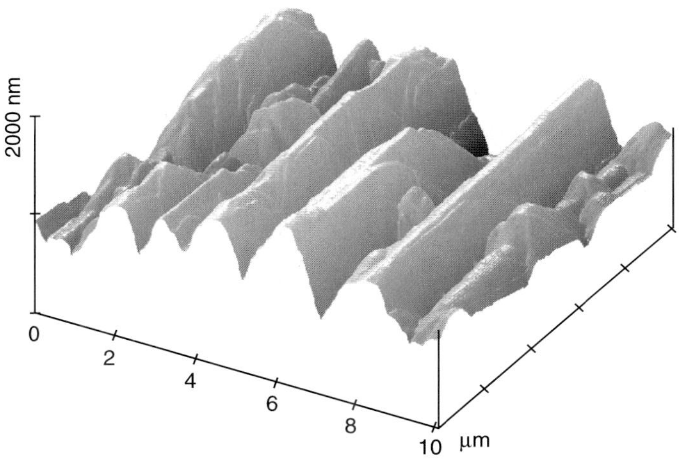

Fig. 1.56. AFM topology on the turbid main face of 1-cyano-5,6-dimethylbenzimidazol after crystallization from methanol, showing huge uniform prism features

It is also possible to create straight three-sided parallel prisms of considerable height by crystallization of a 1-cyanobenzimidazol derivative [77] (Fig. 1.56). One obtains a highly corrugated surface of the type that is stable toward methylamine gas (no surface change).

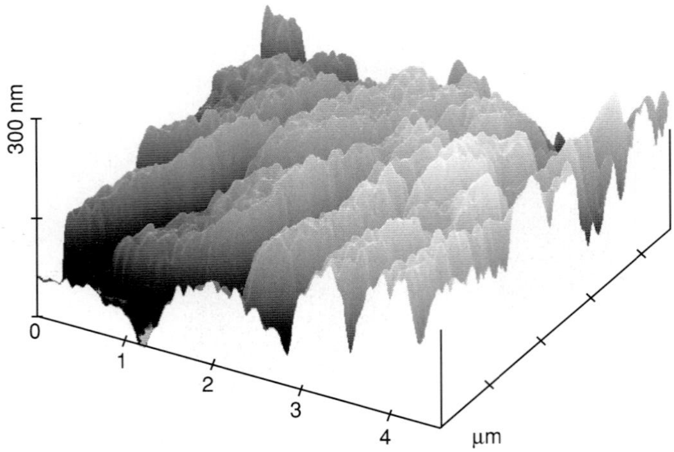

Fig. 1.57. AFM topology on (110) of thiohydantoin ($P2_1/c$) after treatment with methylamine gas

1.9.6 Heights and Valleys

The heights and valleys type differs from the prism type by less regular and extended topology. Usually, segments subdivide the gross features. The directions may be parallel or variant. Figure 1.57 exhibits almost parallel directions of the heights and valleys. The segments are of the volcano type although largely merged with their neighbors [78]. Actually, cleavage planes end at that surface of the original crystal. They are inclined by 66° in the example of Fig. 1.57.

While chemical interaction of tetraphenylethene with gaseous nitrogen dioxide on (10−1) produced rim-type flat craters upon tetranitation (Fig. 1.47), the reaction with bromine gas gives tetrabromination [85] with formation of heights and valleys (Fig. 1.58). This indicates that the crystal structure of the starting material does not alone determine the surface modeling in the case of multiple solid-state cascade reactions. A difference in the chemical reaction mechanisms is autocatalysis by the byproduct hydrobromic acid only in the bromination case. Nevertheless, the feature-forming process can be related to the crystal packing of tetraphenylethene ($P2_1$), which has molecular rows in three almost orthogonal directions [87].

The heights and valleys feature type is not restricted to parallel alignment. The gas–solid chlorine addition in Fig. 1.59 exhibits valleys in three directions. And they are not all connected to each other. Furthermore, the heights consist of very densely packed or merged volcanoes of considerable height [79, 97]. Unfortunately, no crystal structure determination of the starting material is available for a more detailed analysis.

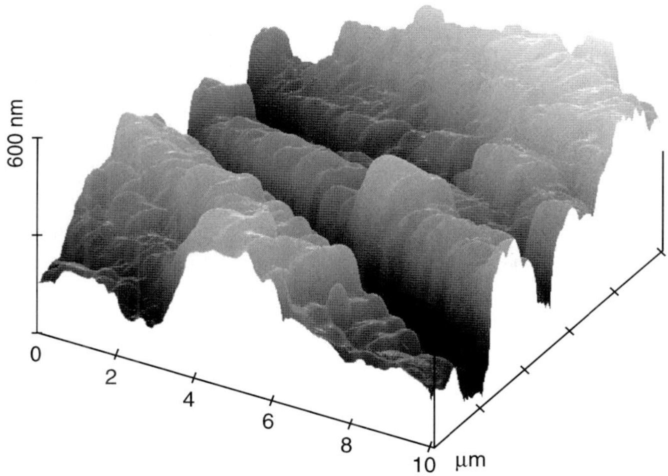

Fig. 1.58. AFM topology on (10–1) of tetraphenylethene after 10 min application of bromine vapor to give tetra-*p*-bromophenylethene and hydrogen bromide

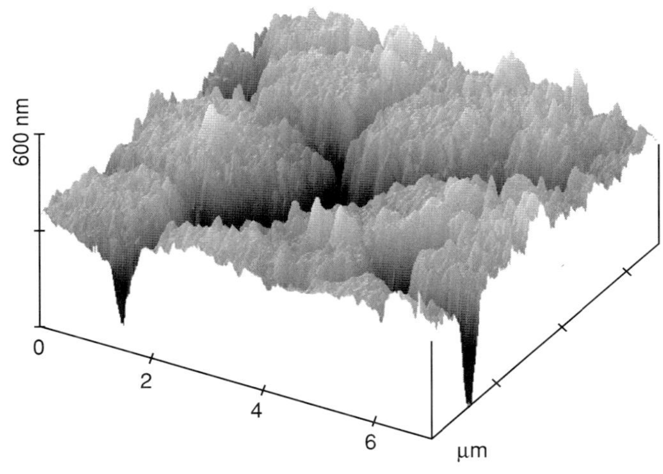

Fig. 1.59. AFM topology on the main face of 5-benzylidene-thiohydantoin after exposure to 1 ml Cl_2 from a syringe in 3 cm distance in air

It is not always easy to differentiate the nonparallel heights and valleys type from the craters and volcanoes type (Sect. 1.9.3, Fig. 1.45). But the features in Fig. 1.60 tend more to the shape of this section. These were produced by huge surface changes in the phase transformation step of a gas–solid diazotization [97]. The preceding gradually increasing features were volcanoes (not shown). Again merging segments are recognizable.

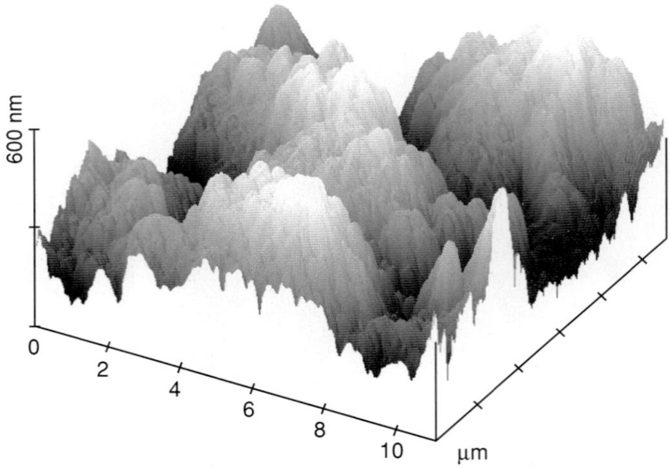

Fig. 1.60. AFM topology on (101) of 4-aminobenzoic acid after extended (10 ml) application of gaseous nitrogen dioxide to give the corresponding diazonium nitrate; the features shown are the result of a sudden phase transformation event

The heights and valleys type may also be recognized in diamond-knife microtome cuts of adhesive tooth–biomaterial interfaces with various cements in dentistry. These determine the lifetime of tooth restoration and are the result of the alignment of composite materials [98]. They are described in more detail in Sect. 1.11.1.

1.9.7 Fissures

Fissures occur in various appearances, usually along cleavage planes. Dibenzylidenecyclopentanone has such a cleavage plane and its direction is shown in the crystal packing diagram of Fig. 1.61. The vertical cleavage plane ends at the (001) crystal surface.

This cleavage direction along the *b*-axis is clearly present in the AFM topology on (001) after photochemical reaction of the crystal. However, a mixed feature type occurred in the photolysis reaction, which gave three different products [99]. The original surface had only corrugations within a few molecular layers. The fissures align along the *b*-direction of the original crystal. The nonparallel fissure in Fig. 1.62 (top right) is probably the result of a crystal defect. Evidently, there must be overall shrinking upon reaction that enforces the fissures in nearly constant distances. Additionally, molecules erupt from the beginning through the surface and release internal pressure that is created by change in molecular shapes as a result of two different dimerization reactions and one competing trimerization reaction [99].

Parallel fissures on a highly corrugated surface are also observed upon more extended irradiation of α-cinnamic acid (Fig. 1.63). In that case there are also

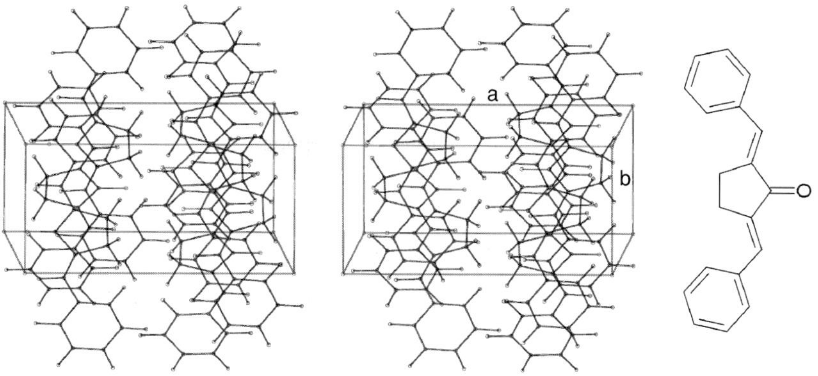

Fig. 1.61. Stereoscopic packing diagram of dibenzylidenecyclopentanone ($C222_1$) on (001) rotated around x by 5° for a better view, showing molecular layers and the slightly interlocked cleavage plane ending there

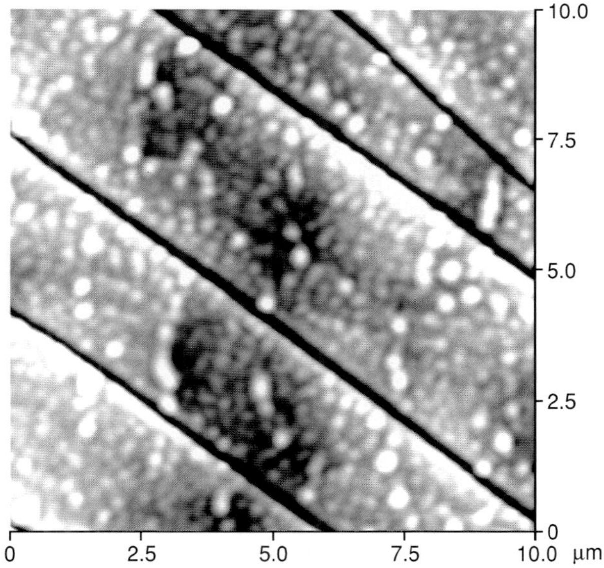

Fig. 1.62. AFM topology on (001) of dibenzylidenecyclopentanone after 15 min irradiation through a Solidex filter; the z-scale is 200 nm

ridges at one side of the fissures along the direction of the vertical (10–1) cleavage plane that cuts the c-axis at an angle of 40° [76]. This is already the phase transformation stage of reaction following the gradual build-up of volcanoes such as in Fig. 1.30.

Fissures may also appear in less regular form as on an aged single crystal of 2-aminophenol in Fig. 1.64. These fissures are broader and not uniform in length and width. This measurement stresses again the necessity to look for

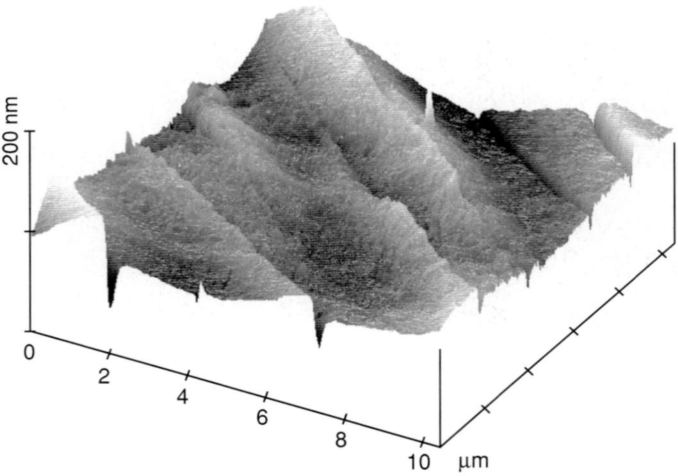

Fig. 1.63. AFM topology on (010) of α-cinnamic acid after 50 min irradiation at 365 nm; the features are the result of the phase transformation step of reaction

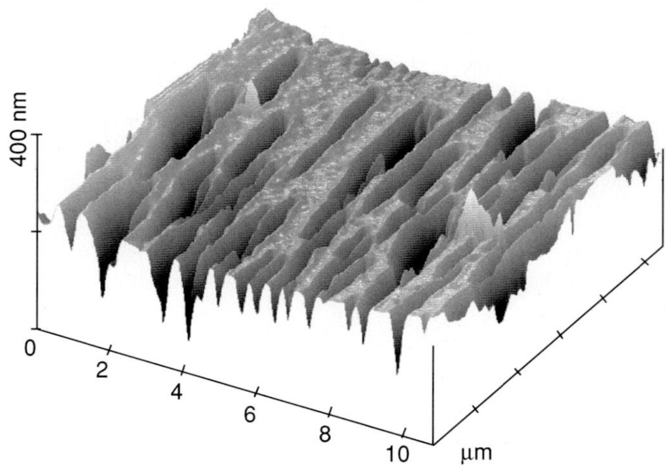

Fig. 1.64. AFM topology on (001) of an aged single crystal of 2-aminophenol, showing broad fissures

submicroscopic features with contact AFM prior to any other surface investigation.

1.9.8 Bricks

Brick-like structures with sharp rectangular edges are another feature type of surface modeling with solid-state chemistry. Thus, the impressive blocks in Fig. 1.65 are built upon a photochemical *cis–trans* isomerization on a crystal

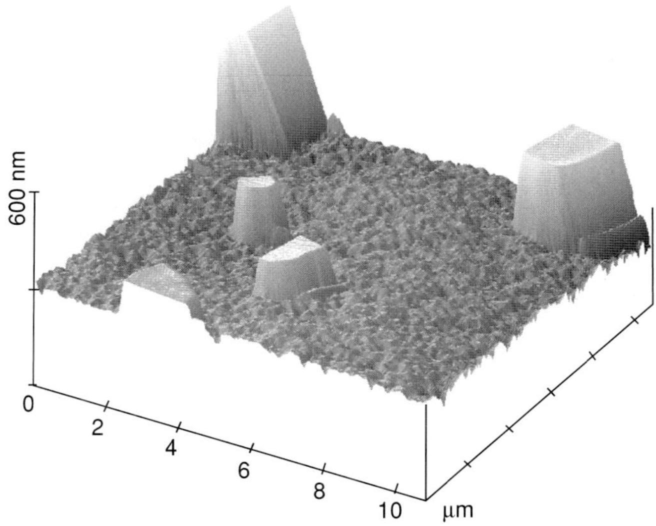

Fig. 1.65. AFM topology on (010) of *cis*-benzylidene butyrolacton after 45 min irradiation at 365 nm (6 mW cm^{-2}) to form the *trans*-isomer

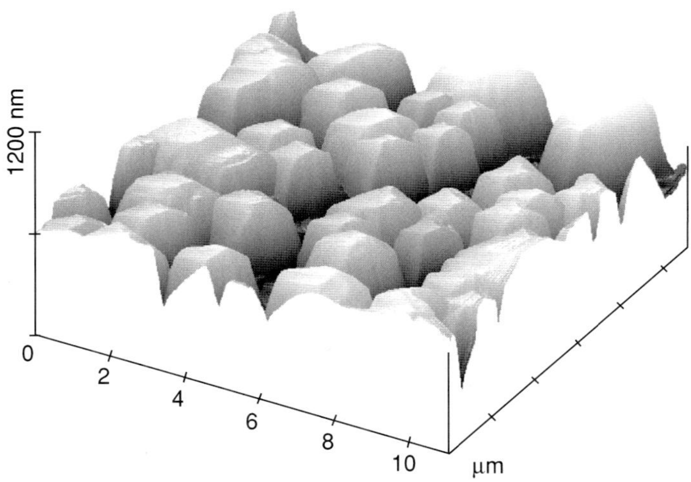

Fig. 1.66. AFM topology on (001) of 2-naphthol (*Cc*) (ref. [100]) with a 4-bromobenzenediazonium nitrate crystal on it at about 1 cm distance after 4 h

without melting [86]. This is the phase transformation state after formation of the product phase from the distorted rebuilt phase.

The sharp-edged bricks in Fig. 1.66 were produced by a solid-solid azo coupling between 2-naphthol and a diazonium salt crystal on it [83]. AFM scanning was performed at about 1 cm distance from the edge of the diazonium

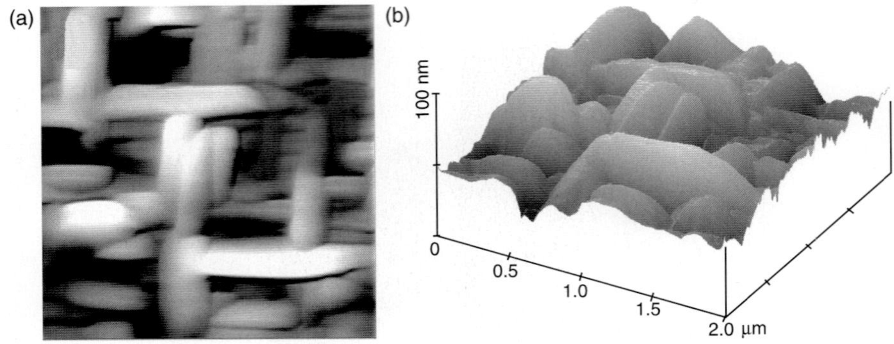

Fig. 1.67. AFM topology on (010) of sulfanilic acid monohydrate after exposure to 0.8 ml NH_3 on the AFM stage; (**a**) top view; (**b**) surface plot, showing submicron brick-like features forming a network at right angles

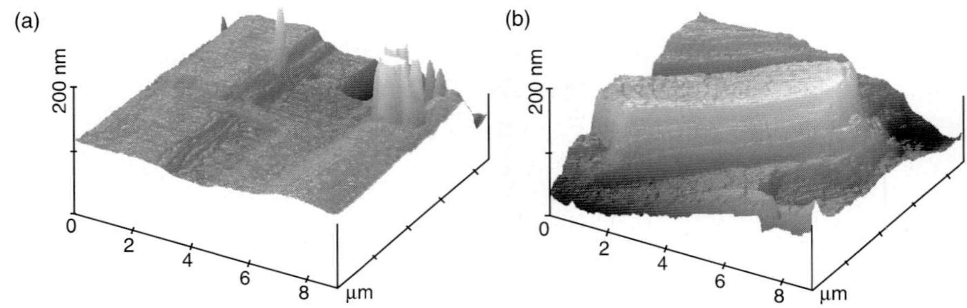

Fig. 1.68. AFM topology on (010) of monolayered diphenylmethanol ($P2_12_12$) (ref. [101]); (**a**) fresh, (**b**) after slow application of NO_2 gas (six portions) with a syringe from 2 cm distance

salt crystal. This surface modeling provides high blocks that are not really "islands" and neither are they "heights and valleys." We must ensure that these features are not artificial in the sense described for small cubes in [12], because they are much larger than the tip dimension.

Brick-like features may interpenetrate at right angles in a gas–solid salt formation. This is shown in a top view and a surface plot in Fig. 1.67, even though the edges are somewhat rounded here. But contact AFM has no problems with that scan. It should be noted that the reagent and the product also play their role in the shape of the surface change. This is the reaction of sulfanilic acid monohydrate with ammonia. If nitrogen dioxide is applied to sulfanilic acid monohydrate, it results in the islands in Fig. 1.37a or large and high islands. Huge bricks can be the result of phase transformation upon gas–solid reactions. Figure 1.68 shows the result of the chemical interaction of nitrogen dioxide with the smooth but obviously not very stable natural

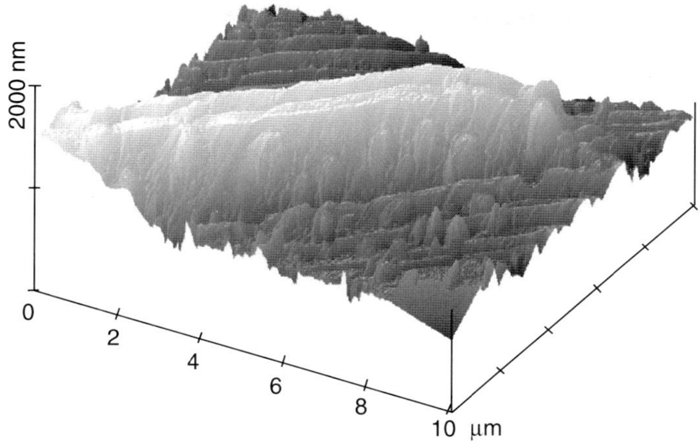

Fig. 1.69. AFM topology on (001) of *rac*-3-carboxy-2,2,4,4-tetramethylpyrrolidine-oxyl after five applications of XeF_2 vapor for 30 s each, forming the corresponding nitrosonium fluoride; scan direction is front to back

surface of a single crystal that is totally changed and easily scanned with contact AFM [79]. This is a result of the phase transformation step.

A gas–solid single electron oxidation of a stable nitroxyl radical with xenon difluoride provides a similar but about 10 times higher large brick after repeated applications until the phase transformation stage occurred following the phase rebuilding stage with volcanoes along the major crystal axis (Fig. 1.69) [87]. Thus, similar features can be constructed with totally different reagents.

More frequently, bricks appear as conglomerates of differently shaped species as in composite crystals or at head sites of large prismatic crystals with well-developed prism faces. Also such important details can be successfully scanned by contact AFM if proper mount to the AFM stage succeeds (Fig. 1.70). These are, however, not the aim of purposeful surface modeling.

All of the reported examples of surface modeling by self-organized solid-state chemical reactions are strictly guided by the crystal packing on the corresponding faces and this has important consequences for the dynamics in molecular crystals. They are now exhaustively studied with organic molecular crystals and a few inorganic salts. Similar feature forming with polymers have also been addressed. Furthermore, it is expected that proper inorganic crystals with their cleavage planes will behave similarly in solid-state reactions. The patterned surfaces might be used for nanoscopic molds when fixed by metallization or related vapor deposition techniques for reproduction purposes in technical applications. The use of crystal packing for self-assembling is very efficient and reliable.

quartz waveguides with core diameters of 3.4 or 5.7 μm and 125 μm step index coating at lengths of 60–120 cm are suitable for sharp pulling. The tested optical fibers were coaxial and did not preserve polarization. As-sharp-as-possible pulling is performed by applying a commercially available micropipette puller (e.g., from Sutter Instruments). End radii of curvature <10–20 nm with a very high aspect ratio are achieved by proper use of pulling parameters. Such a tip is depicted in Fig. 2.2c. The mechanical resonance frequency of the probe is between 100 and 400 kHz, the Q-factor 70–160. The probes are driven to a free oscillation amplitude of 5 nm, which is set to 30–70% damping depending on the sample for maintaining constant shear-force distance. Less damping lacks in sensitivity, more damping will rapidly abrade the sharp tip and decrease the resolution. It should be possible to use etched metal wires as in STM and metal or silicon nanowires in shear-force AFM, and these would provide still higher aspect ratio and more flexibility. The thus-gained higher stability is highly desirable. Flexural stiffness characteristics of nanowires have been investigated [36, 106].

Damping means reduction of the oscillation amplitude due to interaction between the tip and the sample. The mechanism of damping is still a matter of debate. It is certainly not sensitive to surface corrugations as initially proposed [15] but relies on frictional lateral forces that are called shear forces. Various influences are at work. Van der Waals' attraction will certainly influence the vibration amplitude of the tip. In ambient air gas molecules surrounding the tip in the shear-force gap will be limited in their motions and this effect is called hydrodynamic damping. However, these influences do not seem to be sufficient for (very) rough surfaces, which require stronger effects if the vibrating tip is to survive at reasonably rapid scan rates. The van der Waals' forces alone appear insufficient. As the tip comes too close to the surface similar to contact AFM the tapered vibrating tips will break upon approach or abrade upon scanning even if these have a water layer. Usable shear-force damping forces are provided by water layers on polar surfaces (including oxide layers on metals or thiol monolayers on gold), and merging of these layers with the water layer on the silica tip may be important for the mechanism. The water layer on the silica tip may also be of importance on surfaces without such layer requiring closer distance for the same damping (Table 2.1, Sect. 2.5). Also dangling bonds on organic molecular crystals (for example by free side groups or adsorbed gases) may add to the mechanism. A strong indication for these mechanisms is the sudden appearance at a definite distance and good performance only at ambient conditions [107]. The actual distance in the shear-force gap is preferably in the 5 nm ranges and can be changed by the preset damping percentage (30–80%). Fortunately, sufficient shear force is available in most samples because lack of water layers is not very common. We do not discuss here the special problems of shear-force damping in ultrahigh vacuum (UHV) [108].

There are great similarities between shear-force AFM and contact AFM. The reachable lateral and vertical resolutions are better than 1 nm and 0.1 Å, respectively. The resolution of molecular steps on corrugated surfaces is also possible in shear-force mode (e.g., Fig. 2.34a,b). The lateral resolution according to the Rayleigh criterion on rough surfaces is better than 1 nm with shear-force tips of 10 nm and commercial contact tips of 5 nm radii of curvature. However, such lateral resolution capabilities are often not used in large-scale (>1 μm) scans on very rough surfaces, and atomic resolution is not the aim of common shear-force AFM. Shear-force mode may also provide tip imaging if the surface features are taller and sharper than the tip apex, but this rarely happens with very sharp uncoated tips. Such tip imaging by surface features becoming the actual probes is more frequent with blunt, abraded, or broken tips. For example, Fig. 1.73 displays the imaging of a badly broken tip by nanoparticles of 100–200 nm width. It is clearly seen that the tip shape occurs repeatedly and in the same orientation. The different sizes relate to different height position of the nanoparticles. Proof for tip imaging is given by the change of feature orientation with changed scan direction but not if the sample is turned while maintaining the scan direction. It would also change in the case of true feature imaging. Further examples for tip imaging with differently shaped defective tips are shown in Figs. 2.18a, 2.21a, and 2.22a (Chap. 2). There is no difference in behavior with contact AFM in that respect and good measurements are only possible with very sharp probe tips.

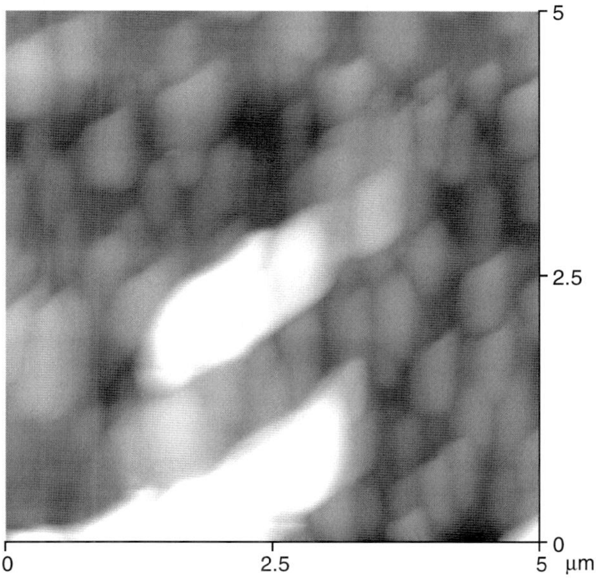

Fig. 1.73. Shear-force AFM (z-scale: 200 nm) imaging of a badly broken tip by nanoparticles with diameters of 100–200 nm that partly aggregate and swim up in a solid resin with 5 nm coverage; the tip shape repeats in shape and orientation giving much larger features than the nanoparticles

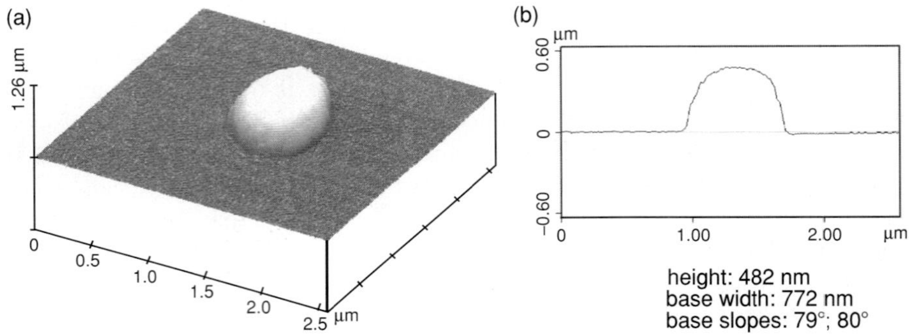

Fig. 1.74. Shear-force AFM image of a 480 nm diameter polystyrene-*co*-acrylamide bead on mica at equal x, y, z scales showing good performance of the probe tip; (**a**) surface plot; (**b**) cross section through the center of the bead

The main advantages of shear-force AFM are the ease and economy of tip pulling and the very high aspect ratio. Therefore very steep slopes and deep structures can be better scanned. This can be shown by the measurement of beads and porous silicon. Figure 1.74 illustrates the performance of the shear-force AFM with a polystyrene-*co*-acrylamide latex particle on mica with nominal radius of 240 nm [109, 110].

Figure 1.74a shows the surface plot with a uniform x, y, and z scale. The section in Fig. 1.74b has the correct height and the circle form on top. Of course, the baseline is enlarged and the base angle is not 90° but is very good at 79° and 80° due to tip–sample convolution with the vibrating tip with about 10° opening angle. Clearly, the final slope of the convolute would be expected at about 45° or smaller with an Si_3N_4 pyramidal tip (cf. Fig. 1.13). The excellent performance of shear-force AFM despite the tip vibration can be favorably and highly economically used for improved measurement of extremely deep steep features.

Porous silicon is an important object for AFM characterization. For example, porous n-type and p-type silicon has been anodically etched (100 mA cm^{-2}, 20 min and 5 mA cm^{-2}, 30 min, respectively) in 48% HF and 95% ethanol (50:50 by volume) to obtain porous silicon measured in ambient air by contact AFM [111]. The features were not very high. The n-type sample resulted in shallow depressions 1 µm wide and "pores" of 130–170 nm width with walls of 40 nm width, the total z-range being 85 nm. The p-type sample gave hills not higher than 2 nm and corrected smallest width of 1.5 nm [111]. Unfortunately, the usually very rapid formation of a silica layer in air was not addressed, neither was the "porosity" evident with such small features.

We were interested in measurements on rough porous silicon with defined layers on it. The investigated sample was coated with 50 nm silica on silicon. A demonstration of the superiority of shear-force AFM for such task is available by comparing with the result of standard contact AFM. Clearly, AFM mea-

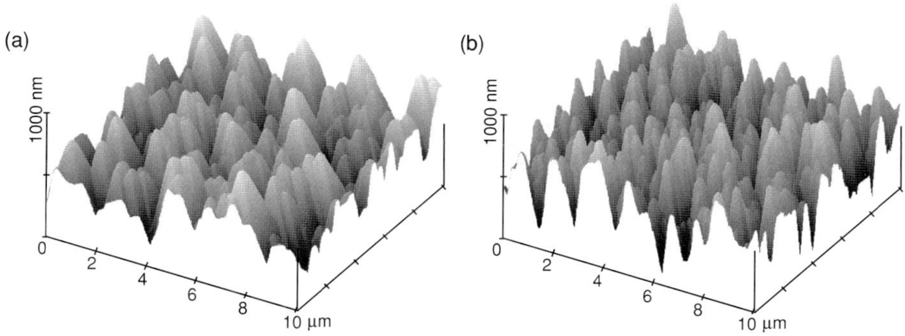

Fig. 1.75. AFM surface images of a sample with 50 nm silica on porous silicon; (**a**) contact AFM using a Si_3N_4 tip; (**b**) shear-force AFM using a vibrating silica fiber tip with high aspect ratio at a scan rate of $50\,\mu m\,s^{-1}$; the contact AFM image is totally artificial (only tip imaging), whereas the shear-force image largely depicts topology with penetration depths exceeding 1,000 nm, but with artifacts in the depths

surement of extremely rough surfaces cannot be achieved with silicon nitride tips in contact mode, as tip imaging occurs at the smaller, densely packed features. The pyramids with rather sharp tops are clearly imaged in the surface plot of Fig. 1.75a. It even shows the inclination of the tip (10–15°) under the tilted cantilever. A much better result is obtained when using a pulled silica tip with about 15 nm "radius of curvature" and very high aspect ratio in shear-force mode, despite the horizontal tip vibration (Fig. 1.75b). The column-type features stand vertical and look the same from all directions. There are also artifacts in the depths, of course. This could also be demonstrated by an inverted image (not shown). Again, perspective surface plots are essential for a realistic interpretation that could not be obtained by top view projections.

Even more impressive is the shear-force AFM on a sample of porous silicon (nominal pore depth about 2 μm and 1–2 μm diameter) coated with 50 nm SiO_2 and 30 nm Si_3N_4 (the sample was kindly provided by Dr. M.J. Schöning of Forschungszentrum Jülich). Two images taken at a scan rate of $70\,\mu m\,s^{-1}$ demonstrate the reliability. The time between the measurements was 3 min in Fig. 1.76a,b. Differences are hardly recognizable in the two images (a) and (b). Thus, the sharp tapered tip of fused silica was not abraded or broken by the measurement under these extreme conditions and the maximal heights match with the pore depth. Such high efficiency is the result of the very rapid shear-force response by the light-sensing control that could not be obtained by any of the other available techniques. Still sharper stiff probes would scan the very depths even better in shear-force mode.

Numerous steep (45–55°) and high (>1 μm) features have also been imaged with shear-force AFM in Fig. 2.32a on irradiated α-cinnamic acid. There is a high level of requirement for shear-force AFM in composite materials with very

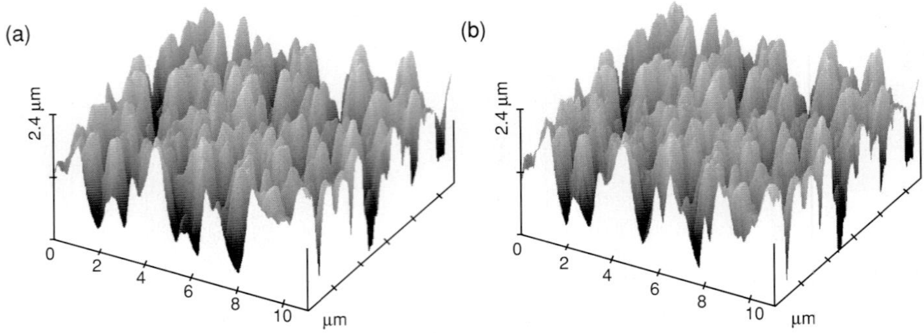

Fig. 1.76. Shear-force AFM on porous Si that is coated with 50 nm SiO_2 and 30 nm Si_3N_4; (**a**) first measurement with height variation of 2.38 µm; (**b**) second measurement with height variation of 2.52 µm, both showing deep descent at the very high roughness

distant surface levels as these may easily occur in microtome-cut preparations due to different elasticities of the different media. Examples that could not have been scanned with standard Si_3N_4 tips are given in Chap. 2 (Figs. 2.70–2.73). The dyed fabric fibers are so much down under the resin surface level that the cantilever would have slid over the rim similar to Fig. 1.9 (Sect. 1.5).

Further development of shear-force tips with high stability and extreme aspect ratio for very deep structures appears possible for the demands of depth gauging and AFM topology at the bottom. Piezos for 200 µm × 200 µm scans with height capabilities of 15 µm are commercially available (for example, the DualScope™ Scanner DS 95–200). Their height capabilities can be profited from by the simply pulled silica tips but not by common cantilever tips. These capabilities are required if living cells as a whole are to be studied with high resolution. For example, there is the long available possibility for the reliable and repeatable imaging of a living somatic cell in culture (therein better with hydrophobized tip) during mitosis that requires a height compensation range of about 20 µm as had been done in [42] seven years ago. Importantly, shear-force AFM with the possibility for very high aspect ratios is much simpler and appears able to favorably compete with the stressed metal probe contact AFM even when it could be adapted to the Milliscope (Sect. 1.3). Large range and very high AFM capabilities will become very powerful tools in many fields such as cytology, histology, microelectronics, micromechanics, micropatterns, etc.

1.11 Further Fields of Practical Application

AFM on (very) rough surfaces has important applications in biosciences and pharmacy. Numerous examples are also reported in Chap. 2 (Sect. 2.10). For example, microtome cuts of tissues are usually very rough and can be studied by shear-force AFM in combination with SNOM for cytological and histological investigations, including cancer prediagnosis. Further fields of application are in polymers, ceramics, mineralogy and geology, metallurgy and corrosion, catalysis, forensics, ancient history, to name a few.

We concentrate here on the high-topology aspects, but not on atomic resolution and other aspects that are not directly related with topology.

1.11.1 Biology and Medicine

AFM imaging of live and fixed cells in culture remains an important field of research [112]. Cell increase or change in shape under the action of drugs has been successfully scanned with contact AFM also under physiological conditions. For example, red blood cells change from a discoid-like shape to a very irregular echinocyte with a lot of ledges under the action of sodium chloride solutions. Similarly, the surface morphology changed upon application of a hypertension drug [113]. Further drugs have been applied to red blood cells and their morphological changes monitored with AFM suggesting a new strategy to describe and understand blood pathologies [114]. Acrosome-intact and acrosome-reacted human sperm were differentiated on a cover slip by contact AFM. Head length at half-maximum height was significantly decreased from 3.56 to 2.99 µm in acrosome-reacted sperm [115]. The AFM was used to investigate the lamellipodial extension as triggered by the application of epidermal growth factor (EGF). An increasing height from about 500 nm to typically 600–800 nm together with a decrease of the elastic modulus by a factor of 1.4 was observed [116]. In a similar AFM study the total cell volume of epithelial canine kidney cells increased after blockade of the Ca^{2+}-sensitive K^+ channel by 17%, while its activation by intracellular elevating of the Ca^{2+} concentration decreased the cell volume by 19%. Hence, local effects by the polarized distribution of K^+ channel activity can be traced [117]. Living endothelial cells of bovine aorta increased their volume upon application of the hormone aldosterone [118]. Related studies deal with the topological imaging of the destruction of bacteria by antimicrobial agents. For example, cell lysis could be imaged by AFM. Staphylococcus aureus NCTC 4163 was exposed to ovine-derived antimicrobial peptides [119]. The action of the antimicrobial peptide PGLa on live Escherichia coli was followed by AFM. It proceeded in two stages: by loss of bacteria topologic features and further damage with total cell rupture, but Mg^{2+} ions partially inhibited these events [120]. The antibiotic daptomycin changed the topology of Bacillus cereus by flattening and shrinking of cells and leakage of cytoplasm through the membrane. Furthermore, the destabilization of flagella was also observed [121]. A detailed image of the amyloid aggregation forming plaques in Alzheimer disease is emerging due to topologic AFM investigations for a better understanding of the pathogenesis holding promise for the future testing of potential therapeutic drugs for interfering the fibril assembly [122–125].

Diamond-knife microtome cuts of adhesive tooth-biomaterial interfaces with various cements in dentistry have been studied to judge the lifetime of tooth restoration for various materials [98]. AFM is the tool of choice for elucidation of the interfacial interaction between adhesive resin, hybrid layer, resin tag, and unaffected dentin. Two photopolymerizable methacrylate resin-based adhesives and one resin-modified glass–polyalkenoate cement were applied to horizontally exposed dentin surfaces of freshly extracted human third molars following the manufacturer's instructions [98]. The 15 μm scans covered z-ranges of 5 μm and showed heights and valleys between the different zones of varied hardness. Furthermore, enlarged scans of the hybrid layer (z-scale: 500 nm) exhibited collagen fibril arrangement (either parallel or loosely organized). The collagen fibrils were found thicker than the normal 80–100 nm. Resin tags were formed in the dentin tubules. The latter firmly attached the resin tag to the tubule orifice wall and also formed an adequate seal of the tubule. After three months' storage in water the hybrid layer became detached from the unaffected dentin, resulting in a gap. Low hybridization can guarantee satisfactory retention and sealing of dentin, and there were differences in the various brands studied. An AFM image of an untreated human tooth microtome is imaged in Fig. 2.51a (the enamel–dentin boundary).

Bone structures and healing of broken bone can be analyzed by AFM directly and after various preparations [126,127]. Topologic AFM also revealed the morphological changes in the demineralization of dentin [128].

1.11.2 Pharmacy

A review for drug characteristics with AFM has appeared [129]. Pharmaceutical tablets are highly composite materials. Therefore, contact AFM is combined with microthermal analysis by replacing the standard AFM tip with a Wollaston wire, thereby allowing the probe to act additionally as a thermistor and temperature probe. Hence, it is possible to map the topology and also the thermal conductivity of the sample. Thermal analysis on highly localized regions of the sample is possible by applying a heating signal to the material. Thus, pharmaceutical tablets can be exhaustively studied [130].

Most pharmaceutical applications of AFM are in the improvement of drug delivery by excipient optimization. Shapes of the particles, sizes, dimples, or pores, dissolution rates (in liquid) are the points of interest [131–133]. Fractal analysis is used to describe three-dimensional AFM surface roughness of pharmaceutical solid particles for quality control. Granules, powders, and freeze-dried powders were analyzed, and a computer program was written to implement the variation method [134].

Furthermore, contact AFM is the tool of choice for the characterization of the performance of pharmaceutical vials out of borosilicate glass that suffer structural and chemical modification upon use [135].

1.11.3 Polymers

Roughness on pristine and aged or reacted polymer surfaces is certainly of enormous practical importance for polymer application. Fig. 1.33 gave an example of a roughening surface reaction of PMMA. Such roughness is of major importance if adhesion of functional coatings is to be applied and improved, for example with optical components that have to be protected from mechanical and environmental atmospheric wear. This subject has been fully covered in various directions in a report [24]. AFM on rough surfaces is of enormous importance in polymer industries also for other reasons. However, there is a tendency to neglect the topology at the expense of phase imaging. Unfortunately, tapping-mode AFM is less reliable with topology, but it is frequently preferred as phase contrast can be recorded that is usually applied for determining physical and chemical differences in composites. Reviews are available [136, 137]. For example, AFM helps in improving bulk properties such as toughness, impact strength, and wear. Nevertheless, contact AFM (without scraping the surface, cf. Fig. 1.14) and shear-force AFM would provide highly useful additional information in most cases. Of particular importance in polymer applications are studies of various phase transitions using a high-temperature accessory. However, viable approaches for detecting stress-induced phase transitions are nanoindentation (Chap. 3) and even better nanoscratching (Chap. 4), and their use is emerging. Micro- and nanowear topologic studies are a continuous subject for AFM on rough surfaces in particular for industrial automotive polymers [138], mostly in combination with nanoscratching tests (Chap. 4). The performance of antithrombogenic polymers in the field of cardiovascular prostheses, artificial hearts, and other devices is related to surface topology. Suitable heparin-like polymers that catalyze the inhibition of thrombin by antithrombin III have been successfully scanned with AFM next to chemical analysis by SEM-EDXA. Functionalization increased the roughness of the surface and its area [139]. Biostability and biocompatibility of polymer dialysis membranes were topologically studied with AFM before and after use [140]. Degradation and ductile-to-brittle transitions occurred upon use and suggested reinforcement with biocompatible alumina–zircon composites.

1.11.4 Ceramics

Contact AFM is generally used for securing smoothness or roughness of ceramic (ultra)thin films. In most cases topology is an undesired property in ceramic layers and low roughness is favorable for applications. Examples are diamond-like carbon coatings for hard disk drives [141], or protecting coatings for turbine blades, diesel engine parts, gas turbine components at very high temperatures and in adverse corrosive environments, etc. [142, 143]. Similarly, wear or polishing on ceramic surfaces is also routinely monitored by contact

AFM [144,145]. Bulk ceramic is usually rich in topologic features. AFM imaging of ceramics topology was facilitated as stylus measurements were common in that field before [146].

Surface roughness of ceramics is of high importance for industrial quality control. Contact AFM provides the basis for applications such as friction, bonding, gluing, etc. This helps in the construction of integrated circuits [147, 148]. Further examples are the topologic characterization of ferromagnetic $BaFe_{12}O_{19}$ sintered powders [149], the particle size distribution of TiO_2 films suitable for technical applications [150], superconductive adhesion coated continuous and flexible round wire [151], and application of glass ceramics out of cordierite and spodumene [152,153].

Ceramic membranes for ultrafiltration or microfiltration have been characterized by AFM. The materials were γ-Al_2O_3 or α-Al_2O_3 [154] and ferroxane [155]. Medicinal applications of ceramics are manifold and AFM is the tool of choice for the analysis of their suitability and performance. A starch thermoplastic/hydroxyapatite composite was exposed to body fluid and AFM monitored the changes in roughness indicating degradation of the polymer matrix and nucleation of calcium phosphate and finally dense uniform covering with it. The results suggest the great potential of the composite for a range of temporary applications in which bone-bonding ability is a desired property [156]. The three-dimensional AFM surface topology of ceramic orthopedic joint prostheses is very important for the performance life of the joint replacement systems [157]. Titanium implants have been coated with alumina and a top layer of porous calcium phosphate under AFM control to form a biocomposite. Thus, an excellent combination of bioactivity and mechanical integrity has been achieved [158]. Fluoroapatite glass has been sputtered on roughened surfaces of Ti_6Al_4V for possible future use as medical implants. An AFM was used to measure the roughness as this greatly affects cell attachment [159].

1.11.5 Mineralogy and Geology

High topologies subject to AFM study occur in geology. Both the texture and the dissolution of minerals are important for geological processes and even in astrophysics. For example, mineral-surface heterogeneity in soils and the outcome of leaching processes deserve topological study by AFM [160]. The shape of the particles helps in the identification of the mix of minerals as detected by X-ray diffraction data.

It is necessary to visualize and characterize pore geometries in rocks by contact and eventually shear-force AFM for accurate estimation of underground flow systems in view of radioactive waste disposal. These studies provide three-dimensional observation of internal rock structure with high resolution for the permeability estimation [161]. Contact AFM was used to study fragments of the Martian meteorite ALH 84001. Viewing of the fragments demonstrated the presence of structures, previously described as nanofossils by NASA representatives who used SEM imaging after gold coating. Careful AFM imaging of the fragments revealed that the observed structures were not an artifact introduced by the coating procedure. However, examination of fracture surfaces near the fusion crust of ALH 84001 with AFM revealed structures also found in Antarctic cryptoendolithic communities. The similarity points to terrestrial origin of the nanofossils and terrestrial microbial activity within a meteorite from the Antarctic blue ice fields. These structures do not bear any resemblance to those postulated to be martian biota, "although they are a probable source of the organic contaminants previously reported in this meteorite" [162].

In the search of aqueous habitats on Mars direct proof of (ancient) flowing water is still lacking, although remote sensing has provided indications of young fluvial systems. To demonstrate that such proof can be provided, surface marks on recent terrestrial sand grains of known origin were examined by AFM. A quantitative three-dimensional analysis can numerically distinguish between aeolian and aquatic transport mechanisms in sedimentary deposits on Earth. Natural grains of quartz, olivine, feldspar pyroxene, and monazite yielded three-dimensional maps of the mineral surface. Fully automated analysis of distribution patterns of the structural elements that constitute the grain surfaces shows characteristic differences in wind-transported and water-transported grains. These can be used as diagnostic fingerprints for the type of transport. This technique is promising for application in future missions to Mars [163].

Hillock growth of several hundreds of nm on (100) of orthopyroxene has been observed by AFM and this observation in combination with AES depth profiles is beneficial for the understanding of the mechanism for weathering and water–rock reactions [164]. The microstructure of palygorskite (accompanied by smaller amounts of smectite, illite, and kaolinite) in Southern Georgia has been studied by AFM. The results suggest that there was an evolving and complex mineralogical and geochemical system during and after its deposition [165]. Contact-mode AFM is able to differentiate various vermiculite samples from different geological origins. The topological differences are attributed to their variable Mg-interlayer contents and the presence of biotite-vermiculite mixed-layer additions in the absence of noticeable reconstruction [166]. Topologic features such as grain distortion, dislocation, stacking faults, grain boundaries, and dislocation walls were observed by AFM on Burma Jadeite. High-quality jadeite can be synthesized by imitating the conditions of metamorphism thus disclosed [167].

Calcite is a subject of biomineralogy. Its crystallization may be biologically controlled. Unicellar algae form calcite shields called coccoliths with associated polysaccharides. AFM studies with Emiliania huxleyi reveal that the latter regulate crystal morphology by enhancing the precipitation of specific faces, a crucial aspect of the biomineralization process. Elaborate shapes of calcite crystals are produced [168]. Biogenic carbonate minerals can have important consequences for the biogeochemical cycle of carbon. Calcite surfaces were hung at discrete depths on a sediment trap array line for a three-day deployment period in subtropical North Pacific waters. AFM revealed the changes in surface morphologies. Organic films developed on the surfaces while carbonate minerals dissolved [169]. Similarly, AFM was the method of choice in studying the dissolution of Al-substituted goethite particles by Pseudomonas mendocina var. bacteria [170]. Dissolution of minerals is an important subject in weathering of sediments and rocks. The basics are crystal dissolution kinetics that gained an additional tool by contact AFM providing high-resolution topologic information [171]. AFM revealed development of microtexture for field-weathered peristeritic feldspar. Rates were calculated and compared to other field rates [172]. Orthoclase dissolution was studied by in situ AFM on (001). There were terrace roughening at pH = 1.1 and step motion at pH = 12.9. Contrasting dissolution mechanisms are inferred for low- and high-pH conditions. No coating was observed under high fluid flow conditions. The rates on (001) are comparable to those derived from steady-state powder dissolution rates and improve the understanding of alkali feldspar-weathering processes [173]. Well and poorly crystallized kaolinites show differences in their AFM topologies. These influenced the rates of dissolution in oxalic acid and inorganic acids, but they were not very different, suggesting that the fundamental structure of kaolinite rather than the specific surface details exerts the greatest influence in that case [174].

Surface modeling techniques in mineralogy were performed under AFM control. Interestingly, photoelectrochemical etching of TiO_2 (rutile) along the c-axis produced a nano-honeycomb porous single crystal structure on (100) with large specific surface of the n-type semiconductor, suitable for photoelectronic devices such as a photocatalyst and a dye-sensitized solar cell [175]. Electrochemical anisotropic hydride growth on Ge (110) and Ge (111) was judged according to feature shapes with an AFM. Substantial insight into the mechanism of hydride growth was obtained, even though the features were not very high (below 100 nm) [176].

1.11.6 Metallurgy and Corrosion

An important field for AFM on rough surfaces is opened in metallurgy and corrosion. Powder metallurgy mills, presses, sinters, and anneals for obtaining substrates that require quality control by AFM for roughness measurement [177,178]. The topology and performance of protective covers or coatings must be most reliably determined by contact AFM [179]. Similarly, the performance of zincation treatments is a matter for AFM characterization [180]. Texture, lattice imperfections, and corrosion resistance of zinc alloy electrodeposits have also been studied with contact AFM [181]. Particularly rich nano- and microstructures were topologically analyzed on steel as an example for multiphase systems [182]. A review covering quality control by AFM analysis in metallurgy is available [183]. Metal corrosion is a severe technical problem that requires detailed topological study by contact AFM, sometimes combined with STM. Broad possibilities have been reviewed [184]. Topologic features play their role from nanoscopic nucleation to microscopic pitting and cracking. A review paper [185] lists and exemplifies various mechanisms. These include corrosion in solution (growth and pitting), intergranular corrosion, stress corrosion, crack growth in solution and in air, environmental influences, and nanofractography with various materials. Improvements of pitting resistance for example in resulfurized stainless steel are to be controlled by AFM measurements [186]. However, intergranular corrosion attack increases with increasing sulfur content in stainless steel [187].

Fatigue phenomena require topologic investigation by AFM for an understanding of the surface deformation [188, 189]. This includes the detection of slip bands, protrusions, crack nucleation, and growth in strain-controlled fatigue tests at various strain amplitudes. These components of the fracture mechanics are very important for the development of macrocracks and failure. AFM led to the development of formalism for the quantitative characterization of the surface deformation of cyclically loaded structures. An in situ system has been described [190]. The corrosion of brass in oxygen-containing water was also studied by AFM [191]. Furthermore, the corrosion and passivation of Ti grains has been topologically and frictionally studied [192]. Also electrochemical corrosion is an obvious subject for AFM as high topologies ensue [193, 194].

More details about the local chemistry of corrosion can be obtained by electrochemical atomic force microscopy (ECAFM) [195]. Examples are the anodic dissolution of copper [196] or inhibition of copper corrosion [197], the surface structure of a lead electrode during redox process [198], corrosion of steel [199, 200], or the uranium (IV)-oxide dissolution and remineralization [201]. A medicinal application of ECAFM studied the oxidation of the implant metals Ti and $TiAl_6V_4$ by H_2O_2 to form oxide layers that should be biocompatible to surgical wound environments. Topological changes occurred. These are related to oxidative mechanisms that take place on the titanium oxide surface and in wound environments. The study provides a foundation for the porous oxide surface model on commercially pure titanium exposed to hydrogen peroxide [202].

The local chemistry of corrosion can also be studied with optical near-field spectroscopy. Electrochemical corrosion of dental alloys with shear-force AFM and SNOM assessment is described in Chap. 2, Sect. 2.9.1.

1.11.7 Catalysis

There are numerous atomically resolved AFM measurements on flat metal surfaces catalyzing gas-phase reactions. Such model studies are not the subject of this book. Heterogeneous catalysts generally owe their efficiency to roughness. It is therefore not surprising that AFM was applied to them [203, 204]. The catalytic surfaces were characterized as valleys, ridges, crevices, dislodged plates, wide pits and craters, and narrow slits. Furthermore, the changes of the surface topology of common fluid cracking catalysts upon operation in a typical refinery have been studied [205]. Pore diameters, pore size distributions, deep trenches, valleys, craters, and short stacks of plates with voids were the features of interest. AFM images allowed the distinction of old and young FCC fractions and coke deposits could be identified. Their burning in air at 600°C did not alter structure and activity. Such characterization of catalyst surfaces is of highest industrial interest and results are mostly not disclosed. AFM is also useful in enzymatic catalysis. The activity of a horseradish peroxidase assembly was related to the surface topology as detected by AFM [206].

1.11.8 Forensics

Topologic measurements with the AFM have found increasing applications in forensic evaluations, and this is not restricted to DNA analysis [207,208]. The human hair is an interesting object for forensics due to characteristic cuticle step height, tilt angle, cuticle diameter [209,210]. Further information about cosmetics and medical diagnoses is also obtained, in particular if lateral force is applied to treated hair [211].

Another topic is forensic examination of line crossings in documents. It is largely facilitated by AFM due to the ambient conditions [212]. Also detection of gunshot residues deposited on the bullet and on the firing hands collected using double-sided tape has been studied by means of AFM in combination with other techniques for considerably increasing the confidence level of such detection [213].

Also a forensic problem may be the determination of the radiation exposure history of common materials, including organic crystals. Exposure to radioactive material leaves a permanent record, which can be read for forensic (and dosimetric) purposes by using AFM [214].

1.11.9 Historical Objects

Both prehistoric fossils and historic antiques deserve AFM investigation, while comparatively scarce applications have been published. Organic-walled Precambrian fossils and microscopic spheromorph acritarchs, permineralized in 650-million-year old cherts of the Chichkan Formation of southern Kazakhstan, have been studied by AFM in combination with laser Raman spectroscopic analysis of individual specimens. The walls of these petrified fossils are composed of stacked arrays of about 200 nm wide angular platelets of polycyclic aromatic kerogen [215]. Thus, AFM helps in the investigation of geochemical maturation of ancient organic matter and in the distinction of true fossils from pseudofossils. Furthermore, a fossil bacterium in dolomite has been analyzed by shear-force AFM and SNOM in Chap. 2, Sect. 2.10.1. Late Cretaceous avian bone tissues from Argentina demonstrate exceptional preservation in a medium-grained sandstone matrix, indicating unusual perimortem taphonomic conditions. AFM topology and AFM pulling curves are required in addition to immunohistology to prove integrity and functionality of retained collagen [216]. Furthermore, mineralized human tissues have been investigated with AFM [217].

The anticorroding property of the Chinese ancient bronze black mirror Hei–Qi–Gu has been elucidated by AFM and supporting techniques. SnO_2 and SiO_2 particles with very small particle sizes are particularly stable and protect the bronze alloy successfully [218]. AFM, XPS, and SEM with XPMA were used to characterize the morphology of the surface of glazes from ancient Kiev and Vizantia, discovered on archeological excavations. Differences in the morphologies point to original manufacture of glazes in ancient Kiev as distinguished from the Byzantine glazes [219]. A more general paper investigated the surfaces of medieval stained glass and ancient copper alloys both chemically and topologically with AFM [220]. Luster is one of the most interesting ancient ceramic decorations. Luster reproductions were compared with medieval luster surfaces by chemical analysis and AFM morphology. It was concluded that Cu and Ag were introduced into the glaze by ion exchange with alkalies from the glaze, rather than forming a superimposed layer on top of the glaze. Surface roughness resulted by the nanocrystals' growth inside the glassy matrix. Silver particles appeared heterogeneously distributed and had a smaller size and higher roughness than copper particles on ancient luster [221].

1.12 Electromagnetic AFM

A new development in AFM uses electromagnetic actuators instead of piezos for distance control and affordable Nanosurf easyScan 2 or Nanosurf Mobile S instruments are already available. These portable instruments are simple in operation and exhibit a high degree of automation. They allow for most AFM techniques with lateral resolutions from 1 nm. Surface areas with very high topology and up to several 100 μm wide can be routinely scanned. These features are of particular interest for real-world surfaces. Figure 1.77 demonstrates the capabilities with a reflector for bicycles. The enormous height resolution is extraordinary and easily achieved. There are many advantages of electromagnetic actuators for AFM when compared to piezo AFM. The range from molecular resolution to the light microscopy scales, but with very high topographic capability (see also Sect. 2.13), is spanned in a low-size portable instrument. It is expected that this technique will find widespread use in applied AFM, teaching, research, and industry.

1.13 Conclusions

Topologic AFM is a mature technique. It is not just a means for atomic or molecular resolution; high topology measurements require different expertise and provide another breakthrough with most widespread practical

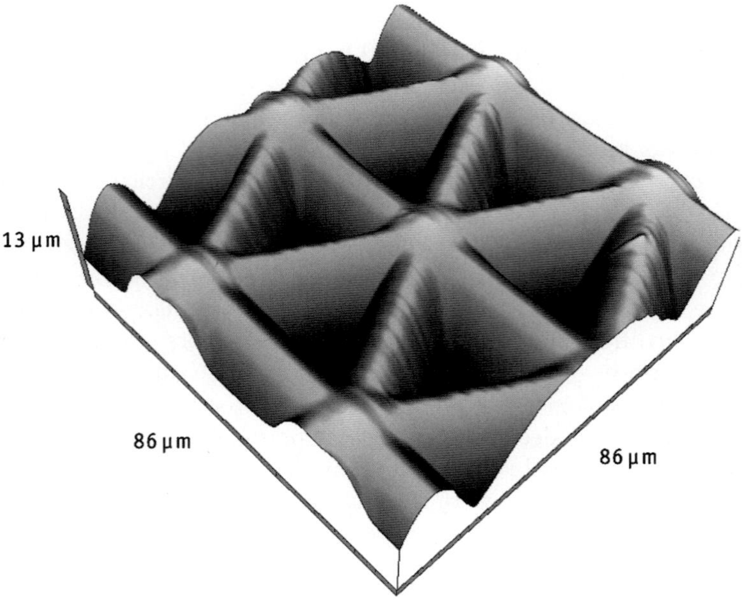

Fig. 1.77. Large-scale AFM image of a technical retroreflector surface with well-resolved 13 µm height differences in an $86 \times 86\,\mu m^2$ area, showing the topographic capabilities of the electromagnetic Nanosurf easyScan instrument; courtesy of Dr. L. Scandella, Nanosurf AG, Liestal/Switzerland

applications in almost all branches of life and sciences. The idea was to scan real-world samples at ambient conditions without change in treatment and evacuation. Another aim was the control of self-assembled high surface feature generation by chemical reaction under control by AFM with classification of feature types and purposeful surface modeling. This has been demonstrated from the early start in 1990s, when AFM was used in the ranges that were typically 10 µm wide and up to 3 µm high. Such measurements had to be experimentally explored without scraping the surface. The success is clearly seen in the high increase in the number of applications since about 1998, which is indicated in Sect. 1.11 preferably with technically useful examples. Unfortunately, there is still widespread reservation in acknowledging the capability of reliably scanning very high and steep surface features without scraping the surface, even if the expertise has been long available. Therefore high topology is frequently not mentioned in the abstracts of relevant papers. Hence, some important papers have almost certainly been missed in the author's literature retrieval efforts. Fortunately, the most robust and least complicated techniques of contact AFM and shear-force AFM provide the most direct and reliable information of topologic surface properties that are so important for practical applications. Numerous spectroscopic techniques can be added for better knowing what the surface features are chemically (e.g., SNOM, friction force, thermoconduction, thermoanalysis, magnetic force, electrostatic force, adhesion, chemical force, etc.). Dynamic AFM such as tapping mode became quite popular, although the topologic information is indirect and less reliable. It is therefore frequently neglected in publications at the expense of phase imaging, the interpretation of which is less rewarding for rough surfaces, which are the rule for real-world samples. These dynamic techniques were not included as they had been the subject of recent review books. There was, however, no review with specific concern of topology as offered for the first time in this book. The success in feature generation and control and further applications of highly resolved topology in life sciences, including medicine and pharmacy, and in chemistry (solid-state syntheses, polymers, ceramics, catalysis, metallurgy, corrosion, etc.), mineralogy/geology, astrophysics, forensics, ancient history, etc., are promising for increased future application in all fields that relate to solid surfaces in academic and industrial research and application.

References

1. G. Kaupp: Photochemie auf neuen Wegen: AFM bei Festkörperphotolysen. Lecture at Arbeitstagung Photochemie, Waldbärenburg (close to Dresden, Germany), March 20–22 (1991)
2. G. Kaupp: Atomic force microscopy in organic gas/solid reactions: How do the new phases build up? Lecture at ICCOSS X, 10th International Conference on the Chemistry of the Organic Solid State, OC 8, Vancouver, July 9–12 (1991)
3. G. Kaupp: New advent in photochemistry: Atomic force microscopy. Lecture at XVth International Conference on Photochemistry, VI-2, Paris, 28.7.–2.8. (1991)
4. G. Kaupp: AFM with nonconducting organic crystals. Lecture at STM'91, International Conference on Scanning Tunneling Microscopy, Interlaken, PC/5, 12.8.–16.8. (1991)
5. G. Kaupp: Photodimerization of cinnamic acid in the solid state: New insights on application of atomic force microscopy. Angew. Chem. Int. Ed. Engl. **31**, 592–594 (1992)
6. G. Kaupp: Photodimerization of Anthracenes in the Solid State: New Results from Atomic Force Microscopy. Angew. Chem. Int. Ed. Engl. **31**, 594–597 (1992)
7. G. Kaupp: Atomic force microscopy in organic gas/solid reactions: How do the new phases build up? Mol. Cryst. Liq. Cryst. **211**, 1–15 (1992)
8. S.A.C. Gould, B. Drake, C.B. Prater, A.L.Weisenhorn, S. Manne, H.G. Hansma, P.K. Hansma, J. Massie, M. Longmire, V. Elings, B. Dixon Northern, B. Mukergee, C.M. Peterson, W. Stoeckenius, T.R.Albrecht, C.F. Quate: From atoms to integrated circuit chips, blood cells, and bacteria with the atomic force microscope. J. Vac. Sci. Technol. A **8**, 369–373 (1990)
9. J.H. Hoh, P.K. Hansma: Atomic force microscopy for high-resolution imaging in cell biology. Trends Cell Biol. **2**, 208–213 (1992)

10. M.C. Goh, P. Markiewicz: The Atomic Force Microscope. Chem. Ind. 1992, 687–691.
11. T. Miyamoto, S. Miyake, R. Kaneko: Wear resistance of C^+-implanted silicon investigated by scanning probe microscopy. Wear **162–164**, 733–738 (1993)
12. U.D. Schwarz, H. Haefke, P. Reimann, H. J. Güntherodt: Tip artefacts in scanning force microscopy. J. Microsc. **173**, 183–197 (1994)
13. G. Binnig, C.F. Quate, C.H. Gerber: Atomic force microscope. Phys. Rev. Lett. **56**, 930–933 (1986)
14. E. Betzig, P. L. Finn, J. S. Weiner: Combined shear force and near-field scanning optical microscopy. Appl. Phys. Lett. **60**, 2484–2486 (1992)
15. P.C. Yang, Y. Chen, M. Vaez-Iravani: Attractive mode atomic force microscopy with optical detection in an orthogonal cantilever/sample configuration. J. Appl. Phys. **71**, 2499–2502 (1992)
16. Q. Zhong, D. Innis, K. Kjoller, V. B. Elings: Fractured polymer/silica fiber surface studied by tapping mode atomic force microscopy. Surf. Sci. Lett. **290**, L688–L692 (1993)
17. M. Dreier, D. Anselmetti, H. J. Güntherodt: Dynamic force microscopy in liquids. J. Appl. Phys. **76**, 5095–5098 (1994)
18. A. Schirmeisen, B. Anczykowski, H. Fuchs: in *Applied Scanning Probe Methods*, (eds. B. Bushan, H. Fuchs, S. Hosaka), 3–39, Springer, Berlin (2004)
19. *Noncontact Atomic Force Microscopy*, S. Morita, R. Wiesendanger, E. Meyer (eds.), Springer, Berlin (2002)
20. *Applied Scanning Probe Methods*, B. Bushan, H. Fuchs, S. Hosaka (eds.), Springer, Berlin (2004)
21. *Scanning Probe Microscopy Analytical Methods*, R. Wiesendanger (ed.), Springer, Berlin (1998)
22. S.N. Magonov, M.H. Whangbo: *Surface Analysis with STM and AFM*, VCH, Weinheim, 1996
23. S. Magonov: Visualization of polymer structures with atomic force microscopy. pp. 207–251 in [20]
24. G. Kaupp, J. Schmeyers, M. Haak, T. Marquardt, M. Plagman, M. Juniel-Kramer: Erforschung von Systemlösungen für transparente Verschleißschutzschichten auf Kunststoffen für die Optik: Organische Schichten und organische Haftvermittlungsschichten mit kovalenter Anbindung. *BMBF Forschungsbericht* 13N6250, 1997, 1–63; Available at docdel@tib.uni-hannover.de
25. G. Binnig, D.P.E. Smith: Single-tube three-dimensional scanner for scanning tunneling microscopy. Rev. Sci. Instrum. **57**, 1688–1689 (1986)
26. C.J. Chen: Electromechanical deflections of piezoelectric tubes with quartered electrodes. Appl. Phys. Lett. **60**, 132–134 (1992)
27. U. Heider, O. Weis: Distortion-free, calibrated $LiNbO_3$ piezoscanner for probe microscopes with atomic resolution. Rev. Sci. Instrum. **64**, 3534–3537 (1993)
28. T. Akiyama, U. Staufer, N.F. Rooij, D. Lange, C. Hagleitner, O. Brand, H. Baltes, A. Tonin, H.R. Hidber: Integrated atomic force microscopy array probe with metal–oxide–semiconductor field effect transistor stress sensor, thermal bimorph actuator, and on-chip complementary metal–oxide–semiconductor electronics. J. Vac. Sci. Technol. B **18**, 2669–2675 (2000)
29. T.R. Albrecht, S. Akamine, T.E. Carver, C.F. Quate: Microfabrication of cantilever styli for the atomic force microscope. J. Vac. Sci. Technol. A **8**, 3386–3396 (1990)

30. O. Wolter, T. Bayer, J. Greschner: Micromachined silicon sensors for scanning force microscopy. J. Vac. Sci. Technol. B **9**, 1353–1357 (1991)
31. M. Yamaki, T. Miwa, H. Yoshimura, K. Nagayama: Efficient microtip fabrication with carbon coating and electron beam deposition for atomic force microscopy. J. Vac. Sci. Technol. B **10**, 2447–2450 (1992)
32. H. Ximen, P.E. Russell: Microfabrication of AFM tips using focused ion and electron beam techniques. Ultramicroscopy, **42–44**, 1526–1532 (1992)
33. C.V. Nguyen, K.J. Chao, R.M.D. Stevens, L. Delzeit, A. Cassell, J. Han, M. Meyyappan: Carbon nanotube tip probes: Stability and lateral resolution in scanning probe microscopy and application to surface science in semiconductors. Nanotechnology **12**, 363–367 (2001)
34. A. Marquez-Lucero, J.A. Gomez, R. Caudillo, M. Miki-Yoshida, M. Jose-Yacaman: A method to evaluate the tensile strength and stress–strain relationship of carbon nanofibers, carbon nanotubes, and C-Chains. Small **1**, 640–644 (2005)
35. H. Kado, K. Yokoyama, T. Tohda: Atomic force microscopy using ZnO whisker tip. Rev. Sci. Instrum. **63**, 3330–3332 (1992)
36. W. Lee, H.G. Sohn, J.M. Myoung: Prediction of the structural performance of ZnO nanowires grown on GaAs (001) substrates by metalorganic chemical vapor deposition (MOCVD). Mater. Sci. Forum **449–452**, 1245–1248 (2004)
37. S. Akamine, R.C. Barrett, C.F. Quate: Improved atomic force microscope images using microcantilevers with sharp tips. Appl. Phys. Lett. **57**, 316–318 (1990)
38. T. Hantschel, E.M. Chow, D. Rudolph, D. Fork: Stressed metal probes for atomic force microscopy. Appl. Phys. Lett. **81**, 3070–3072 (2002)
39. T. Hantschel, E.M. Chow, D.K. Fork: Scanning probe system with spring probe. US Pat. Appl. Publ. US 20030182993 (2003)
40. T. Hantschel, E.M. Chow, D.K. Fork, M.A. Rosa, D. De Bruyker: Scanning probe system with spring probe and actuation/sensing structure. US Pat. Appl. Publ. US 20030183761 (2003)
41. T. Hantschel, E.M. Chow, D. Rudolph, C. Shih, L. Wong, D.K. Fork: Stressed-metal NiZr probes for atomic force microscopy. Microelectron. Eng. **67–68**, 803-809 (2003)
42. T. Mariani, C. Frediani, C. Ascoli: A three-dimensional scanner for probe microscopy on the millimetre scale. Appl. Phys. A **66**, S861–S866 (1998)
43. T. Mariani, C. Ascoli, P. Baschieri, C. Frediani, A. Musio: Scanning force images through the 'Milliscope' – a probe microscope with very wide scan range. J. Microsc. **204**, 53–60 (2001)
44. G. Kaupp, J. Schmeyers, U. Pogodda, M. Haak, T. Marquardt, M. Plagmann: AFM for the imaging of large and steep submicroscopic features, artifacts and scraping with asymmetric cantilever tips. Thin Solid Films **264**, 205–211 (1995)
45. D. Keller: Reconstruction of STM and AFM images distorted by finite-size tips. Surf. Sci. **253**, 353–364 (1991)
46. D. Keller, F. S. Franke: Envelope reconstruction of probe microscope images. Surface Science **294**, 409–419 (1993)
47. L. Montelius, J. O. Tegenfeldt: Direct observation of the tip shape in scanning probe microscopy. Appl. Phys. Lett. **62**, 2628–2630 (1993)
48. G. Kaupp: Atomic force microscopy in the photochemistry of chalcones. J. Microsc. **174**, 15–22 (1994)

49. P. Dietz, K.H. Herrmann, O. Inacker, H.D. Lehmann, P.K. Hansma: Scanning force microscopy of synthetic membranes in air and under water: surfaces, cross-sections and fouling. Proc. SPIE **1639**, 186–197 (1992)
50. A.K. Fritzsche, A.R. Arevalo, A.F. Connolly, M.D. Moore, V. Elings, C.M. Wu: The structure and morphology of the skin of polyethersulfone ultrafiltration membranes: A comparative atomic force microscope and scanning electron microscope study. J. Appl. Polymer Sci. **45**, 1945–1956 (1992)
51. H. Kamusewitz, M. Keller, D. Paul: Membrane characterization by means of pneumatic scanning force microscopy. Thin Solid Films **264**, 184–193 (1995)
52. D.J. Shuman: Scanning tip orientation adjustment method for atomic force microscopy. US Pat. Appl. Publ. US 2004200261 (2004)
53. G.C. Gazzadi, P. Luches, S. F. Contri, A. di Bona, S. Valeri: Submicron-scale patterns on ferromagnetic–antiferromagnetic Fe/NiO layers by focused ion beam (FIB) milling. Nucl. Instrum. Methods Phys. Res., B **230**, 512–517 (2005)
54. G. Kaupp, A. Herrmann, M. Haak: Review Commentary – Artifacts in scanning near-field optical microscopy (SNOM) due to deficient tips. J. Phys. Org. Chem. **12**, 797–807 (1999)
55. P. Maivald, H.J. Butt, S.A.C. Gould, C.B. Prater, B. Drake, J.A. Gurley, V.B. Elings, P.K. Hansma : Using force modulation to image surface elasticities with the atomic force microscope. Nanotechnology **2**, 103–106 (1991)
56. J.E. Houston: Interfacial force microscopy; Selected applications, pp. 41–73 in [20]
57. V. Scherer, M. Reinstädtler, W. Arnold: Atomic force microscopy with lateral modulation, pp. 75–115 in [20]
58. B. Bhushan: Micro/nanotribology studies using scanning probe microscopy, pp. 171–205 in [20]
59. G. Kaupp: Solid-state molecular syntheses: complete reactions without auxiliaries based on the new solid-state mechanism Cryst. Eng. Commun. **5**, 117–133 (2003)
60. G. Kaupp: Nanostructures or submicrostructures from gas/solid reactions as probed by atomic force microscopy. J. Vac. Sci. Technol. B **12**, 1952–1956 (1994)
61. G. Kaupp, M. Plagmann: Atomic force microscopy and solid state photolyses: Phase rebuilding. J. Photochem. Photobiol. A **80**, 399–407 (1994)
62. B. Bushan: Nanotribologic scratching wear, local deformation, pp. 147–168 in [20]
63. J. M. Helt, J. D. Battaes: Wear of Mica under Aqueous Environments: Direct Observation of Defect Nucleation by AFM. Langmuir **21**, 633–639 (2005)
64. R. Kaneko, E. Hamada: Microwear processes of polymer surfaces Wear **162–164**, 370–377 (1993)
65. G. Kaupp: AFM and STM in photochemistry including photon tunneling. Adv. Photochem. **19**, 119–177 (1995)
66. G. Kaupp: AFM on chemically reacting crystals. Mol. Cryst. Liq. Cryst. **242**, 153–169 (1994)
67. G. Kaupp: Unpublished results
68. G. Kaupp, A. Herrmann: Unpublished results
69. G. Kaupp: Supermikroskopie bei organischen Kristallreaktionen. Chem unserer Zeit, **31**, 129–139 (1997)

70. P. E. Hillner, A.J. Gratz, S. Manne, P.K. Hansma: Atomic-scale imaging of calcite growth and dissolution in real time. Geology **20**, 359–362 (1992)
71. T. Maeda, G.H. Lee, T. Ohnishi, M. Kawasaki, M. Yoshimoto, H. Koinuma: Molecular layer-by-layer growth of $SrTiO_3$ and $BaTiO_3$ films by laser molecular beam epitaxy. Materials Science & Engineering B **41**, 134–137 (1996)
72. M. Yoshimoto, T. Ohnishi, G.H. Lee, K. Sasaki, H. Maruta, H. Koinuma: Atomic-scale analysis of laser MBE growth of oxide thin films by in situ RHEED and CAICISS. Mater. Res. Soc. Symp. Proc. **502**, 227–236 (1998)
73. A. Brazdeikis, A.S. Flodstroem: Surface morphology and nucleation of infinite-layer $(Sr, Ca)CuO_2$ films on atomically flat $SrTiO_3$ (100) substrates studied by atomic force microscopy. Appl. Surf. Sci. **107**, 95–101 (1996)
74. G. Kaupp, U. Pogodda, J. Schmeyers: Gas/Solid Reactions with Acetone. Chem. Ber. **127**, 2249–2261 (1994)
75. J. Uilk, S. Bullock, E. Johnston, S.A. Myers, L. Merwin, K.J. Wynne: Surface science of elastomeric coatings prepared from α,ω-dihydroxypoly (dimethylsiloxane) and the ethoxysiloxane mixture "ES40". Macromolecules **33**, 8791–8801 (2000)
76. G. Kaupp, M. Haak: Phase rebuilding of α-cinnamic acid in tail irradiations. Mol. Cryst. Liq. Cryst. **313**, 193–198 (1998)
77. G. Kaupp, J. Schmeyers, J. Boy: Quantitative gas–solid reactions with ClCN and BrCN: Synthesis of cyanamides, cyanates, thiocyanates, and their derivatives. Chem. Eur. J. **4**, 2467–2474 (1998)
78. G. Kaupp, J. Schmeyers: gas–solid reactions of aliphatic amines with thiohydantoins: Atomic force microscopy and new mechanisms. Angew. Chem. Int. Ed. Engl. **32**, 1587–1589 (1993).
79. G. Kaupp, J. Schmeyers: Unpublished results
80. A. Herrmann, G. Kaupp, T. Geue, U. Pietsch: Atomic force microscopy and grazing incidence diffraction investigation of the gas–solid diazotization of sulfanilic acid. Mol. Cryst. Liq. Cryst. **293**, 261–275 (1997)
81. G. Kaupp, J. Boy: Unpublished results
82. S. Nakatsuji, A. Takai, M. Mizumoto, H. Anzai, K. Nishikawa, Y. Morimoto, N. Yasuoka, J. Boy, G. Kaupp: Preparation and magnetic properties of novel CT complexes derived from organic stable radicals. Mol. Cryst. Liq. Cryst. **334**, 177–184 (1999)
83. G. Kaupp, A. Herrmann, J. Schmeyers: Waste-free chemistry of diazonium salts and benign separation of coupling products in solid salt reactions. Chem. Eur. J. **8**, 1395–1406 (2002)
84. G. Kaupp, J. Schmeyers, F. Toda, H. Takumi, H. Koshima: Mechanism of solid-solid resolution of pantolactone. J. Phys. Org. Chem. **9**, 795–800 (1996)
85. G. Kaupp, A. Kuse: 100% yield in 300 gas–solid reactions covering 22 reaction types: A benign option for industrial solid-state production. Mol. Cryst. Liq. Cryst. **313**, 361–366 (1998)
86. G. Kaupp, M. Haak: E/Z isomerization in crystals – phase rebuilding on photolysis with long-wavelength radiation. Angew. Chem. Int. Ed. Engl. **35**, 2774–2777 (1996)
87. G. Kaupp, J. Schmeyers: Gas/solid reactions with nitrogen dioxide. J. Org. Chem. **60**, 5494–5503 (1995)
88. M.A. Rodriguez, S.D. Bunge: Diphenylamine. Acta Crystallogr. E **59**, o1123-o1125 (2003)

89. Y.Y. Wei, B. Tinant, J.P. Declercq, M.V. Meerssche: Structures of Four *para*-substituted 1,1-diphenylethylene derivatives: 4,4'-ethylenedibenzonitrile (I); 4,4'-ethylenedianisole (II); *p*-(1-phenylethylene)anisole (III) and *N*,*N*-Dimethyl-4-(1-*p*-Nitrophenylethylene)aniline (IV). Acta. Crystallogr. C **43**, 86–89 (1987)
90. G. Kaupp: Supermicroscopy in Supramolecular Chemistry: AFM, SNOM, and SXM, in *Comprehensive Supramolecular Chemistry,* Vol. **8**, 381–423 + 21 color plates (ed. J.E.D. Davies), Elsevier, Oxford (1996)
91. G. Kaupp: Solid-state reactivity/gas–solid and solid–solid reactions in *Encyclopedia of Supermolecular Chemistry,* Taylor & Francis, E-ESMC-120024356 (2005)
92. G.M.J. Schmidt: Photodimerization in the solid state. Pure. Appl. Chem. **27**, 647–678 (1971)
93. E. Heller, G.M.J. Schmidt: Topochemistry XXXIII. Solid-state photochemistry of some anthracene derivatives. Israel J. Chem. **9**, 449–462 (1971)
94. J.R. Green, D.R. Wheeler: X-ray investigation of some plastic crystals. II. Density of vacancies in cyclohexane, cyclohexanol, and dl-camphene. Mol. Cryst. Liq. Cryst. **6**, 13–21 (1969)
95. G. Kaupp, J. Schmeyers, J. Boy: Waste-free solid-state syntheses with quantitative yield. Chemosphere **43**, 55–61 (2001)
96. J. Schmeyers, F. Toda, J. Boy, G. Kaupp: Quantitative solid–solid synthesis of azomethines. J C.S. Perkin 2 **1998**, 989–993; addendum **2001**, 132
97. G. Kaupp, J. Schmeyers, M. Haak, T. Marquardt, A. Herrmann: AFM in organic solid state reactions. Mol. Cryst. Liq. Cryst. **276**, 315–337 (1996)
98. Y. Yoshida, B. Van Meerbeek, J. Snauwaert, L. Hellemans, P. Lambrechts, G. Vanherle, K. Wakasa, D.H. Pashley: A novel approach to AFM characterization of adhesive tooth–biomaterial interfaces. J. Biomed. Mater. Res. A **47**, 85–90 (1999)
99. G. Kaupp: Atomic force microscopy on photoreactive organic solids: Nanostructures. Mol. Cryst. Liq. Cryst. **252**, 259–268 (1994)
100. B. Marciniak, E. Rozycka-Sokolowska, V. Pavlyuk: 2-Naphthalenol. Acta Crystallogr. E **59**, o52–o53 (2003)
101. S. Ebbinghaus, D. Abeln, M. Epple: Diphenylmethanol, CCDC, deposit no. 108059
102. R. Toledo-Crow, P.C. Yang, Y. Chen, M. Vaez-Iravani: Near-field differential scanning optical microscope with atomic force regulation. Appl. Phys. Lett. **60**, 2957–2959 (1992)
103. R.D. Grober, T.D. Harris, J.K. Trautman, E. Betzig: Design and implementation of a low temperature near-field scanning optical microscope. Rev. Sci. Instrum. **65**, 626–631 (1994)
104. K. Karrai, R.D. Grober: Piezoelectric tip–sample distance control for near-field optical microscopes. Appl. Phys. Lett. **66**, 1842–1844 (1995)
105. R. Brunner, A. Bietsch, O. Hollricher, O. Marti: Distance control in near-field optical microscopy with piezoelectric shear-force detection suitable for imaging in liquids. Rev. Sci. Instrum. **68**, 1769–1772 (1997)
106. Y. Isono, T. Namazu, T. Tanaka: AFM bending testing of nanometric single crystal silicon wire at intermediate temperatures for MEMS. IEEE International Conference. On Micro-Electro-Mechanical Systems, Technical Digest, 14th, Interlaken, Switzerland, pp. 135–138; published by Institute of

Electrical and Electronics Engineers, New York, NY (2001); Chem Abstr. **135**: 100824 (2001)

107. M.A.D. Roby, G.C. Wetsel, C.Y. Wang: Scanned probe lateral-force determination of fluid-dynamic effects near a solid surface in air. Appl. Phys. Lett. **69**, 130–132 (1996)
108. M.J. Gregor, P.G. Blome, J. Schöfer, R.G. Ulbrich: Probe-surface interaction in near-field optical microscopy: The nonlinear bending force mechanism. Appl. Phys. Lett. **68**, 307–309 (1996)
109. E. Teixeira-Neto, G. Kaupp, F. Galembeck: Latex particle heterogeneity and clustering films. J. Phys. Chem. **107**, 14255–14260 (2003)
110. E. Teixeira-Neto, G. Kaupp, F. Galembeck: Spatial distribution of serum solutes on dry latex sub-monolayers determined by SEPM, SNOM, SC microscopy. Coll. Surf., A **243**, 79–87 (2004)
111. M. Sassaki, R.A. Douglas, M.U. Kleinke, O. Teschke: Structure imaging by atomic force microscopy and transmission electron microscopy of different light emitting species of porous silicon. J. Vac. Sci. Technol. B **14**, 2432–2437 (1996)
112. K. Sinniah, J. Paauw, J. Ubels: Investigating live and fixed epithelial and fibroplast cells by atomic force microscopy. Cur. Eye Res. **25**, 61–68 (2002)
113. M. Targosz, P. Czuba, M. Szymonski: Erythrocyte surface modification due to the pharmacological treatment studied with atomic force microscopy. Opt. Appl. **32**, 275–281 (2002)
114. M. Girasole, A. Cricenti, R. Generosi, A. Congiu-Castellano, G. Boumis, G. Amiconi: Artificially induced unusual shape of erythrocytes: An atomic force microscopy study. J. Microsc. **204**, 46–52 (2001)
115. A. Mai, W. Weerachatyanukul, M. Tomietto, D.D.M. Wayner, G. Wells, R. Balhorn, A. Leader, J.L. Cyr, N. Tanphaichitr: Use of atomic force microscopy for morphological analyses of acrosome intact and acrosome-reacted human sperm. Mol. Reproduct. Dev. **63**, 471–479 (2002)
116. C. Rotsch, K. Jacobson, J. Condeelis, M. Radmacher: EGF-stimulated lamellipod extension in adenocarcinoma cells. Ultramicroscopy **86**, 97–106 (2001)
117. S.W. Schneider, P. Pagel, C. Rotsch, T. Danker, H. Oberleithner, M. Radmacher, A. Schwab: Volume dynamics in migrating epithelial cells measured with atomic force microscopy. Pflug. Archiv: Eur. J. Physiol. **439**, 297–303 (2000)
118. S.W. Schneider, Y. Yano, B.E. Sumpio, B.P. Jena, J.P. Geibel, M. Gekle, H. Oberleithner: Rapid aldosterone-induced cell volume increase of endothelial cells measured by the atomic force microscope. Cell Biol. Int. **21**, 759–768 (1997)
119. R.C. Anderson, R.G. Haverkamp, P.L. Yu: Investigation of morphological changes to Staphylococcus aureus induced by ovine-derived antimicrobial peptides using TEM and AFM. FEMS Microbiol. Let. **240**, 105–110 (2004)
120. A. da Silva, O. Teschke: Effects of the antimicrobial peptide PGLa on live Escherichia coli. Biochimica et biophysica acta **1643**, 95–103 (2003)
121. P.C. Braga, D. Ricci, M. Dal Sasso: Daptomycin morphostructural damage in Bacillus cereus visualized by atomic force microscopy. J. Chemotherapy **14**, 336–341 (2002)
122. M. Stolz, D. Stoffler, U. Aebi, C. Goldsbury: Monitoring biomolecular interactions by time-lapse atomic force microscopy. J. Struct. Biol. **131**, 171–180 (2000)

123. N.M. Kad, N.H. Thomson, D.P. Smith, D.A. Smith, S.E. Radford: Beta(2)-microglobulin and its deamidated variant, N17D from amyloid fibrils with a range of morphologies in vitro. J. Mol. Biol. **313**, 559–571 (2001)
124. H. Lin, R. Bhatia, R. Lal: Amyloid beta protein forms ion channels: implications for Alzheimer's disease pathophysiology. FASEB J: Off. Publ. Federat. Am. Soc. Exper. Biol. **15**, 2433–2544 (2001); erratum in FASEB J. **16**, 759 (2002)
125. R. Liu, C. McAllister, Y. Lyubchenko, M.R. Sierks: Residues 17–20 and 30–35 of beta-amyloid play critical roles in aggregation. J. Neurosci. Res. **75**, 162–171 (2004)
126. V. Baranauskas, I. Garavello-Freitas, Z. Jingguo: Observation of the bone matrix structure of intact and regenerative zones of tibias by atomic force microscopy. J. Vac. Sci. Technol. A **19**, 1042–1045 (2001)
127. H.F. Knapp, G. C. Reilly, A. Stemmer, P. Niederer, M.L. Knothe Tate: Development of preparation methods for and insights obtained from atomic force microscopy of fluid spaces in cortical bone. Scanning **24**, 25–33 (2002)
128. N. Nakabayashi, A. Watanabe, K. Igarashi: AFM observation of collapse and expansion of phosphoric acid–demineralized dentin. J. Biomed. Mater. Res. A **68**, 558–565 (2004)
129. E. Yonemochi: Assessment of drug characteristics by atomic force microscopy (AFM). Yakuzaigaku **63**, 129–132 (2003)
130. P.G. Royall, D.Q.M. Craig, D.M. Price, M. Reading, T.J. Lever: An investigation into the use of microthermal analysis for the solid state characterization of an HPMC tablet formulation. Int. J. Pharm. **192**, 97–103 (1999)
131. A. zur Muehlen, E. zur Muehlen, H. Niehus, W. Mehnert: Atomic force microscopy studies of solid lipid nanoparticles. Pharm. Res. **13**, 1411–1416 (1996)
132. A. Ringqvist, L.S. Taylor, K. Ekelund, G. Ragnarsson, S. Engstrom, A. Axelsson: Atomic force microscopy analysis and confocal Raman microimaging of coated pellets. Int. J. Pharmaceutics **267**, 35–47 (2003)
133. L. Mu, M.M. Teo, H.Z. Ning, C.S. Tan, S.S. Feng: Novel powder formulations for controlled delivery of poorly soluble anticancer drug: Application and investigation of TPGS and PEG in spray-dried particulate systems. J. Controlled Release **103**, 565–575 (2005)
134. T. Li, K. Park: Fractal analysis of pharmaceutical particles by atomic force microscopy. Pharm. Res. **15**, 1222–1232 (1998)
135. M.S. Schwarzenbach, P. Reimann, V. Thommen, M. Hegner, M. Mumenthaler, J. Schwob, H.J. Guentherodt: Topological structure and chemical composition of inner surfaces of borosilicate vials. J. Pharmac. Sci. Technol. **58**, 169–175 (2004)
136. A. Karbach, D. Drechsler: Atomic force microscopy – a powerful tool for industrial applications. Surface and Interface Analysis **27**, 401–409 (1999)
137. G.K. Bar, G.F. Meyers: The application of atomic force microscopy to the characterization of industrial polymer materials. MRS Bull. **29**, 464–470 (2004)
138. G.S. Blackman, L. Lin, R.R. Matheson: Micro- and Nano-Wear of Polymeric Materials. ACS Symp. Ser. **741**, 258–269 (2000)
139. M.C. Porte-Durrieu, C. Aymes-Chodur, N. Betz, C. Baquey : Development of heparin-like polymers using swift heavy ion and gamma radiation. I. Preparation and characterization of the materials. J. Biomed. Mater. Res. **52**, 119–127 (2000)

140. B.A. Konduk, A.H. Ucisik: A study on the characterization and biostability of used and virgin dialysis membranes and biocompatibility of the composite biomaterials. J. Aust. Ceram. Soc. **37**, 63–81 (2001)
141. P. Lemoine, J.M. Mc Laughlin: Continuity and topography of ultrathin diamond-like carbon films characterized by scanning electron microscopy/energy dispersive X-ray analysis and atomic force microscopy. J. Vac. Sci. Technol. A **17**, 176–182 (1999)
142. A. Portinha, V. Teixeira, J. Carneiro, M.F. Costa, N.P. Barradas, A.D. Sequeira: Stabilization of ZrO_2 PDV coatings with Gd_2O_3. Surf. Coat. Technol. **188–189**, 107–115 (2004)
143. A. Portinha, V. Teixeira, A. Monteiro, M.F. Costa, N. Lima, J. Martins, D. Martinez: Surface analysis of nanocomposite ceramic coatings. Surf. Interf. Anal. **35**, 723–728 (2003)
144. K. Yamana, M. Miyamoto, S. Nakamura, K. Kihara: AFM observations of ceramic surfaces. Key Eng. Mater. **111–112**, 57–70 (1995)
145. B. Baretzky, B. Reinsch, U. Taeffner, G. Schneider, M. Ruehle: Continuous microscopy of ceramic materials with atomic force microscopy. Z. Metallkd. **87**, 332–340 (1996)
146. M.G. Gee, N.J. McCormick: 3D topography measurements of ceramics. Nat. Phys. Laboratory (United Kingdom) Report CMMT(A) **35**, 1–13 (1996)
147. J. Burger, G. Dietler, M. Binggeli, R. Christoph, O. Marti: Aspects of the surface roughness of ceramic bonding tools on a nanometer scale investigated with atomic force microscopy. Thin Solid Films **253**, 308–310 (1994)
148. C.E. Rabke: Precision failure analysis and QC inspection. Adv. Mater. Processes **156**, 32–34 (1999)
149. S. Palomares-Sanchez, S. Ponce-Castaneda, F. Ruiz, M. Mirabal-Garcia, J.R. Martinez: Topographic characterization of barium hexaferrite powders using AFM. Scr. Mater. **34**, 1195–1200 (1996)
150. P. Falaras, A.P. Xagas: Roughness and fractality of nanostructured TiO_2 films prepared via sol–gel technique. J. Mater. Sci. **37**, 3855–3860 (2002)
151. A.E. Rokhvarger, L.A. Chigrinsky: Unconventional nanoparticle technology of superconductor ceramic articles. Mater. Res. Soc. Symp. Proc. **776**, 49–54 (2003)
152. S. Salgado, C. Perez, C. Diaz, J.G. Banuelos, J.M. Saniger: Residues as alternative raw materials in glass ceramic, Environment Conscious Materials: Ecomaterials. Proceedings International Symposium Ottawa, Aug. 20–23, 2000, 309–314, Publisher: Canad. Institute of Mining, Metallurgy and Petroleum, Montreal (2000)
153. B. Karmakar, G. Heide, G.H. Frischat: Comparative atomic force and scanning electron microscopic studies disclosing nanocrystallinity in cordierite glass–ceramics: Examples of surface modification. Indian J. Phys. A **78**, 47–51 (2004)
154. A. Bottino, G. Capannelli, A. Grosso, O. Monticelli, O. Cavalleri, R. Rolandi, R. Soria: Surface characterization of ceramic membranes by atomic force microscopy. J. Membrane Sci. **95**, 289–296 (1994)
155. M.M. Cortalezzi, J. Rose, G.F. Wells, J.Y. Bottero, A.R. Barron, M.R. Wiesner: Ceramic membranes derived from ferroxane nanoparticles: A new route for the fabrication of iron oxide ultrafiltration membranes. J. Membrane Sci. **227**, 207–217 (2003)

156. I.B. Leonor, A. Ito, K. Onuma, N. Kanzaki, R.L. Reis: In vitro bioactivity of starch thermoplastic/hydroxyapatite composite biomaterials: An in situ study using atomic force microscopy. Biomaterials **24**, 579–585 (2002)
157. L. Blunt, X.Q. Jiang: Three dimensional measurement of the surface topography of ceramic and metallic orthopaedic joint prostheses. J. Mater. Sci. Mater. Med. **11**, 235–246 (2000)
158. L.P. He, Y.W. Mai, Z.Z. Chen: Effects of anodization voltage on CaP/Al_2O_3-Ti nanometer biocomposites. Nanotechnology **15**, 1465–1471 (2004)
159. J. K. Bibby, N.L. Bubb, D.J. Wood, P.M. Mummery: Fluorapatite-mullite glass sputter coated Ti_6Al_4V for biomedical applications. J. Mater. Sci. Mater. Med. **16**, 379–385 (2005)
160. C. Liu, X. Li, F. Xu, P. M. Huang: Atomic force microscopy of soil inorganic colloids. Soil Sci. Plant Nutr. (Tokyo) **49**, 17–23 (2003)
161. T. Hirono, M. Takahashi, W. Lin: Pore space visualization in unlithified sediments and rocks. Gensh Bakk Kenk. **8**, 179–189 (2002)
162. A. Steele, D.T. Goddard, D. Stapleton, J.K. Toporski, V. Peters, V. Bassinger, G. Sharples, D.D. Wynn-Williams, D. McKay: Investigations into an unknown organism on the martian meteorite Allan Hills 84001. Meteorit. Planet. Sci. **35**, 237–241 (2000)
163. A. Kempe, F. Jamitzky, W. Altermann, B. Baisch, T. Markert, W.M. Heckl: Discrimination of aqueous and aeolian paleoenvironments by atomic force microscopy – a database for the characterization of Martian sediments. Astrobiology **4**, 51–64 (2004)
164. J. Xue, Y. Pan, G. Luo, H. Lu: Chemical components and microtopography of (100) surface for orthopyroxene. Chinese Sci. Bull. **48**, 931–934 (2003)
165. M.P.S. Krekeler, S. Guggenheim, J. Rakovan: A microtexture study of palygorskite-rich sediments from the Hawthorne formation, Southern Georgia, by transmission electron microscopy and atomic force microscopy. Clays Clay Miner. **52**, 263–274 (2004)
166. C. Marcos, I. Rodriguez, L.C. DeRenno, J.I. Paredes: Vermiculite surface structure as imaged by contact mode AFM. Europ. J. Mineral. **16**, 597–607 (2004)
167. Y. Gio, Z. Pan, G. Song: Metamorphic plastic deformation origin of jadeite jade and its simulating synthesis. Guisuan. Tong. **20 (Suppl.)**, 4–8 (2001)
168. K. Henriksen, S.L.S. Stipp, J.R. Young, M.E. Marsh: Biological control of calcite crystallization: AFM investigation of coccolith polysaccharide function. Am. Miner. **89**, 1709–1716 (2004)
169. P.J. Troy, H. Yuan, F.T. Mackenzie: Changes in surface morphology of calcite exposed to the oceanic water column. Aqua. Geochem. **3**, 1–20 (1997)
170. P.A. Maurice, Y.J. Lee, L.E. Hersman: Dissolution of Al-substituted goethites by an aerobic Pseudomonas mendocina var. bacteria. Geochim. Cosmochim. Acta **64**, 1363–1374 (2000)
171. A. Luettge: Crystal dissolution kinetics studied by vertical scanning interferometry and Monte Carlo simulations. A brief review and outlook. Nanoscale Structure and Assembly at Solid–Fluid Interfaces **1**, 209–247 (2004) (eds. X.Y. Liu, J.J. De Yoreo) Kluwer Academic Publishers, Norwell, MA. (2005)
172. M.A. Nugent, P. Maurice, S.L. Brantley : The field dissolution rate of feldspar in a Pennsylvania (USA) spodosol as measured by atomic force microscopy. Proceedings 9th International Symposium on Water–Rock Interaction, Taupo, N. Z. 387–390 (eds. G.B. Arehart, J.R. Hulston), Balkema Publ. Rotterdam, Neth. (1998)

173. H.H. Teng, P. Fenter, L. Cheng, N.C. Sturchio: Resolving orthoclase dissolution processes with atomic force microscopy and X-ray reflectivity. Geochem Cosmochim Acta **65**, 3459–3474 (2001)
174. S.H. Sutheimer, P.A. Maurice, Q. Zhou: Dissolution of well- and poorly crystallized kaolinites: Al specification and effects of surface characteristics. Am. Miner. **84**, 620–628 (1999)
175. T. Sugiura, T. Yoshida, H. Minora: Microstructural Observation of Photoelectrochemically Tailored Nano-Honeycomb TiO_2. Electrochemistry (Tokyo) **67**, 1234–1236 (1999)
176. K.D. Kepler, A.A. Gewirth: In situ AFM and STM investigation of electrochemical hydride growth on Ge(110) and Ge(111) surfaces. Surf. Sci. **303**, 101–113 (1994)
177. Y.X. Zhou, R. Naguib, H. Fang, K. Salama: Development of cube textured Ni–W alloy tapes using powder metallurgy along with high energy ball milling for HTS coated conductors. Superconductor Sci. Technol. **17**, 947–953 (2004)
178. K. Hanada, N. Nakayama, M. Mayuzumi, T. Sano, H. Takeishi: Tribological properties of Al–Si–Cu–Mg alloy-based composite-dispersing diamond nanocluster. Diamond and Related Materials **11**, 749–752 (2002)
179. I. Calliari, M. Dabala, G. Brunoro, G.M. Ingo: Advanced surface techniques for characterization of Cu-based artifacts. J. Radioanal. Nucl. Chem. **247**, 601–608 (2001)
180. T.M. Ko, W.C. Ng, W.T. Chen: Zincation treatments for electroless nickel under-bump metallurgy in the flip-chip packaging. Proc. Electrochem. Soc. **99-9**, 340–351 (1999)
181. R. Ramanauskas: Structural factor in Zn alloy electrodeposit corrosion. Appl. Surf. Sci. **153**, 53–64 (1999)
182. T.R. Yanez, Y. Houbaert : Characterisation of multiphase steel surface topography by atomic force microscopy. Sonderb. Praktisch. Metallograp. **32**, 79–82 (2001)
183. K.H. Koch: Analysis and metallurgy. Part 2. CLB Chem Labor Biotech. **52**, 128–133 (2001)
184. J. Flis: Scanning tunneling microscopy (STM) and atomic force microscopy (AFM) and their application in corrosion studies. Ochr. Przed Koroz. **41**, 34–39 (1998)
185. K. Minoshima, K. Komai: Nanometric visualization of localized damage by scanning probe microscopy. Mater. Sci. Res. Int. **2**, 209–219 (1996)
186. H. Krawiec, V. Vignal, E. Finot, O. Heintz, R. Oltra, J.M. Olive: Local electrochemical studies after heat treatment of stainless steel. Role of induced metallurgical and surface modifications on pitting triggering. Metal. Mater. Trans. A **35**, 3515–3521 (2004)
187. B. Schluter, G. Barkleit, F. Schneider, K. Mummert: Analysis of intergranular corrosion attack of stainless steels by means of atomic force microscopy and optical microscopy. Part 2: influence of impurity content. Mater. Corros. **51**, 115–119 (2000)
188. L. Cretegny, A. Saxena: AFM characterization of the evolution of surface deformation during fatigue in polycrystalline copper. Acta Mater. **49**, 3755–3765 (2001)
189. K. Kowal, J. DeLuccia, J.Y. Josefowicz, C. Laird, G.C. Farrington: Atomic force and scanning electron microscopy of corrosion and fatigue of an aluminum–copper alloy. Mater. Res. Soc. Symp. Proc. **409**, 201–206 (1996)

190. K. Minoshima, Y. Oie, K. Komai: In situ AFM imaging system for the environmentally induced damage under dynamic loads in a controlled environment. ISIJ Int. **43**, 579–588 (2003)
191. P.A. Zhdan, J.E. Castle: The corrosion of admiralty brass in ultra pure water measured by means of AFM and solution analysis. Proc. Electrochem. Soc. **2001–22**, 822–834 (2001)
192. J.W. Schultze, B. Davepon, F. Karman, C. Rosenkranz, A. Schreiber, O. Voigt: Corrosion and passivation in nanoscopic and microscopic dimensions: The influence of grains and grain boundaries. Corr. Eng. Sci. Technol. **39**, 45–52 (2004)
193. T. Martin, K.R. Hebert: Atomic Force Microscopy study of pit formation during anodic etching of aluminum. Proc. Electrochem. Soc. **99–27**, 671–680 (1999)
194. J. Jakubowicz, H. Jungblut, H.J. Lewerenz: Initial surface topography changes during divalent dissolution of silicon electrodes. Electrochim. Acta **49**, 137–146 (2003)
195. B. Mao, X. Cai, C. Shi, Z. Xie, J. Tang, Y. Chen, Z. Tian, Z. Tian: Application of electrochemical scanning probe microscopes in characterization and modification of solid–liquid interfaces. Fud Xueb., Zir. Kex. **38**, 7–12 (1999)
196. J. Y. Josefowicz, L. Xie, G.C. Farrington: Observation of intermediate CuCl species during the anodic dissolution of Cu using atomic force microscopy. J. Phys. Chem. **97**, 11995–11998 (1993)
197. E. Rocca, G. Bertrand, C. Rapin, J.C. Labrune: Inhibition of copper aqueous corrosion by non-toxic linear sodium heptanoate: mechanism and ECAFM study. J. Electroanalyt. Chem. **503**, 133–140 (2001)
198. K. Uosaki, K. Konishi, M. Koinuma: Electrochemical atomic force microscopy study on the surface structure of a lead electrode during redox processes and surface atomic arrangement of electrochemically formed $PbSO_4$ in H_2SO_4 Solution. Langmuir **13**, 3557–3562 (1997)
199. M.U. Kamachi, Y. Katada: Electrochemical atomic force microscopic studies on passive films of nitrogen-bearing austenitic stainless steels. Electrochim. Acta **46**, 3735–3742 (2001)
200. T.J. Mckrell, J.M. Galligan: Electrochemical atomic force microscopy study of the dissolution kinetics of 304 stainless steel. Mater. Res. Soc. Symp. Proc. **451**, 549–554 (1997)
201. J. Romer, M. Plaschke, G. Beuchle, J.I. Kim: In situ investigation of U(IV)–oxide surface dissolution and remineralization by electrochemical AFM. J. Nucl. Mater. **322**, 80–86 (2003)
202. J.P. Bearinger, C.A. Orme, J.L. Gilbert: Effect of hydrogen peroxide on titanium surfaces: in situ imaging and step-polarization impedance spectroscopy of commercially pure titanium and titanium, 6-aluminium, 4-vanadium. J. Biomed. Mater. Res. A **67**, 702–712 (2003)
203. M.L. Occelli, S.A.C. Gould, G.D. Stucky: The study of the surface topography of microporous materials using atomic force microscopy. Stud. Surf. Sci. Catal. **84 Pt.A**, 485–492 (1994)
204. M.L. Occelli, S.A.C. Gould, B. Drake: Atomic force microscopy examination of the topography of a fluid cracking catalyst surface. ACS Symp. Ser. **571**, 271–293 (1994)

205. M.L. Occelli, S.A.C. Gould: The use of atomic force microscopy (AFM) to study the surface topography of commercial fluid cracking catalysts (FCCs) and pillared interlayered clays (PILCs) cracking catalysts. Stud. Surf. Sci. Catalysis **149**, 71–104 (2004)
206. Y.M. Li, Z.C. Chen, D.S. Lu, X.F. Lin: Topography and electro-catalysis property of horseradish peroxidase assembly active film. Wuli Huaxue Xuebao **19**, 474–477 (2003)
207. T. Nakagawa: Scanning probe microscope for DNA sequence determination. Jpn. Kokai Tok. Koho JP 06289017 A2 (1994)
208. J. Vesenka, E. Morales: Scanning probe microscopy in biology with potential applications in forensics. Proceedings of the 3rd Symposium Atomic Force/Scanning Tunneling Microscopy, Baltimore, May 9–12, 1998, 31–48, Publisher Kluwer Academic, New York (1999)
209. J.R. Smith: A quantitative method for analyzing AFM images of the outer surfaces of human hair. J. Microsc. **191**, 223–228 (1998)
210. S.P. Gurden, V.F. Monteiro, E. Longo, M.M.C. Ferreira: Quantitative analysis and classification of AFM images of human hair. J. Microsc. **215**, 13–23 (2004)
211. R.L. McMullen, S.P. Kelty: Investigation of human hair fibers using lateral force microscopy. Scanning **23**, 337–345 (2001)
212. S. Kasas, A. Khanmy-Vital, G. Dietler: Examination of line crossings by atomic force microscopy. Forensic Sci. Internat. **119**, 290–298 (2001)
213. M. D'Uffizi, G. Falso, G.M. Ingo, G. Padeletti: Microchemical and micromorphological features of gunshot residue observed by combined use of AFM, SA-XPS and SEM + EDS. Surf. Interf. Anal. **34**, 502–506 (2002)
214. J. Sharma, J.P. Teter, R.J. Abbundi, N.A: Guardala. Determination of radiation exposure history of common materials and computer hardware by using atomic (and magnetic) force microscopy. Appl. Phys. Lett. **82**, 2236–2238 (2003)
215. A. Kempe, J.W. Schopf, W. Altermann, A.B. Kudryavtsev, W.M. Heckl: Atomic force microscopy of Precambrian microscopic fossils. Proc. Nat. Acad. Sci. USA **99**, 9117–9120 (2002)
216. R. Avci, M.H. Schweitzer, R.D. Boyd, J.L. Wittmeyer, F.T. Arce, J.O. Calvo: Preservation of bone collagen from the late cretaceous period studied by immunological techniques and atomic force microscopy. Langmuir **21**, 3584–3590 (2005)
217. K.D. Jandt, M. Finke, P. Cacciafesta: Aspects of the physical chemistry of polymers, biomaterials and mineralized tissues investigated with atomic force microscopy (AFM). Coll. Surf. B **19**, 301–314 (2000)
218. C.Z. Fan, S. Minoru, I. Miyoshi, A. Tadahiro, M. Mitsuo, A. Ishikawa, Y.S. Wu, Z.C. Li: Analyses of composition and structure on the ancient bronze black mirror "Hei–Qi–Gu" and search into the treating process of its surface layer. Sci. China, Ser. B **37**, 1421–1428 (1994)
219. V.A. Zazhigalov, S.I. Klimovskii, V.V. Strelko, A.V. Zazhigalov, A. Kowal: Structure and morphology of the surface of glazes from ancient, Kiev Vizantia. Dopovidi Natsional'noi Akademii Nauk Ukraini 120–124 (2001) Chem. Abstr. **137**, 93303 (2002)

220. M. Schreiner, G. Woisetschlager, I. Schmitz, M. Wadsak: Characterization of surface layers formed under natural environmental conditions on medieval stained glass and ancient copper alloys using SEM, SIMS and atomic force microscopy. J. Anal. Atom. Spectrom. **14**, 395–403 (1999)
221. J. Roque, T. Pradell, J. Molera, M. Vendrell-Saz: Evidence of nucleation and growth on metal Cu and Ag nanoparticles in lustre: AFM surface characterization. J. Non-Cryst. Solids **351**, 568–575 (2005)

2

Scanning Near-Field Optical Microscopy

2.1 Introduction

The history of SNOM (scanning near-field optical microscopy) [other acronyms are NSOM (near-field scanning optical microscopy), PSTM (photon scanning tunneling microscopy), and STOM (scanning tunneling optical microscopy) — the same acronyms are used for the near-field microscopes] started with the idea of a very small aperture (smaller than $\lambda/2$ of the light) for the light source. The confined light passing through small apertures may break the Abbe diffraction limit of microscopic resolution. This technique is overwhelmingly used in common SNOM equipment even though the light is collected in the far field. The possibility for apertureless SNOM with uncoated illuminated tips and reflection back to the fiber was first proposed by Courjon et al. [1, 2] who tried to use a phase effect for the reflection at near-field distance ($<\lambda/2$) but such effect could not be usefully applied. It was only a new and apparently unexpected or even denied [3] near-field effect in shear-force distance that allowed for SNOM on rough surfaces in constant distance (more precisely in constant vibration damping gap) with sharp uncoated tapered waveguides [4–9]. It is the strongly enhanced reflection back into the very sharp fiber at shear-force distance that allows for SNOM on "real-world" rough surfaces without topological artifacts up to high slopes for all types of transparent and opaque samples [4–9], where hot and blunt metal-coated tips cannot follow the topology.

Scanning techniques are able to provide the topology of the rough surface and the light interaction in SNOM provides the additional information that can also be imaged upon the topography. Depth of focus has been a severe problem for normal light microscopy at resolutions above the $\lambda/2$ resolution limit. Modern solutions for that problem are confocal microscopy and digital microscopy. The latter technique is also able to obtain 3D topographic images of surfaces with optical information. It overlaps with SNOM at the larger area ranges with remarkable depth of focus capabilities that is briefly mentioned in Sect. 2.13 even if it is not a nanotechnique.

2.2 Foundations of SNOM

The history of SNOM and its relation to further microscopic techniques has been reviewed [10] and is not repeated here. The discussion in [10] is limited to the idea of aperture SNOM, which breaks the diffraction limit with a submicroscopic light source. The light is passed through a submicroscopic aperture of metallized devices. Submicroscopic local resolutions are achievable but only on test samples or samples without topology. This is valid for all different modes of aperture SNOM (illumination mode, collection mode, reflection mode). The major problems are the heat (all three illumination modes) and bluntness (emission and all three collection modes) of metal-coated tips that cannot reasonably be operated in shear-force distance of less than 10 or 5 nm to a surface. Therefore, measurements are mostly restricted to constant height techniques over very flat surfaces necessarily with interference errors and topologic artifacts if topology is present. Furthermore, shadowing effects are difficult to treat. The technical problems in the manufacture of suitable tips are manifold, because rather thick metal coatings are required for eliminating light leakage. The smaller the aperture is, the more heated is the blunt tip with flat end to very high temperatures ($100 -> 470°C$) [11,12] and that in very close distance to the scanned surface! In sharp completely coated tips the melting point of Al at 660°C is easily reached. The heat production has been used to 'vaporize' the aluminum coating at the tapered tip end for the production of sharp 400 nm far-field apertured SNOM tips that can be used for positive submicron lithography on photostable light absorbing organic crystals by local surface melting [13](Sect. 2.12). Furthermore, use of the heat on metal-coated tips for producing pits has been described in [14]. Light leakage of the metal-coated tips causes artifacts and reduces contrast, and these effects can be only partly overcome by modulation techniques. The propagating modes become evanescent and are strongly damped if the diameter of the waveguide is smaller than $\lambda/2$ (cut-off). Therefore the tapered region must be as short as possible. The radiation intensity scales with the sixth power of the aperture radius, the limit being about 50 nm [10]. Far-field detection of the evanescent light (that has become propagating by the tip–sample interaction) through the transparent sample in transmission mode will be dependent on the nonuniform bulk composition. The latter cannot be assessed from the surface experiment looking for local variations in composition. Various types of apertured tips and the relation to other techniques that break the diffraction limits have been exhaustively described in [10]. These include flat apertures, small quartz or Si_3N_4 tips sticking out of the aperture, metal-coated tetrahedral tips with one edge uncoated leading to a localized excitation of a metal particle at the tip, and hollow metal pyramids. Tetrahedral tips connected to tuning fork distance control have also been realized [15]. However, the scan rates are very slow ($500\,\mathrm{nm\,s^{-1}}$), samples need to be transparent, and heights above 24 nm were not tested. Also active probes are discussed [10]. For example, a submicroscopic diamond crystal that was previously exposed to 2-MeV electrons was attached to a SNOM fiber probe and its fluorescence used for near-field imaging [16]. Alternatively, an apertured metal coated tip may be additionally coated with fluorescent molecules embedded in a polymer such as PMMA and the fluorophore excited by absorption of short wavelength light through the fiber. Problems of bleaching out are less severe with fluorescent nanoparticles that are embedded in the additional coating. For example it is possible to have only one CdSe nanoparticle (core: 3.8 nm diameter; 1 nm ZnSe shell) over the whole 150 nm aperture, and this would pave way for molecular resolution (<10 nm) at room temperature after coupling-in of 458-nm light [17]. It was however not pointed out how the heat problem of apertured tips could be handled in that endeavor. Test samples and flat surfaces have been tested with these techniques, mostly in terms of resolution (>50 nm) and "edge contrast" well below the diffraction limit of classical light microscopy. The various typical difficulties with aperture tips in terms of multiple artifacts, contrast, sensitivity, topology, distance or height regulation, interference (between probe edge and the sample), and resolution have been collected and theoretically discussed [18]. However, the so-called "credibility requirements" for near-field optical images that are claimed ("no correlation between topographic and near-field image" and "correlated structures must be displaced by a constant amount between topology and optical image") contradict the actual goal of SNOM, which is detection of chemical contrast either with or without topology (e.g., metal patterns on or in supports). Such chemical contrast must occur precisely at the site of the chemical species in order to be really credible. Displacements (by some casual asperity at the metal-coated tip or for other reasons) are also a type of artifact (Sect. 2.6.5). Noninterpretable images are of little practical value.

At this point it appears suitable to compare the various principle techniques of SNOM that have been used. They are grouped in Fig. 2.1. Most of them require very flat surfaces such as monolayers or molecular species on mica or graphite but are not usable on rough "real-world" surfaces. The exception is method C, where very sharp tapered tips are used apertureless, that is without metal coating.

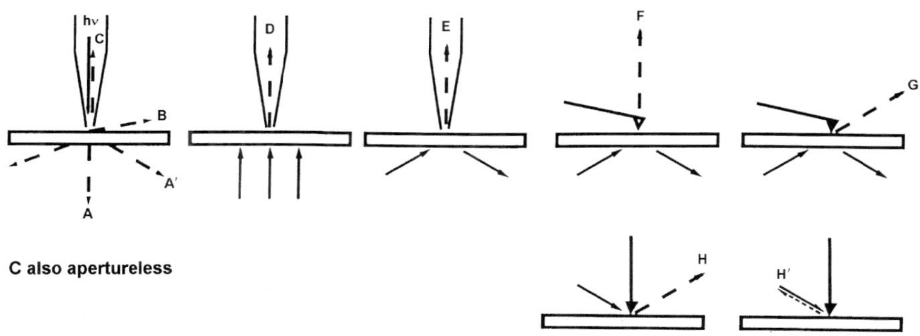

Fig. 2.1. The various principles of SNOM techniques; aperture SNOM (A, A', B, C) with metal-coated light sources, D and E with separate source; apertureless SNOM by light tunneling (F, G), scattering with double modulation (H, H′), and under shear-force regulation with uncoated tips according to the C geometry

Apertureless SNOM is still less popular but possible according to four principally different techniques.

Photon tunneling (PSTM) relies on a collection of forbidden light illuminated through the sample at angles within the range of total reflection by a sharp transparent (F) or opaque tip (G) on the flat surface [19]. Interference occurs between the tip field and the background field, and the fringes may obscure the SNOM image [20]. Furthermore, topology and local changes of the refractive index combine to the optical signal. Separations have been tried by a distance modulation technique [21].

The second technique SIAM (scanning interferometric apertureless microscopy) uses double modulation (vertical vibration of the tip and lateral of the sample) with reflection of the light from an interferometer back to the interferometer (H'). The phase shift due to scattering at an AFM tip at the other side of a transparent plate with reference to a second spot without such scattering is scanned. The modulations decrease the background signal. Resolutions down to 1 nm are claimed [22, 23]. However, the origin of the high-resolution features remained unexplored (microscope cover slip) [22] or the image is recognizably artificial (dark/bright edge contrast of mica and droplets is seen in Figs. 2.2, 2.3 of [23]). Few samples of interest are clear transparent glasses and permissible surface topology must be very low. The liftoff artifact of the tip pulling away from the surface (it goes out of laser focus) can only be corrected for samples with topology variations less than 20 nm.

The third apertureless technique uses near-field scattering. It focuses laser beams to the apex of metal tips as scattering devices that are in tapping mode to the surface, in order to obtain field enhancement under the metal tip (H). The dither serves the demodulation of the signal at higher harmonics for reducing the background of scattering from both the illuminated probe shaft and the sample [24, 25]. The optical near-field interaction signal is extracted by the demodulation procedure. Clearly, this technique is restricted to very flat surfaces. No improvements provide excited single fluorophores at the tip end [26], including FRET (fluorescence resonant energy transfer) in the near-field [27]. For example, a 50 nm gold step on glass gave a much-extended artificial stripes contrast (cf. Sect. 2.6.2) without discontinuity at the edge site over more than 10 μm when a *p*-polarized beam scattered at a silicon tip [28]. Coherent imaging of plasmon patterns used a carbon nanotube tip bundle as the scatterer with heterodyne detection and second or third harmonic demodulation [29]. The 91 nm wide gold disks (20 nm height on glass) revealed dipolar oscillation, however, both phase and amplitude are structured and therefore not valid materials contrast even if they closely relate to the tip-broadened 140 nm topology. Still, the near-field scattering theory does not seem to cover the complete physics: far infrared ($\lambda = 150\,\mu m$) near-field imaging (150 nm resolution) relies on ac coupling of the tip-surface system instead of scattering, and this mechanism may also play a role at near infrared and perhaps even visible imaging [30]. More detailed descriptions of the theory, various experimental setups, and artifacts of the two types (amplitude and phase or intensity) of scattering apertureless SNOM are given in [31]. Simulations of the near-field around silver tips have been published [32].

The fourth apertureless choice is illumination through and reflection back to the very sharp uncoated tapered fiber in shear-force distance at a horizontal dither at 100–400 kHz, which uses a new physical effect for the local light collection in the very close near-field: enhancement of reflectivity into sharp dielectric tips in the shear-force gap [4–9]. The sharp uncoated tip is used both as an emitter and a detector in that technique. This apparently unexpected new experimental knowledge will certainly help to better understand the phenomenon of shear-force in very close distance that is still subject to debate, despite enormous theoretical efforts [33]. The favorable technique uses crossed polarizers for illumination and detection in order to efficiently suppress undesired primary stray light and light reflected from the front end of the fiber. However, the polarization is not maintained in the apertureless SNOM for "real-world" samples. The technique has therefore nothing in common with depolarization SNOM using uncoated tips at constant height (instead of constant distance) and a polarization manipulator [34]. Edge contrast but not chemical contrast is obtained in depolarization SNOM.

As previous theoretical modeling efforts did not predict the experimentally found internal reflection enhancement [4–9] we do not refer to all of these studies here. Even the constructors of the commercial instrument and the supplier (DME, Copenhagen) were apparently not aware of the enhancement necessity [35]. In fact, the now experimentally secured reflection enhancement in the shear-force gap is the only reason for the good performance of the apertureless shear-force SNOM mode, excluding topologic artifacts. It does not help that simulation of the intensity on the basis of the vector Helmholtz equation in the strict regime of linear optics claimed the possibility for near-field resolution for the reflected intensity orthogonal to the incident polarization for an uncoated tip with 60 nm diameter at the apex and 30 nm distance [36]. These parameters are absolutely unsuitable for the experimental conditions of artifact-free SNOM, which works only under the conditions of enhanced reflectivity that requires end radii of <20 nm and distances in the range of 5 nm (see below). Furthermore, the objections against constant distance SNOM in terms of topologic artifacts [37] do not apply under the favorable conditions of the new near-field enhanced reflection back to the sharp fiber, a fact that has been largely overlooked in the past, when blunt (etched) and therefore easily broken tips were used.

The difference between aperture and apertureless tapered tips in technique C (Fig. 2.1) may be demonstrated with an on-scale comparative sketch, which also indicates the shear-force condition (distance <10 nm, typically 5 nm) (Fig 2.2). Clearly, hot metal-coated tips require thermal stability of the sample and absence of topology, while ambient sharp tips can easily follow the topology with slope angles up to the tip gradient with shear-force distance control even on delicate organic or biological surfaces. Experimental evidence shows that the sharp tip must have an end radius of <20 nm for being a suitable acceptor for strongly enhanced near-field light within the shear-force gap.

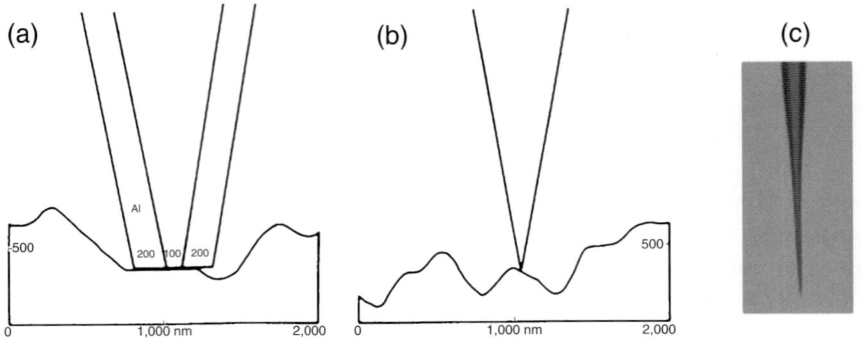

Fig. 2.2. Sketch of (**a**) a typical metal-coated tip, (**b**) a sharp uncoated tip, and (**c**) photo of an uncoated tapered tip indicating the high aspect ratio without resolution of the end radius, of course

Furthermore, very sharp tips have no problem with interference phenomena or very steep temperature gradients. However, the illuminated spot under the uncoated tip is rather large (about 1 µm in diameter). This means we need the large difference in far-field and near-field coupling of the reflected light intensity back to the tip if near-field resolution can be achieved.

We are able to experimentally demonstrate the new near-field effect with great ease: very sharp tips collect reflection within the near-field gap much more efficiently than far-field stray light [4–9]. An XY plotter in the approach procedure for the SNOM experiment monitors such enhanced reflectivity (Fig 2.3). Approach of tapered laterally vibrating tips (near the mechanical resonance at frequencies of 100–400 kHz with 5 or 10 nm free amplitude) for shear-force AFM is subdivided into fast coarse approach with retraction by 1 µm and stop as triggered by the damping control and a slow close approach from that waiting position. This works both with dark and illuminated waveguides. Figure 2.3 depicts the reflected intensity during coarse approach of the laterally vibrating illuminated tip in its left part. When it feels the start of damping the shear force control triggers the immediate retraction and stops in order to prevent penetration into the surface and breaking there. At the same time a short flash of reflected light is recorded at the surface point. The right part of Fig. 2.3 traces the close approach. From the waiting position cautious approach is started down to the preset damping efficiency (usually 30–60%), and the scanning of SNOM under shear-force distance control (simultaneously AFM) is started thereafter at again largely increased optical signal in shear-force distance. Figure 2.3 clearly exhibits a constant far-field stray light signal at the detector of the reflected light level through the fiber that is independent of the tip surface distance. Only at the surface point and during scan when the tip is in shear-force distance is the remarkable signal

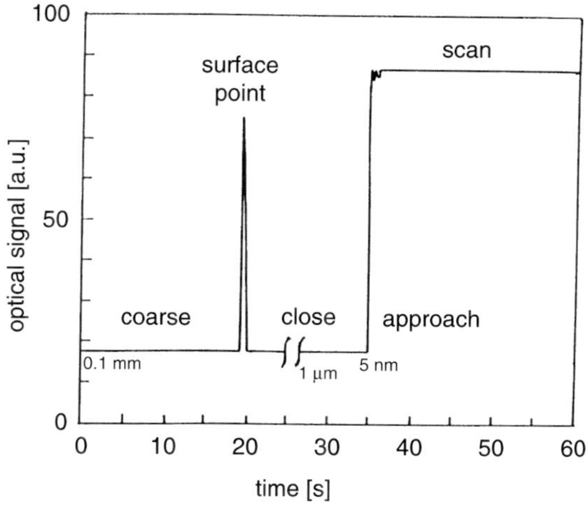

Fig. 2.3. Approach of a sharp illuminated multimode tapered quartz glass tip to a typical surface while monitoring the signal level showing highly enhanced light intensity level while the tip is in shear distance

enhancement seen. Thus, the scan in constant shear-force distance monitors only variations in the near-field enhanced signal, and the constant far-field part does not contribute to the contrast. Due to this unexpected strong enhancement effect in the shear-force gap we get near-field resolution and avoid "crosstalk" with the AFM detection of the topology. Hence, our near-field effect in the shear-force gap secures that topology does not influence the optical contrast (Fig. 2.3). Clearly, it works only in shear-force distance and only with sharp tips (final radius <20 nm), and the optical resolution is in the range of the tip radius or somewhat better.

The sharpness requirements are severe (<20 nm apex radius). Thus, the "new probes for scanning near-field optical microscopy" uncoated at lengths from 15 to 150 µm and with apex radii from 200 to 2000 nm as offered by LovaLite (Troyes, France) are totally unsuitable for our technique with shear-force enhancement of the reflection: neither the enhancement will occur nor the topology (Fig. 2.2) can be followed.

Actually, our near-field enhancement effect for the collection efficiency in the shear-force gap does not work for blunt or most (exception in Fig. 2.16) broken tips. These do not exhibit the enhancement but may even decrease the signal level with respect to the far-field level while in shear-force distance. This is clearly shown by the approach behavior of a typical broken tip (similarly with blunt tips) in Fig. 2.4. Here the intensity decreases at the surface point and during alleged shear-force scanning. In the absence of the enhanced coupling of shear-force gap reflected light no SNOM signal is obtained but only far-field artifacts. These are presented in Sect. 2.6. As the enhancement is absolutely necessary (the SNOM signal level should be more

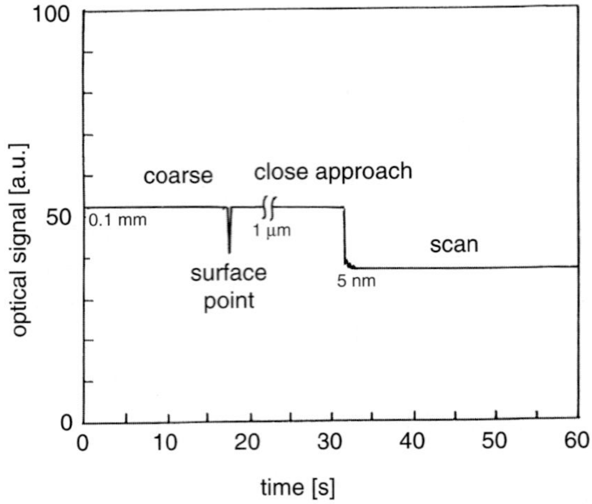

Fig. 2.4. Approach of a broken multimode tapered quartz glass tip to a typical surface while monitoring the signal level; only artificial far-field artifacts are obtained under such unsuitable conditions

than at least twice the far-field level and may reach factors of up to >50) every single measurement must control the signal level. Tips may be blunt from the beginning, or they may abrade and break upon handling, approach, or measurement.

Sharp fused quartz tips can be routinely pulled (as sharp as possible) to end radii between 20 and <10 nm (detected by SEM), and are the more stable the smaller the radius is. Etched tips [14] are blunt (end radius about 60 nm) and do not reach the required sharpness. These experimental requirements may be the reason for previous failures or topological artifacts with this apertureless technique that is easier and more versatile than any other SNOM technique (Fig. 2.1) and can be used for the more important applications to (very) rough "real-world" surfaces. Resolution limits are as yet slightly below 10 nm, topological limits are encountered at extremely steep (depends on the taper angle) or extended vertical slopes and overhangs that are not suitable for any scanning probe technique. Impossible topology will give artificial contrast next to valid SNOM contrast in "normal" regions of the sample.

Applications of SNOM to rough and very rough surfaces that are most common in nature and industry require a stable technique and the easy and cheap access to the probe tips. The chemical (materials) contrast in SNOM requires a chemically or physically non-uniform surface. It is more diverse than the contrast in conventional microscopy (illuminated topology, absorption, fluorescence, phase, polarization, Rayleigh, Raman). Topologic errors are absent in shear-force gap reflection-back-to-the-fiber SNOM if the tip is sharp and not broken and if a considerable enhancement of the reflected intensity is realized. The enhancement factor (>2 to >50 times the far-field level) depends on the substrate and the instrumental alignments; it needs to be recorded during the scanning. The chemical contrast differentiates very sensibly. It contains the microscopic contrast features and also the variations of the shear-force efficiency between the composite partners and electrostatics (constant damping at different distances) and the specific enhancement qualities that vary between the different composites in surprisingly large ranges. Rankings are difficult due to so many different influences, but highly reflecting materials (metals, salts, polar substances) tend to produce higher enhancement factors. There are also enormous differences in the penetration depth of the light, and the influence of surface layers on the shear-force distance at a preset damping is very decisive. But all of that belongs to the chemical contrast. Even closely related organic compounds in a composite can be easily differentiated. The reason for the close-near-field contrast remains largely undetermined. Fluorescence and Raman SNOM contrast can be locally analyzed by diffraction and spectroscopic analysis. The so-called "edge contrasts" (these are not materials contrast!) are of little practical interest as edges, protrusions, and ditches are more easily characterized by AFM. Only chemical (materials) contrast is of practical use in SNOM.

2.3 Equipment

Aperture SNOM equipment has been well described in [10] and is not repeated here. We also do not repeat the unusually complicated setups of scatter-type apertureless SNOM. Very sharp uncoated monomode or better multimode tapered quartz glass waveguides with end radii of 10–20 nm are required in the shear-force reflection back to the fiber technique. Both mono- or multimode-fused quartz waveguides with core diameters of 3.4 or 5.7 and 125 μm-step index coating at lengths of 60–120 cm are suitable for sharp pulling. The tested optical fibers were coaxial and did not preserve polarization. They are easily and highly economically pulled with a well-adjusted laser puller such as the P-2000 micropipette puller (Sutter Instruments, Navato, CA) at the proper settings. The more stable are the pulled tips with respect to breakage, the smaller the radius that may approach 10 nm or below (typically 15 nm). There is the risk of breakage in shear-force distance particularly with sticky materials or in the absence of a reasonable shear-force mechanism (e.g., on platinum). Therefore, etched tips [14] that are more brittle and blunt (end radius typically 60 nm) are unsuitable also due to their high tendency for breakage (breakage does not apply if these are metal coated for the fabrication of aperture tips). Figure 2.2 is an on-scale sketch for a typical situation of shear-force AFM and distance regulation. It indicates that successful use of apertured "tips" (irrespective of their enormous heat in shear-force distance) cannot be substantiated also for geometric reasons, even if there were a small protrusion at the end.

The various functions of the robust commercially available RASTERSCOPE SNOM 4000 of Danish Micro Engineering A/S (DME), Copenhagen are depicted in Fig. 2.5. A detailed description of its construction is given

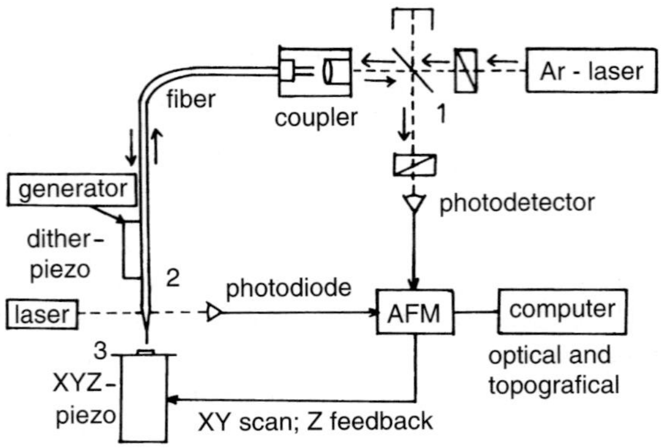

Fig. 2.5. Block diagram of an apertureless reflection-back-to-the-fiber SNOM setup with shear-force distance control and cross-polarization to minimize straylight; **1**: beam splitter and crossed polarizers; **2**: shear-force arrangement; **3**: sample mount on a piezo stage

in [38]. This setup uses light-controlled constant damping control (2) for maintaining the shear-force distance of the horizontally vibrating tip (100–400 kHz) (distance variations at different surface components are part of the chemical contrast). The He/Ne laser beam is shadowed by the edge of the vibrating fiber and provides the feedback signal for maintaining the preset amplitude (cf. Fig. 1.72, Sect. 1.10). The damping level may be chosen. Typical settings are between 30% and 60%. The shear-force mechanisms are described in Chap. 1, Sect. 1.10. The fast light control allows for scan speeds of up to 200 μm s^{-1} following high and varied surface topology in large scans with shear-force feedback. This is highly superior to the slow mechanical (piezo) or acoustic (tuning force) control. Any preference of the apparently more popular slow techniques at the expense of the most rapid trigger for the adjustment of the vibrational amplitude has to be paid with enormous difficulties as the tips are at risk, the sensitivity decreases, and extremely slow scan rates are the consequence. Importantly, shear force control also provides a simultaneous topologic image in addition to the optical image without crosstalk. All variations in the enhanced near field light are chemical contrast due to all effects that determine its efficiency. The free lateral oscillation amplitude is 5 or 10 nm. Any artifacts (deviations from true chemical contrast) are easily recognized [9] and covered in detail in Sect. 2.6.

The light source is a high-quality laser (e.g., an Ar^+ laser, emitting at 488 nm) focused to a coupler through a beam splitter and a crossed polarizer for effective elimination of straylight that would add to the back coming reflected light (1). The tapered coaxial optical waveguide does not retain polarization. This or the degree of polarization at the end of any polarization conserving tips can be checked with a polarizer in the far-field of the free tip end. The light passes through the uncoated optical waveguide and produces a concentric light spot of about 1 μm diameter on the sample at a piezo table (3) (typical energies are 60–80 μ W). The optically controlled feedback is fast enough so that scan rates between 10 and 120 μm s^{-1} can be achieved at a resolution of 512 × 512 pixels. Fast scanning may be required for minimizing loss of contrast by heating of strongly light absorbing surfaces [5]. The feedback dynamics has been checked with porous silicon samples at a rate of 50 μm s^{-1} in Fig. 1.75 or with 70 μm s^{-1} in Fig. 1.76 (Sect. 1.10). The light that is reflected back through the fiber consists of a constant far-field background and the enhanced reflection in the shear-force gap (typical energy 1 μ W). It is decoupled, separated from the primary straylight, detected by a photodetector with photomultiplier, and processed to an optical image by a computer, which also processes the AFM information in an independent channel. An energy meter must constantly control the signal level. Suitable edge filters or interference filters may also separate primary light from fluorescence light or the reflected light may be coupled to an optoacoustic spectrometer with waveguide optics for spectral analyses (local fluorescence and Raman). The whole setup is very robust and easily installed as shown in Fig. 2.6.

2.4 Optical Resolution

The reflection-back-to-the-fiber SNOM that works on the basis of the new close-near-field effect is able to achieve subwavelength resolution on rough and very rough surfaces comparable to or even better than the concomitant shear-force AFM resolution, due to the strongly enhanced near-field collection in the shear-force gap. However, maximal resolution is not usable in mostly used large scans of 10 μm × 10 μm with a pixel size of 18 nm. That range is, however, more important or typical for practical applications on "real-world" surfaces. However, there are also important technical questions with nanoparticles, and maximal resolution of apertureless SNOM has to be assessed. As described in Sect. 2.2, the end radius of the tapered uncoated tip varies from <10 to 20 nm, depending on the pulling result.

A first indication of the high resolution power is obtained from fluorescing nanoparticles that swam up in a resin. The depth of the resin cover on top was judged to be 5 nm as the AFM width was consistently found 10 nm larger than the fluorescence SNOM contrast [39]. In another experiment an autoxidized skew (slope 10–15°) (110)-surface of anthracene had formed structured features of anthraquinone with a pronounced chemical contrast. The peak-to-peak distance of the optical contrast obtained with a sharp uncoated tip was measured in several instances to <20 nm [6]. This is an excellent resolution for apertureless SNOM that cannot be obtained with any other SNOM technique on such sample. Similarly, a polycarbonate substrate gave protrusions on the polymer with bright contrast and sub-20 nm resolution as shown in Fig. 2.7. A similar resolution result (15 nm) was claimed in the literature [38].

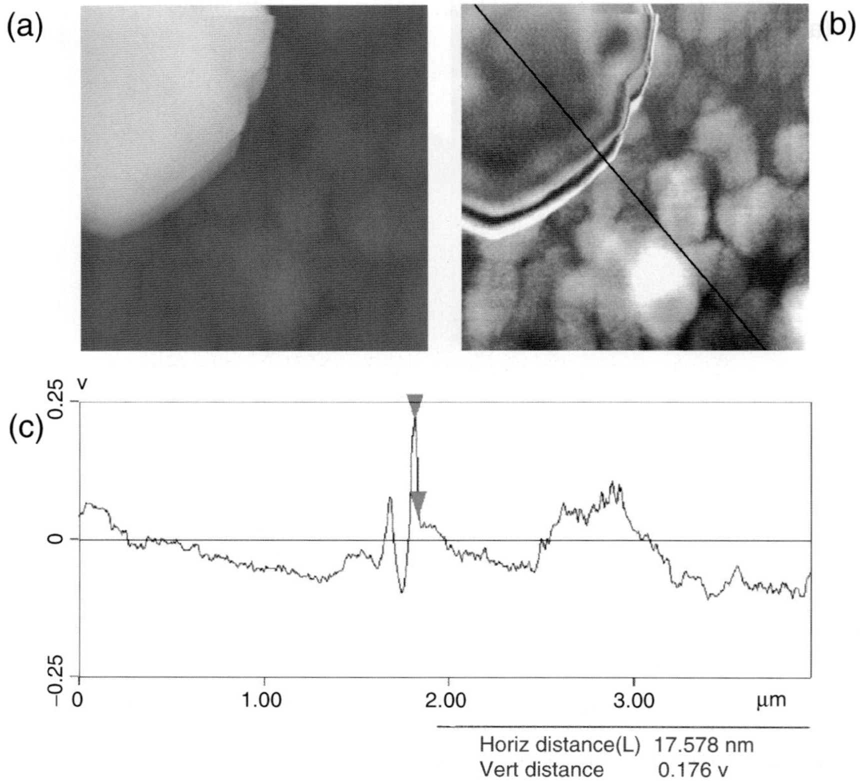

Fig. 2.11. Simultaneous shear-force apertureless AFM (**a**) and SNOM (**b**) at a 1 µm vertical edge of a sodium chloride crystal on glass; the optical image at the edge contains only artificial but not materials contrast; the section (**c**) shows the high resolution of the stripes

shear-force measurements. But these obstacles do not usually occur on "real-world" natural or technical surfaces. Sticky surfaces are difficult.

2.5 Dependence of the Reflectance Enhancement from the Shear Force

The efficiency of the shear-force damping ($100\,d/d_0$, where d is the set amplitude and d_0 the free amplitude of vibration) of the horizontally vibrating quartz glass tapered tips is dependent on the surface material in question. Thus, for example 50% damping will be obtained at different distances on different materials. This feature already is an important part of the chemical contrast, which is highly distance dependent. Also the optimum for the apertureless SNOM measurement has to be found for every different sample as a

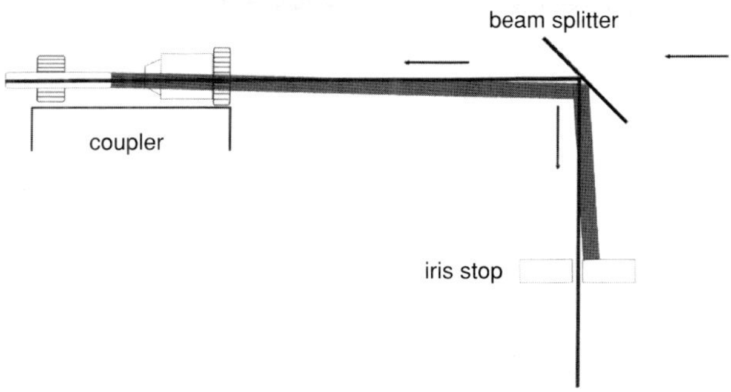

Fig. 2.12. Iris stop as an additional tool for the elimination of primary straylight

compromise between depth of optical contrast and risk of breaking the tip at too close a distance. Hence it is both of theoretical and practical importance to know more about reflectance enhancement and damping, which is related to distance.

The enhancement factor F (factor $F = I/I_0$, where I is the total intensity and I_0 the far-field background intensity) is strongly dependent on the exclusion of primary straylight that should be further diminished by an iris stop as sketched in Fig. 2.12. This feature increases the optical contrast by removing the reflected light from the waveguide cut end, which is never perfectly orthogonal to the incoming beam direction.

The enhancement factors F are different for different materials and can reach values of up to $F \geq 50$ (found on Al with ambient Al_2O_3 and water layer on it), while factors of $F \geq 2$ are sufficient for pure optical contrast at poorly near-field reflecting compounds. Approach curves to various preset damping values have been carefully measured with various materials. This is exemplified in Fig. 2.13 for two series of experiments for gold surfaces that had been treated with hydrogen sulfide in order to achieve a shear-force mechanism with the shortest possible self-assembled sulfide monolayer and in Fig. 2.14 for finely polished dolomite. These curves are only obtainable in a certain range of damping [40]. Too low settings will not properly engage while too high settings will abrade the tip and thus decrease the efficiency of the enhanced reflection. The sensitivity of the process is obtained from the linear part of the curves, and all points of a curve have been obtained with the same tip by successively increasing the preset damping value, up to final decline of the factor F. The tips were abraded then and sometimes broken. Independent tests secure that tips that have been used only in the linear part retained their sharpness, while those that experienced the horizontal or declining part of the approach were abraded. Clearly, the optimum for the safe SNOM measurement is to be sought in the middle of the linear part of the curves. It is of high interest that the increase of the signal or the factor F is linearly rather than exponentially

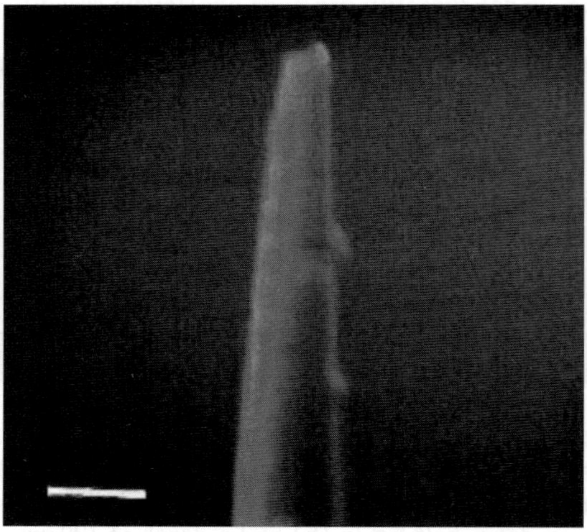

Fig. 2.17. SEM image (30,000 ×; bar: 500 nm) of the broken tip with protrusion at the *right* edge

1.01 μm wide elongated all features in scan direction, and the optical image was identical with the topology image. As we have glass the breakage may also result in a largely irregular torso. Slow continuous abrasion tends to increase the end radius of the tip. Tips may also be broken before measurement or during approach, and the appearance of the artificial images will be very different depending on the shapes of tip and surface.

2.6.2 Stripes Contrast

The artificial stripes contrast might be circular and straight. Figure 2.18 contains both types. Nanoparticles of 100–200 nm size (partly aggregated) that swam up under an afterward hardening resin were scanned with a particular broken tip that produced an AFM topology with sharp tip imaging and gave circular fringes around the small protrusions (up to 40 nm high) or almost straight stripes in the nonenhanced artificial optical image. These types of artifact in the optical image are easily recognized and occur in many published images that were taken with other SNOM techniques either apertureless (e.g., [1]), or when using apertures.

An optical stripes artifact has also been discussed with a very sharp tip at a vertical edge (Fig. 2.11). The reason here is the extreme topology that cannot be followed even by a sharp taper. The stripes contrast is recognized at the alternation of dark and bright parallel zones. This occurs with blunt tips not only at isolated surface protrusions but also at extended positive features. An example is shown with a partly autoxidized crystal of 2-mercaptobenzothiazole on (001), which exhibits islands of the corresponding disulfide that were imaged with a blunt taper at a 10 μm scan with shear-force AFM. The AFM

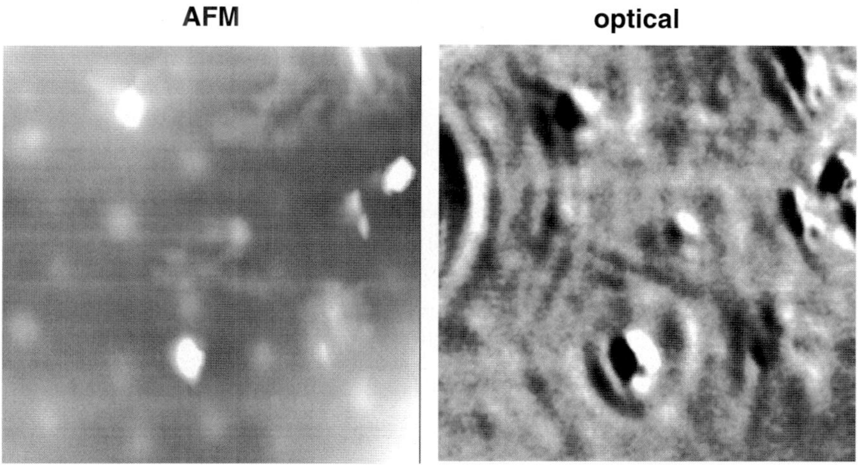

Fig. 2.18. Shear-force AFM (10 µm, $Z = 100$ nm) and optical image of 100–200 nm wide nanoparticles embedded in a resin obtained with a broken tip showing tip imaging and artificial optical features that are related to topology (upward *dark*, downward *bright*, see also Sect. 2.6.3) as circular fringes (there was a particle out of the *left* edge of the image) and some stripes

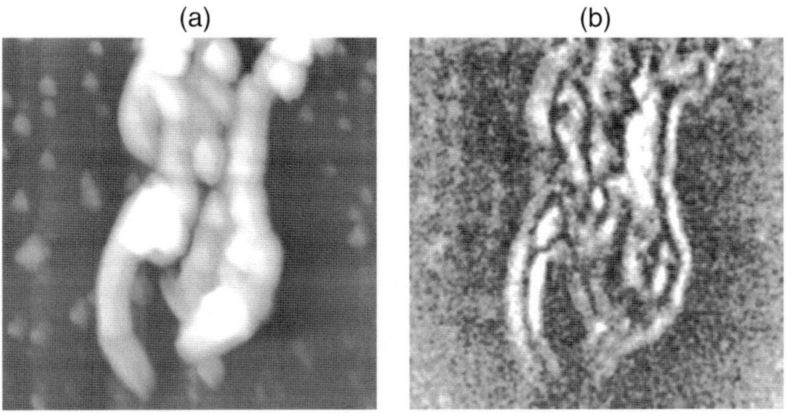

Fig. 2.19. Simultaneous shear-force AFM (**a**) (11 µm, z-range: 800 nm) and artificial SNOM scan (**b**) on partly autoxidized 2-mercaptobenzothiazole with a blunt tip showing some tip imaging and artificial stripes contrast at the sites of the islands

exhibits clear signs of tip imaging (rhombs) and the optical contrast is an artificial stripes contrast (Fig. 2.19), because there is no enhancement to the blunt tip. The topology is easy, the slope angles ranging from 10° to 22°. However, blunt tips experience severe problems at these slopes. In fact, the topology is not extreme for a sharp tip, and a dark contrast of the disulfide with respect to the brighter thiol is obtained under the conditions of enhanced near-field reflection (Sect. 2.7, Fig. 2.28). The poor performance of a 100 nm wide tip probing 282 nm latex spheres is also recognizable in [43] where a stripes contrast is to be seen.

It should be noted again that a steepness contrast is unavoidable even for very sharp tips if the slope angles are larger than the gradient of the tip (for example Fig. 2.11).

Interesting though artificial images are obtained with blunt or broken vibrating tips if they scan along extended slope areas, even if such slopes would not pose any problems to sharp tips. Series of parallel stripes are obtained along the rims and slopes that, of course, do not tell any useful optical properties but those that are artificial. We present here sample images for the classification of the artifact that may be easily avoided by using sharp tips. Fluorescent PET fibers (polyethyleneterephthalate) were embedded in a common resin (Technovit, Kulzer, Wehrheim, Germany) and cut with a microtome. The topologies in Figs. 2.20a,c were obtained using blunt tips. The optical images (b) and (d) at the two different sites did not profit from signal enhancement in shear-force distance and are thus artificial. The five stripes in Fig. 2.20b down right derive from a blunt tip exploring a uniform slope area with a tilt of 32° over a range of 1.5 μm in constant shear-force distance. In Fig. 2.20d the ring stripe artifacts were created at slopes of 40–70°, the steeper areas being to the right in Fig. 2.20c. Such slopes are not troublesome for sharp tips in enhanced near-field fluorescence collection [39] (Sect. 2.11.4). Clearly, the artificial optical response in Figs. 2.20b,d is not a useful fluorescence contrast.

Interference fringes have also been obtained in PSTM imaging of a 91-nm latex sphere in constant height mode making use of image enhancement [44]. Clearly, such optical response is artificial. It appears that similar artifacts should also be discussed in [45] where etched fibers were in use. Figures 2.3, 2.4, 2.6, 2.8, 2.11, and 2.12 there show some resemblance with the present Fig. 2.20 and are likely to be the result of blunt and/or broken tips. The stripes contrast is different from the "edge-contrast" that was described in [34]. The origins of these latter stripes are largely unknown.

2.6.3 Inverted Derivative of the Topology in the Optical Response

If the tip is so severely blunt that small protrusions do not give a reasonable topology in the shear-force AFM the simultaneous optical contrast may be an inverted derivative of the distorted topology. An example is given for a sample of 100–200 nm nanoparticles embedded in a polyvidone resin (Fig. 2.21). The AFM image (a) depicts the tip geometry as scanned by the sharper nanoparticle as the probe (tip imaging) with the cross section given. The optical artifact (b) is not the result of a contrast enhancement differentiation. But a similar image can also be created by highpass filtering of the AFM artifact in Fig. 2.21a followed by inversion. An enhancement factor of only $F = 1.3$ was

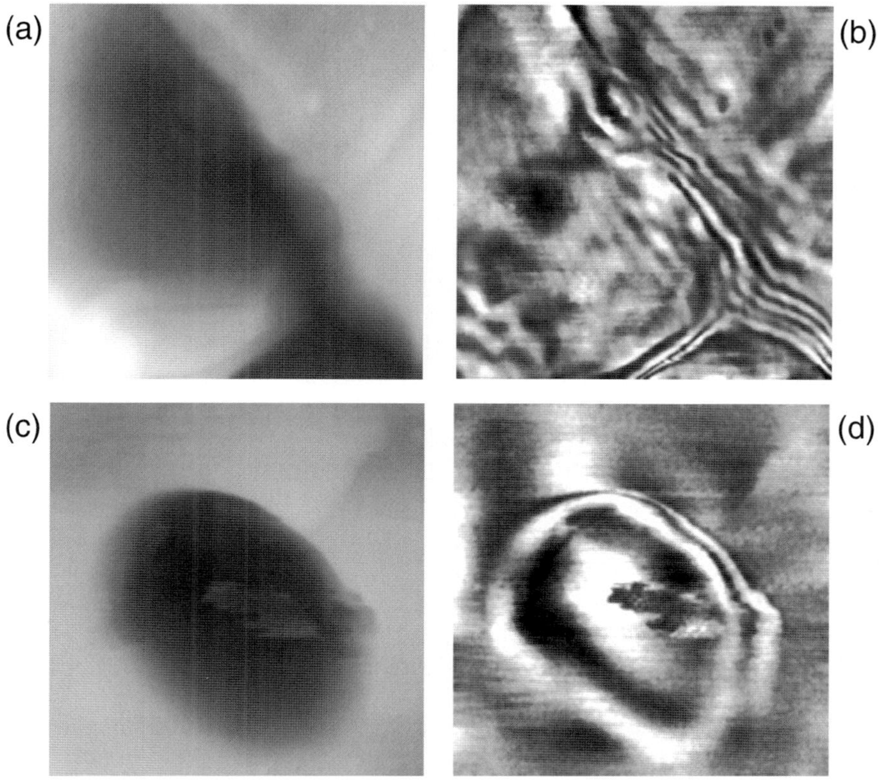

Fig. 2.20. Simultaneous shear-force AFM and SNOM scans on two sites of a microtome cut of a fluorescent PET fiber with a broken tip: (**a**) AFM (10 μm, z-scale: 4 μm); (**b**) artificial optical stripes contrast; (**c**) AFM (2 μm; z-scale: 2 μm); (**d**) artificial optical stripes contrast

recorded probably because the axial symmetric tip had a tiny protrusion on its center. The bright circle in (b) reflects the shoulder ring in the huge distorted topology of (a) [9]. Also Fig. 2.18 has some indication of a negative derivative at the protrusion sites in addition to the stripes artifact in its optical part. A similar optical feature can be also seen in [45] (there Fig. 2.11) reporting a SNOM experiment with an etched (that means blunt) taper. It should now be interpreted as an artifact. Figure 2.21 has nothing in common with the frequently applied technique of contrast enhancing in constant height mode by differentiation as in [44], where PSTM interference fringes surrounding a sphere were contrast enhanced by such technique.

It must be stressed that a sharp tip on the same sample as in Fig. 2.21 gives the correct size of the nanoparticles and the correct fluorescence SNOM contrast (Sect. 2.11.4, Fig. 2.77).

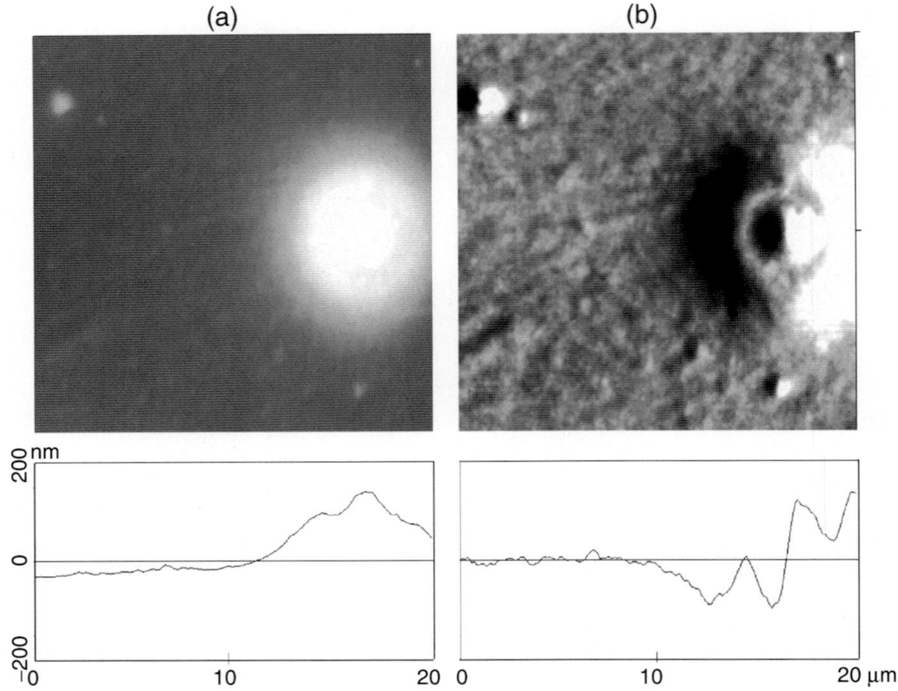

Fig. 2.21. Simultaneous artificial AFM (**a**) (20 µm, z-scale: 400 nm) and artificial SNOM scan (**b**) of fluorescent nanoparticles (100–200 nm) embedded in polyvidone with an axially symmetric broken tip showing the optical image like an inverted derivative of the diffuse AFM response also in five small features; the sections for (**a**) and (**b**) are horizontal through the centers of the large features

2.6.4 Inverted Contrast

There are some reports of contrast reversal during SNOM measurements in the literature [3]. The authors used tips with end radii of 50–100 nm. However, these break very readily upon scanning [4–9, 40]. It was proposed that dust particles had adsorbed to the tip and hung from it. With coated tips the interpretation of false contrast invoked a concept of shear-force phase effects [46]. It is our experience that contrast inversion may occur upon tip breakage during scanning, the occurrence of that event being immediately detected by the drop of the energy meter reading for the reflected light. Another report claimed 15 nm optical resolutions at 2–5 nm high Al island steps with 40% of the tips, whereas 60% of the tips provided inverse contrast that cannot be near-field but must be far-field contrast [38]. No reasons for the difference were given and no enhancement phenomena were reported. Again, the reason seems to be quite obvious: the deficient measurements must be the result of blunt and/or broken tips that do not profit from the enhancement in the shear-force

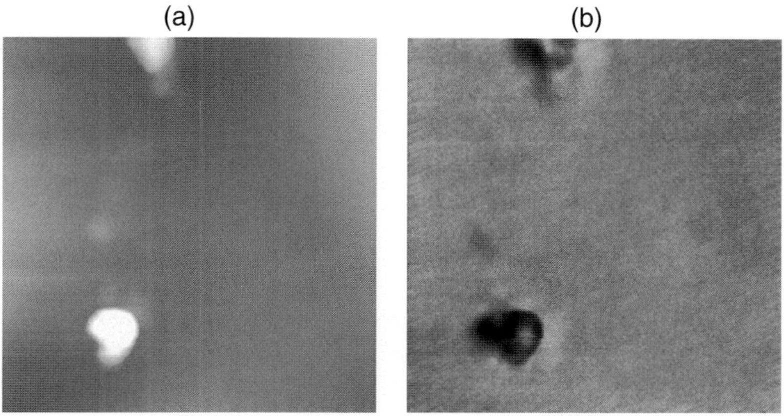

Fig. 2.22. Simultaneous artificial AFM (10 μm, z-scale: 300 nm) and artificial optical fluorescence scan (an edge filter for elimination of the 488-nm light was applied) with an apparently asymmetric broken tip showing tip imaging in the topology (**a**) and split false contrast in the optical image (**b**)

gap but collect far-field light at a much lower level of intensity. The sign of the chemical contrast (brighter or darker) cannot be securely predicted as various influences are at work. However, sign change will be recognized when the tip breaks during measurement (see Fig. 2.15). When starting with a blunt tip a faulty sign of an optical contrast may ensue. Accompanying satellites clearly recognize the artifact. Figure 2.22 gives an example for a fluorescing nanoparticle that must give a positive contrast as the primary 488-nm light was eliminated by a cutoff filter. However, it is inverted and has the diagnostic satellite because of a broken tip. The topologic features in (a) derive from artificial tip imaging. The nanoparticles are only 100–200 nm wide but the brightest feature has a diameter of 1.3 μm and a height of 135 nm. The optical image (b) provides exclusively artificial far-field contrast; the fluorescence is totally obscured. Valid fluorescence SNOM contrast using a sharp tip on the same sample is presented in Sect. 2.11.4 (Figs. 2.76 and 2.77).

Generally, either positive topology or negative topology may give bright or dark optical contrast in valid SNOM if it is chemical contrast. For comparisons a true negative optical contrast of an island structure is added in Fig. 2.23. It was measured with a sharp tip with the indispensable enhancement [8]. Importantly, the coastline of the chemically different island is closely followed in the optical image, and despite the topology there is almost no loss in resolution.

2.6.5 Displaced Optical Contrast

The "golden rule" for valid chemical contrast requires the optical signal precisely at the site of the reflecting species be it in a pit, on a hill, or at

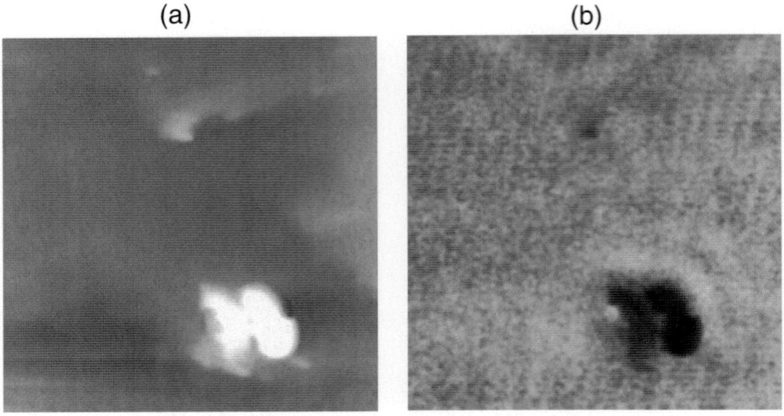

Fig. 2.23. Simultaneous shear-force AFM (**a**) (3 μm, z-scale: 500 nm) and valid SNOM (**b**) images of 9,10-dibromoanthracene islands on anthracene as measured with a sharp tip, showing the negative chemical contrast at the correct sites

a plane. Hence, the most important checks for the validity of any SNOM contrast (also in reflection-back-to-the-fiber mode) are precise site or topology correspondence of the optical contrast. If these conditions are not met one observes artifacts in the optical and very often also in the topologic image. A striking example of image displacement is shown in Fig. 2.24, where a Pt/C test pattern (Kramer test sample) [47] was scanned. This surface does not provide a useful shear-force mechanism but leads to wear of the tip. Abrasion of tips was detected by inspection with a light microscope after several scans. It is useless to judge the edge resolution of our SNOM (Sect. 2.4) with artificial images that become increasingly worse with any further scan.

The SNOM in Fig. 2.24b is indeed artificial: first, the dark features that seem to reproduce the topology are significantly displaced from the topology and second, there are bright areas to the right of seven dark features in Fig. 2.24b, although of a different character than in Fig. 2.21b.

A similar displacement of topologic and optical features is also recognizable in a paper that imaged the surface of an ITO sample using a blunt uncoated tip with 100-nm diameter [43]. A related problem occurred in the photon tunneling through a Si_3N_4 pyramidal tip when a chromosome was scanned in force contact: the optical image was largely displaced and some interference fringes appeared [44]. Clearly, the optical features in Fig. 2.8 of that study should be considered artificial. Also SIAM absorption contrasts after correction for the liftoff artifact reveal this type of artifact (Fig. 2.2 in [48], as yet undiscussed).

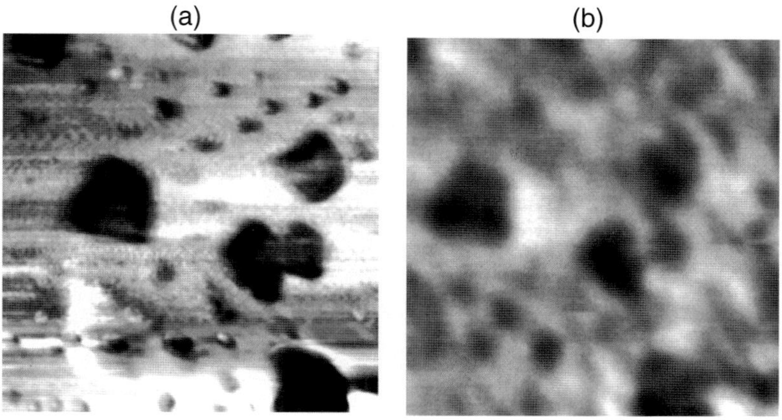

Fig. 2.24. Simultaneous shear-force AFM (**a**) (5 µm) and SNOM (**b**) of a Pt/C test pattern with an abraded tip showing noncoincidence of topologic and optical image; the optical response is artificial and there was no significant enhancement; dark areas in (**a**) are depressions down to the supporting glass of 2 nm height

2.6.6 Local Far-Field Light Concentration

A remarkable artifact may occur on uniform surfaces if a regular pattern concentrates the illuminating light at certain spots. The reflection efficiency will be higher there under regular conditions, but it is not materials contrast that is recorded. A gold-coated double grating with square blocks of 500 nm height offers favorable conditions for such an effect due to the highly reflecting side faces. Figure 2.25 was taken at a rate of 70 µm s^{-1} in order to minimize thermal effects by the light absorption of the gold layer [5]. It shows that a good shear-force resolution of the topology (a) is obtained as well as an optical contrast, which, however, cannot be materials contrast. Rather a far-field effect is imaged in (b). There is increased pickup of light if the tip is down between the steep walls from all sides and the illuminated spot is about 1 µm wide. The thermal effect (decrease of resolution at lower scan rate) is exceptional as gold absorbs about 37% of the incident light and cannot efficiently transport the heat to the insulating support. Excellent resolution that compares with the resolution obtained with nonilluminated tip is obtained at a rate of 100 µm s^{-1} [5].

2.6.7 Topologic Artifacts

A review of SNOM on rough surfaces has to take into account eventual topologic artifacts. These occur at extreme topology that also cannot reasonably be handled with AFM, depending on the tip gradient. Tapers with high aspect ratio may be usable up to 70° or 80° surface slopes if the shear-force control is

Fig. 2.25. Simultaneous shear-force AFM (**a**) (10 μm, *z*-scale: 1 μm) and optical far-field contrast (**b**) of a double grating of gold on an insulating polymer indicating increased pickup of light in the depth between the reflecting sidewalls but not a materials contrast

very rapid. Furthermore, artifacts occur by necessity if laterally vibrating tips are blunt or broken (Sect. 2.2, Fig. 2.4). Valid SNOM images must precisely correlate with simultaneous topology if that is caused by a chemical or physical state variation (Sect. 2.6.5). This requirement is the evident basis for chemical contrast and local spectroscopy. If the rough surface is chemically and physically uniform no contrast occurs in genuine SNOM. Furthermore, distinction of chemically different topologies of comparable heights in the same image is also possible due to their different contrast. If a sample consists only of bulk material there will be no SNOM contrast at its various sites. It has been unduly (by not knowing or not acknowledging the shear-force gap enhancement) claimed that such correlation of "true optical contrast" to surface topology should not exist even for uncoated tips [37,45,49]. Even further, it was unduly claimed "from general principles" that "all SNOM configurations at constant distance mode" contain topologic artifacts, "particularly with highly resolving sharp uncoated tips" [50]. While these considerations might relate to the artificial steepness contrast at (near) vertical steps, they are irrelevant for well-done reflection-back-to-the-fiber SNOM at slopes below the gradient of the tip. The opposite is true: images that do not exhibit correlation to the topology, if the latter is the reason for nonuniformness, are artificial. There are also some cases where chemical or physical change is created locally on a rough surface without significant change of the topology [5,6], or the reflecting species can be embedded closely under the surface (Sect. 2.4; Fig. 2.8). Only such cases exclude the topologic correlation. Of course, topology on chemically and physically uniform surfaces must not give an optical contrast if sharp uncoated tips with high aspect ratios are used (artificial steepness contrast occurs at slopes that approach or exceed 80°) [6, 8, 51] and if local far-field

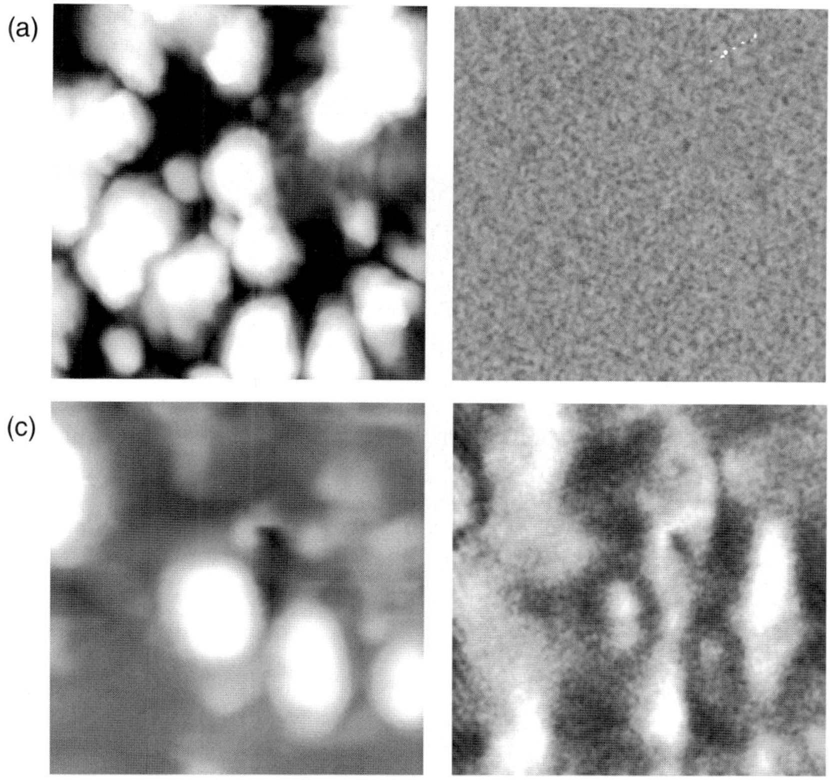

Fig. 2.26. Simultaneous shear-force AFM (**a**, **c**) (10 μm, z-scale: 500 nm) and SNOM (**b**, **d**) of large features on a uniform 4-nitroaniline surface; (**a**) and (**b**) are correct images with a sharp tip (enhanced reflectance) showing complete absence of an optical contrast because of chemical uniformity on the rough surface; (**c**) and (**d**) were measured with a blunt tip in the absence of significant shear-force gap reflectance enhancement and exhibit a bad AFM (tip imaging) and a severe optical artifact that is also recognizable by the absence of correlation with the "topology"

light concentration is excluded [5]. Not every topology is caused by nonuniformity. Thus, topology itself does not give rise to optical contrast in properly done apertureless reflection SNOM using the variations in the shear-force gap enhanced reflected light to uncoated sharp tips [7]. This is valid for both protrusions and pits on the same surface [39] and amply documented [5–8,13,52]. However, the favorable situation is lost if the sharp tip breaks or if blunt tips are used.

Due to the importance of this finding the correct situation is compared to the artificial measurement in a very rough chemically uniform surface in Fig. 2.26. It depicts the AFM and the optical image of a 4-nitroaniline surface on its (100) face after short application of NO_2 gas [53, 54]. The tip was sharp in (a, b) and blunt in (c, d). This difference can already be judged by the difference in sharpness of the AFM topologies as there is more tip–sample convolution in Fig. 2.26c. Figure 2.26d is totally artificial and does not contain any SNOM information as there was no enhancement of the reflectance, while Fig. 2.26b proves the uniformity of the sample surface at all sites, giving no change in enhanced reflectance at constant distance throughout.

It is important to repeat that very high uniform topology (Fig. 2.26a,b) does not give an optical contrast provided the uncoated tip is sharp and experiences the reflectance enhancement in the shear-force gap that is so easily recorded (Fig. 2.3). Furthermore and contrary to the undue cited statements, if the optical measurement on a rough surface is artificial (no enhancement) the optical contrast features do not fit to the topology, as shown in Fig. 2.26c,d. This calls for severe alertness if claims of "valid SNOM" are based on the mere nonequivalence of topological and optical image in the literature. For example, Fig. 2.26d is artificial!

Numerous successful SNOM measurements with precise site or topology correlation of chemical contrast or fluorescence contrast have been published [4–8, 13, 52, 55], whereas deviations from such close correlation or the appearance of optical contrast on chemically uniform surfaces (for example in [45]) indicate the recording of artifacts similar to the ones that are described for broken tips in this chapter.

The collection of the various appearances of artifacts in the optical images at unsuitable conditions of measurement is absolutely necessary for the judgment of SNOM results here and in the literature. There are, however, instances where both valid and artificial contrast occurs in the same image due to single impossible topological sites (e.g., sharp ridges). Therefore, reference to the sample pictures in this section is highly valuable for practical applications of SNOM, the majority of which is concerned with rough surfaces and these include all types of materials (oxides, salts, metals, semiconductors, glasses, organics, medicinal tissues, etc.). All of these in seven sections classified artifacts occur also on flat surfaces without topology under unsuitable conditions of every SNOM technique, and numerous examples from the literature have been cited in this chapter at the appropriate sites.

The following sections deal with successful SNOM measurements and identify artificial sites of the valid images where these cannot be avoided.

2.7 Application of SNOM in Solid-State Chemistry

The chemical contrast of SNOM is an invaluable tool for the evaluation of important mechanistic questions in solid-state chemistry. These are very diverse and must be elucidated in the nanoscopic range.

There exist questions of eminent practical importance. For example: why can crystals of highly oxidizable organic molecules be handled in ambient air? Surprisingly, crystals of mercaptobenzothiazole are remarkably stable toward

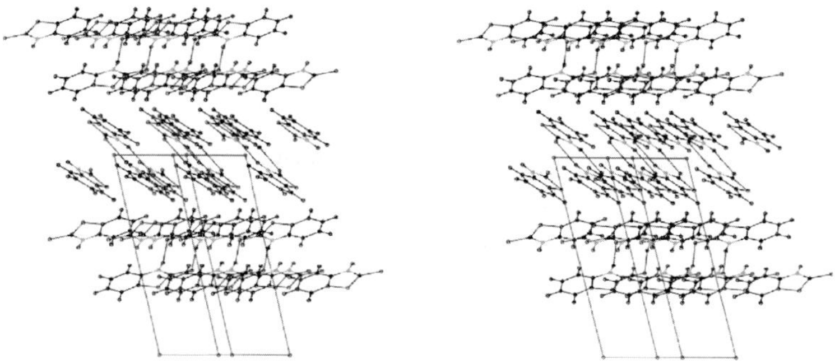

Fig. 2.27. Stereoscopic view of the molecular packing of 2-mercaptobenzothiazole with (001) on *top* showing flat double layers internally hydrogen bridged and thus with all functional groups shielded from chemical attack under (001)

autoxidation in air. They occur as thin plates with most prominent (001) faces [56]. AFM measurements on aged crystal surfaces revealed rare isolated islands indicating extremely impeded reactivity toward oxygen. The interpretation rests on the crystal packing, which exhibits double layers parallel to the surface by connecting single layers in pairs by hydrogen bonds. The hydrogen bonds connect two molecular entities each to give essentially planar eight-ring dimers while forcing all reactive groups to the interior of the double layers (Fig. 2.27). Therefore, autoxidation should be impeded by a single molecular layer, and it is concluded that the oxidation product (the corresponding disulfide) can only start to form at some defect sites and that the reaction spreads from there forming islands. Submicroscopic proof for this mechanism is provided by SNOM.

Figure 2.28 shows both the topology of such islands and by virtue of the dark chemical contrast at the sites of the islands that these are of different material (chemical contrast). Unlike Fig. 2.19 this measurement is valid as there was enhancement of the reflection in the shear-force gap. This clear-cut result is the necessary support for the different chemical composition at the islands and thus for the macroscopic effect of the uppermost single molecular surface layer of the bulk crystal structure. The material is not indefinitely stable, but it may be safely handled in ambient atmosphere for some technically important time.

A similar self-protection against autoxidation at the molecular level is observed with anthracene in its common crystalline appearance as scales with overwhelming (001) faces (Fig. 4.37, Chap. 4), where the molecules stand on their short side and hide the reactive 9,10-positions (stable for >6 months in ambient clean air with AFM precision). Similarly, only the reactive 9-position (but not also the 10-position) are present at the flat {110}-surface in the prism form of anthracene (Fig. 4.35), and the crystal is protected against autoxidation except at crystal faults [6]. Prism crystals of anthracene tend to exhibit

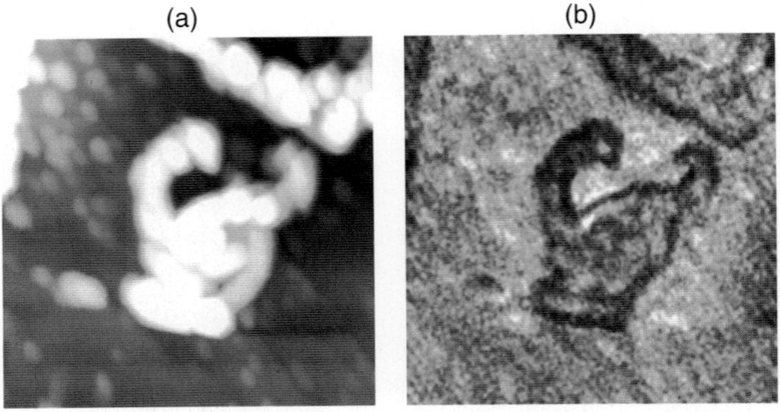

Fig. 2.28. Simultaneous AFM (**a**) (11 µm, z-range: 500 nm) and SNOM (**b**) of a partially autoxidized (001)-face of 2-mercaptobenzothiazole showing islands and negative chemical contrast precisely at the island sites

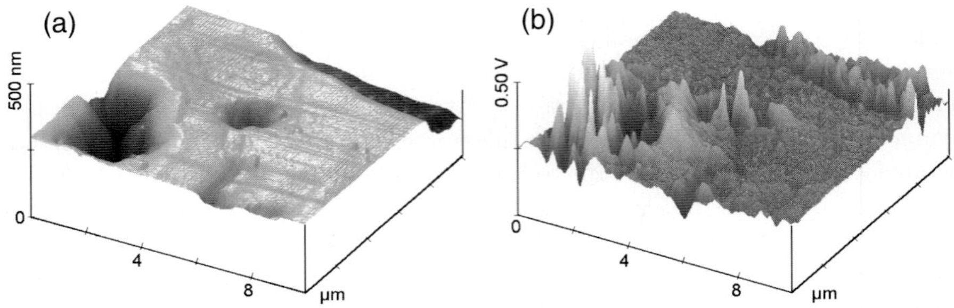

Fig. 2.29. Simultaneous AFM (**a**) (10 µm, z-range: 500 nm) and SNOM (**b**) surfaces of a {110}-face of anthracene with some faults that was pre-exposed to ambient clean air; the slopes rarely exceed 20° (but 30° is also observed); the SNOM image (**b**) indicates three types of contrast: normal at the flat sites and bright (positive) or dark (negative) at the slopes and rims due to different local advance of reaction

craters of up to 400 nm depths on their main surfaces. In that crater slopes both reactive centers are accessible for oxygen attack and oxidation to produce 9,10-anthraquinone can occur. Neither anthracene nor anthraquinone crystals absorb at 488 nm; however, an intermediate product, the charge transfer sandwich complex between both of them, does so heavily, as shown by the pure independently synthesized brown complex (absorption spectrum in [6]). Only SNOM can reveal the complicated situation at the crystal faults on {110} of anthracene prism crystals that were pre-exposed to air and can autoxidize there after an unpredictable induction period of weeks or months. The AFM topology in Fig. 2.29a cannot provide chemical information even though the rims on top of the slopes might point to some reaction with ambient oxygen.

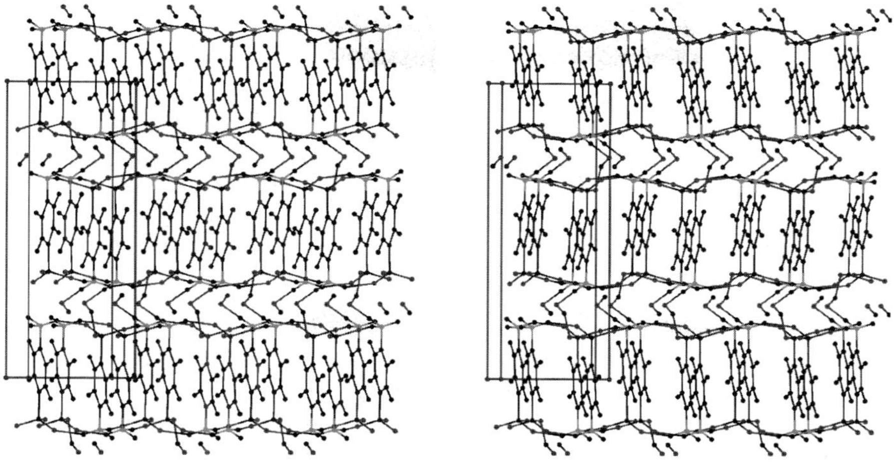

Fig. 2.30. Stereoscopic view of the molecular packing and hydrogen bonding of sulfanilic acid monohydrate along [100] (but rotated around Y by 10° for a better view) with (010) on *top* showing hydrogen-bridged strings of molecules that can only be chemically attacked from the sides after the (010)-surface has all accessible amino groups (any other molecule of the topmost layer) diazotized

However, SNOM proves that autoxidation has started at the slopes and at the rims by providing normal (anthracene), bright (anthraquinone), and dark contrast (light absorbing charge transfer complex). Clearly, no autoxidation is discernable at the flat sites of the surface. Further images are reported in [5]. If the prism form of anthracene crystals is avoided, this important material can be stored in (clean) air for very long time without being autoxidized.

Another practical problem of worldwide importance is the switch to chemical production that proceeds without producing wastes. A large number of unexpected though environmentally benign sustainable gas–solid and solid–solid reactions have emerged recently and have been mechanistically investigated with AFM [57]. The close correlation with the crystal packing and face selectivity endows them with credence, as they require far-reaching molecular migrations within the crystal, which were previously denied even in textbooks claiming the opposite by following Schmidt's topochemistry hypothesis [58]. SNOM must therefore secure the experimentally proven occurrence of waste-free chemistry for facilitation of the suitable industrial investments. The performance of SNOM might be demonstrated with the gas–solid diazotization of sulfanilic acid monohydrate to give the corresponding diazonium salt by the action of gaseous NO_2 [6, 59]. The AFM investigation shows that the gas–solid diazotization of sulfanilic acid monohydrate with gaseous NO_2 comes to a stop on its natural (010) face, whereas attack from the sides of the hydrogen-bonded infinite ribbons occurs very efficiently. This is clearly explained with the packing diagram of Fig. 2.30. It appears impossible to have the molecules migrate up over the (010)-surface, because the hydrogen bonds

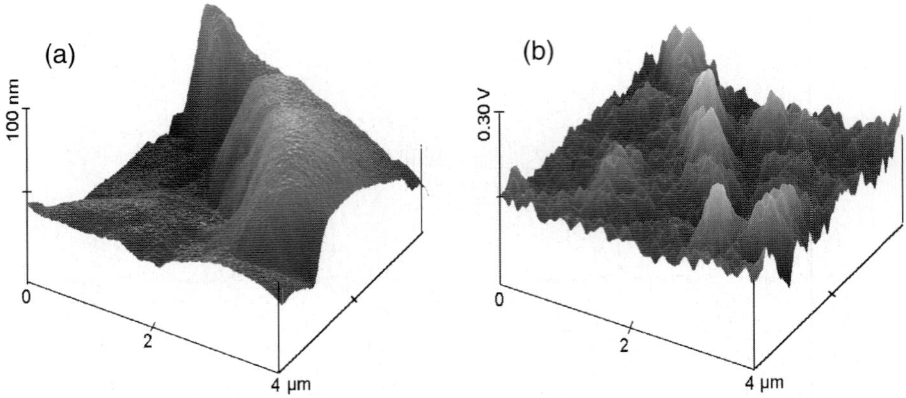

Fig. 2.31. Simultaneous AFM (**a**) and SNOM (**b**) of step sites on the (010)-face of sulfanilic acid monohydrate after short exposure to NO_2 gas (1:4 mixture with air injected from 1 cm distance and left for 1 min), showing bright (positive) chemical contrast precisely at the slope sites in the various directions

keep the molecular strings in their position preventing access to the second layer etc. Conversely, all layers are accessible from the sides of the crystal. This conclusion at the molecular level is unprecedented. It must and can be secured with SNOM at steps or slopes on the (010) surface. Figure 2.31 gives an example. The step site had been searched on the surface by AFM and did not give a SNOM contrast before the application of a small amount of gaseous NO_2 indicating chemical uniformity. Furthermore, the AFM image did not observably change at the chosen scale after the partial diazotization.

The bright contrast is only seen at the slopes. Thus, the surface is indeed only covered with a half layer of diazonium salt in the flat parts, whereas more profound diazotization occurs at slope sites from where the diazonium salt material gives rise to bright reflection. Longer-term diazotization leads to a complete reaction of the whole crystal without melting and without producing wastes, as was shown with powdered material and chemical analysis. Thus, the preparative and AFM evidence and interpretation has been nicely confirmed by SNOM. For the same goal, a controversy in solid-state photochemistry required scrutiny by AFM that had to be complemented by SNOM on rough surfaces. Thus, α-cinnamic acid, the standard of Schmidt's topochemistry (claiming absence of molecular migrations), gives unexpected (and previously denied) submicroscopic features as a consequence of anisotropic molecular migrations upon irradiation when the dimer α-truxillic acid is formed. The features correlate strictly with the crystal packing of α-cinnamic acid, and they therefore definitely prove the long-range molecular migrations within the crystal (Sect. 1.9). In no stage of the reaction could a SNOM contrast be obtained. Therefore, the photoreaction proceeds everywhere in the region of light penetration (down to 1,300 layers or through the whole crystal, depending on the

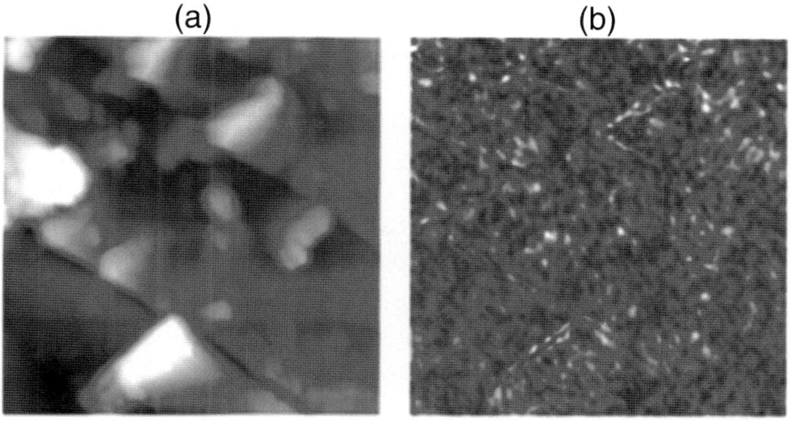

Fig. 2.32. Simultaneous AFM (**a**) (11 μm, z-range: 1.5 μm) and apertureless shear-force SNOM (**b**) of the (010) face of α-cinnamic acid after 15 min irradiation at $\lambda > 290$ nm showing anisotropically grown features in the AFM and only slight local steepness contrast but no chemical contrast in the SNOM image

wavelength for absorption), and the features do not differ in composition at rough sites or at flat parts. As the photodimer (without C=C double bonds) should reflect differently from the cinnamic acid (with a C=C double bond) any possibility for a cluster-by-cluster type photoreaction without continuous molecular migrations before crystal disintegration is additionally excluded as chemical contrast cannot be seen at and around low and μm high features that grow out of the bulk above the initially flat (010) surface. The result of Fig. 2.32, as well as numerous measurements at earlier stages of the photolysis under varied conditions, shows that no chemical contrast is discernible even at very high features. The very reliable performance of the applied apertureless shear-force SNOM is indicated by the very minor steepness contrast at the edges of the very high (>1 μm) and steep (45–55°) features in (b). There is no trace of a chemical contrast; the rough anisotropically structured surface is thus chemically uniform.

Similarly, a missing chemical contrast in SNOM has been used in the discussion of Fig. 1.23 (Sect. 1.7) for the exclusion of so-called "defect site photoreaction" of anthracene (with previous claims of absorbed photon energy migration to the defect sites) in favor of chemical reaction at the bulk sites where the light is absorbed [55].

For industrial purposes it is also important to understand consecutive chemical reactions. While apertureless shear-force SNOM on rough surfaces is able to help in the development of sustainable and waste-free reactions, it is also able to differentiate consecutive chemical reactions. For example, phthalimide, a heterocycle with many polar groups, can both hydrate the surface and chemically react to give phthalic acid monoamide. It is a useful example

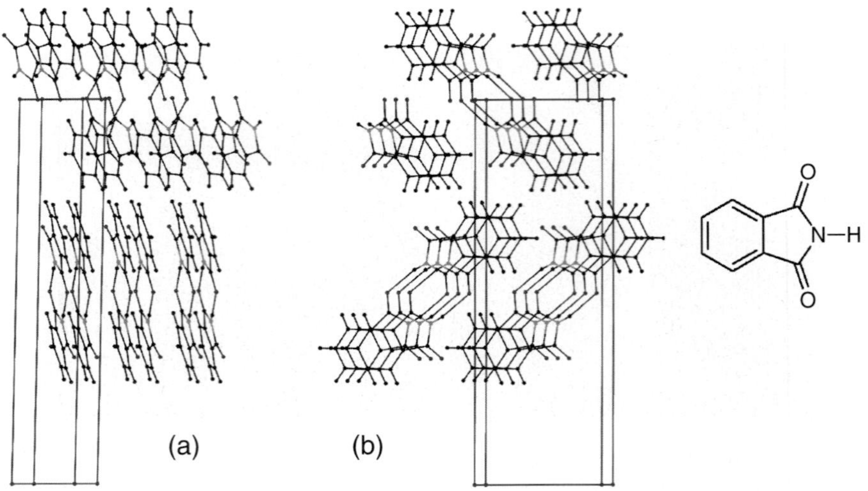

Fig. 2.33. Crystal packing of phthalimide including hydrogen bonds; (**a**) on (010); (**b**) on (101); both images are turned around Y by 10° for better view; the (001) face is on *top* in both images

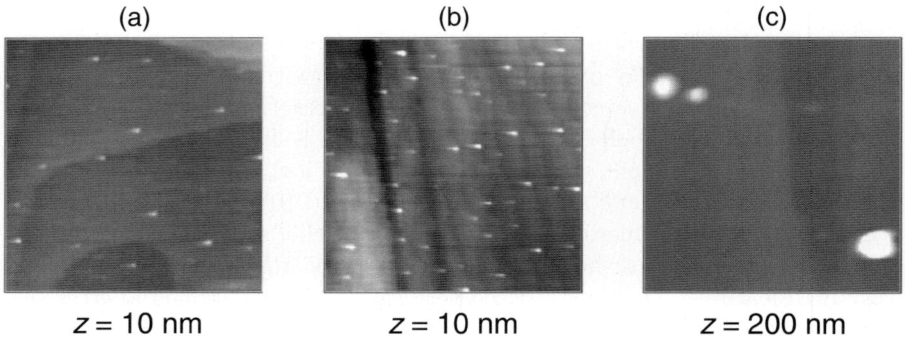

Fig. 2.34. AFM topologies (5 μm) on (001) of phthalimide at room temperature; (**a**) fresh from acetone; (**b**) after 2 d in air at relative humidity of 50–60%; (**c**) after 16 h storage in a vacuum near the saturation pressure of water at room temperature

as the crystal packing on its most prominent (001) face has the functional groups hidden under the hydrophobic benzene rings (Fig. 2.33). However, surface hydration is not excluded, as one of the oxygen atoms per molecule is not efficiently shielded (Fig. 2.33b). A double-layer situation as on (001) in Fig. 2.33a,b should be unreactive. Therefore, an island mechanism can be expected at best on that face.

The AFM analysis reveals a very flat initial (001) surface with molecular steps (bilayers) and some peaks with heights of 2.3–3.3 nm (Fig. 2.34a) that persist in moist air. The surface hydration in wet air is the only process within

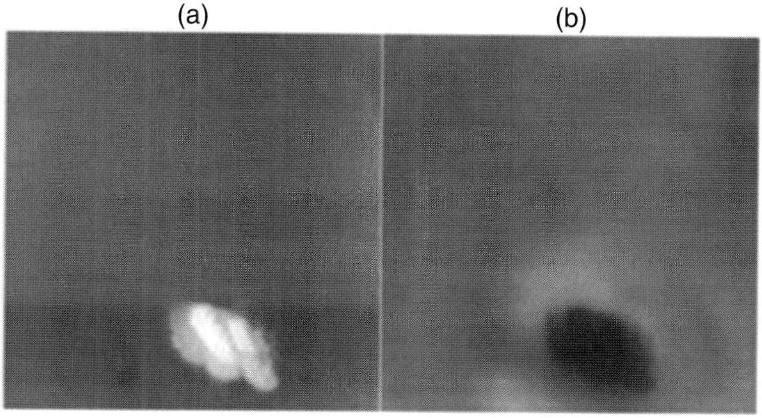

Fig. 2.35. Shear-force AFM (**a**) (5 μm, z-range: 400 nm) and apertureless SNOM (**b**) (488 nm) of phthalimide on (001) after 16 h in almost saturated water vapor

2 days. A uniform change at the 10 nm ranges is obtained (Fig. 2.34b). Numerous bilayer steps and deeper ditches align along the long crystal axis [10] and are most likely hydrated at least at the molecular steps and beyond. No SNOM contrast was obtained at that stage. More extended exposure to moisture (weeks) or exposure to nearly saturated water vapor for 16 h produces islands of considerable height (up to 180 nm) while the direction of preference is retained (the ditch is now ca. 10 nm deep) as shown in Fig. 2.34c. The larger isolated islands are the result of a chemical reaction with long-range molecular migration. Most likely is hydrolytic ring opening to give phthalic acid monoamide only at the islands (?). In view of the minute quantities of material, this conclusion is subject to SNOM scrutiny, which is given in Fig. 2.35.

The simultaneous AFM and SNOM at a different site of the above sample disclose a strong chemical contrast, which is dark in this case and very different from the unreacted hydrated flat plane, in support of the localized chemical reaction (Fig. 2.35). The tip had to go over slopes that never reached 40° except for one region near the top of the island feature. The reliability of the measurement is demonstrated by the precise local correspondence with the topologic image. The broad and very minor positive features do not seem to be part of the chemical contrast in Fig. 2.35b. Similar dark chemical contrast in a more advanced state of hydrolysis has been found with the SNOM using 633-nm light [8, 60].

2.8 Physical-State SNOM Contrast

Different crystal modifications that might coexist will give different reflectivity in the shear-force gap and optical contrast is expected. Also different

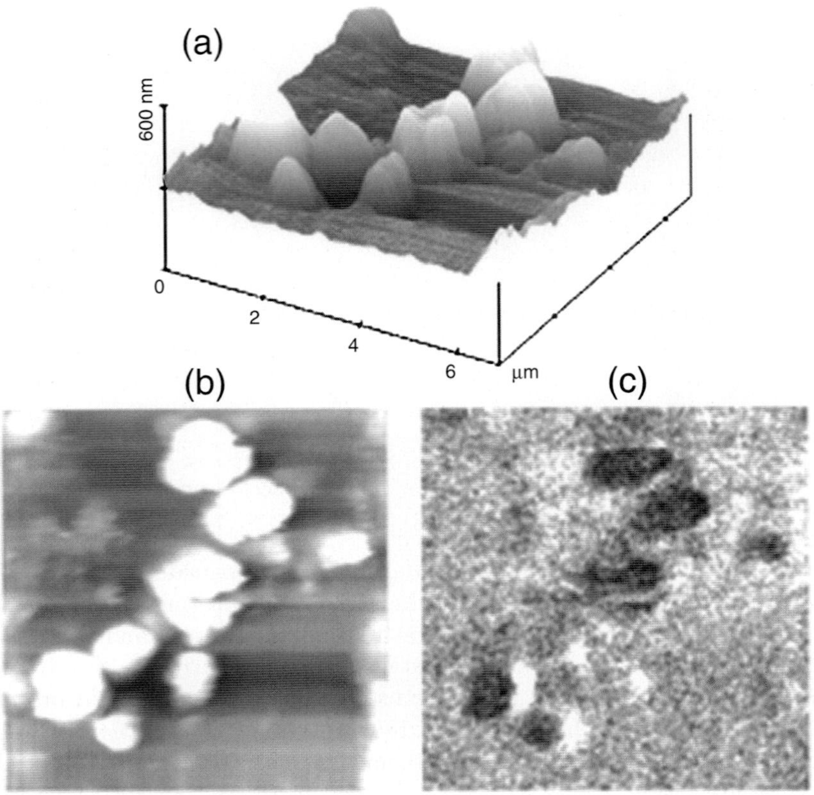

Fig. 2.36. Simultaneous shear-force AFM (**a**, **b**) and apertureless SNOM (**c**) (488 nm) on a fresh 2-(p-dimethylaminobenzylidene)-3-*oxo*-2,3-dihydroindole surface showing microcrystallites at different orientation and different types of physical state contrast in the optical image

orientations of dichroic crystals will give a SNOM contrast due to different absorbance and reflectivity. Pure dichroic crystals of the photostable merocyanine 2-(p-dimethylaminobenzylidene)-3-*oxo*-2,3-dihydroindole (diffuse reflection spectrum: $\lambda_{max} = 445$, 480, 530 sh, 546, 580, 600 sh, 665 sh nm) exhibit well-developed crystallites on their surfaces when crystallized from methanol (Fig. 2.36a,b). They are able to dissipate the heat formed upon radiationless decay and give rise to the physical state contrast of Fig. 2.36c [52]. The crystallites on the AFM surface have different orientation. Three types of optical contrast are seen in the SNOM image: medium at the base and dark or bright or medium at the sites of the crystallites. This is neither chemical contrast and nor topologic artifact, as differently oriented crystallites behave differently due to different molecular orientation.

2.9 Technical Applications

Apertureless SNOM in the reflection back to the sharp tip on rough surfaces is highly application based because it is particularly simple and economic. It should now become state of the art after the basic physical facts and the experimental understanding have been worked out even in advance of successful theoretical modeling. Application possibilities are manifold and very promising. First industrial applications have already succeeded. For example, existing and new dental alloys have to be checked for their corrosion properties in order to prevent harm to the patients. Electrochemical corrosion models are analyzed by chemical analysis. Also submicroscopic analysis of the surface after corrosion tests is very helpful. But SNOM provides a new approach by pointing out local differences in the corrosion process by its chemical contrast. Submicroscopic analysis of the surface after corrosion tests will have to be urgently executed. They are equally important with further implant materials. Actually, a widely used and approved alloy was eliminated from the market immediately after SNOM measurements indicated the formation of galvanic elements, which give rise to pitting corrosion. Furthermore, better alloys that do not exhibit the pitting could be identified as substitutes [61]. Further emerging fields of application are nanoparticles, varnishes, finishes, glazes, dyes, semiconductors, catalysts, and composites, to name a few. Applications in biology and medicine are treated in Sect. 2.10. SNOM on rough surfaces complements confocal microscopy and digital optical 3D microscopy (Sect. 2.13) in various ways due to higher resolution and chemical contrast. Applications of local fluorescence and Raman are treated in Sect. 2.11.

2.9.1 SNOM on Dental Alloys

The study of SNOM on pristine and corroded dental alloys was part of a government project in cooperation with Dr. Liefeith IBA, Heiligenstadt, Germany. The results are published in the final report of the BMBF [61]. The standardized electrochemical corrosion proceeded differently with various dental alloys, ranging from smooth uniform leaching (favorable) to submicroscopic pitting (unacceptable). This is easily detected by shear-force AFM and simultaneous apertureless SNOM. Submicroscopic pitting will occur upon creation of local galvanic elements. Figure 2.37 shows the results of a cobalt-containing abraded alloy in AFM and SNOM before and after high electrochemical load. The degree of abrasion is typical for direct use at the dentist. The edges enforce stripes artifacts in the SNOM image but there is no additional chemical contrast in (b) and nor so in (d), even though the topology (c) has changed. This appears to be a good result if chemical analysis of the leached metal ions is also acceptable. Similarly favorable results were obtained with polished or mold samples. Such judgment is not available by confocal microscopy or XPS, and other techniques.

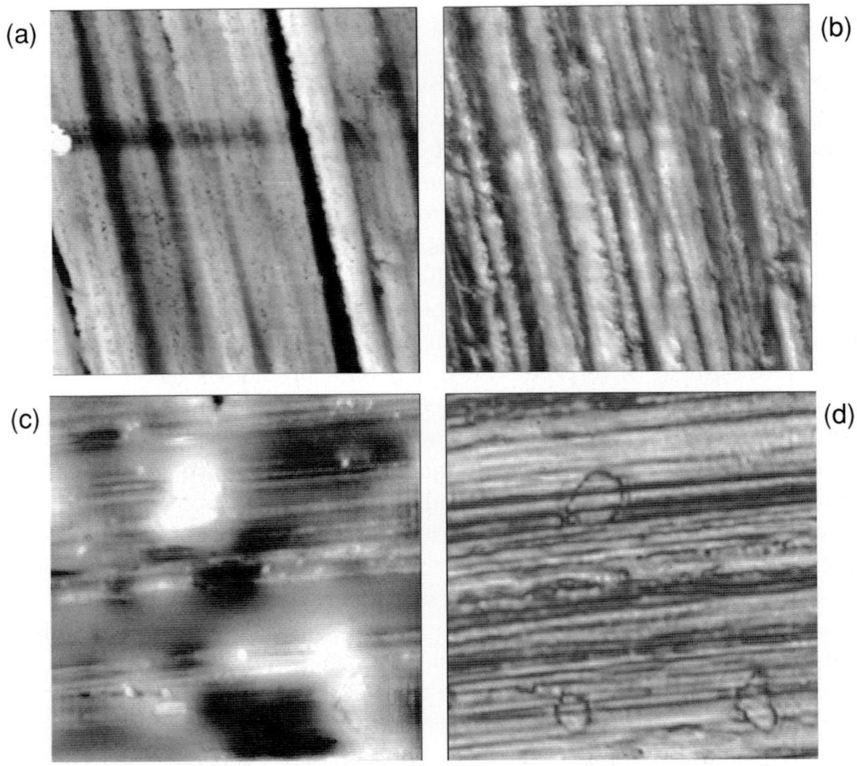

Fig. 2.37. Simultaneous shear-force AFM (**a**) (10 µm, z-range: 1 µm) and (**c**) (z-range: 600 nm) and apertureless SNOM (**b**, **d**) of an abraded cobalt alloy; (**a**) and (**b**) before, (**c**) and (**d**) after electrochemical load

Totally different is the behavior of a widely used nickel-containing alloy that gave the results of Fig. 2.38. The edges in the unused abraded alloy (a) are sharp and give the corresponding stripes contrast with some small dark points that are chemical contrast (b). After electrochemical treatment the edges disappear and several craters are formed (c). Importantly, the SNOM shows chemical contrast (d) indicating different metallic composition at the different sites and that means local galvanic elements. Clearly, submicroscopic pitting occurred and the unacceptable material was immediately withdrawn from the market and is not used further as a dental alloy after these results.

Another example of a smoothly corroding copper-containing dental alloy gives the results of Fig. 2.39. Again we have sharp edges (a) and an artificial stripes contrast in the SNOM (b) at their sites. The additional high feature in (a) is responsible for a strong chemical contrast in (b) that can be clearly identified next to the artificial stripes. At low electrochemical load the additional topology and the corresponding chemical contrast disappear. Only the stripes remain (not shown). Most importantly, after high electrochemical load

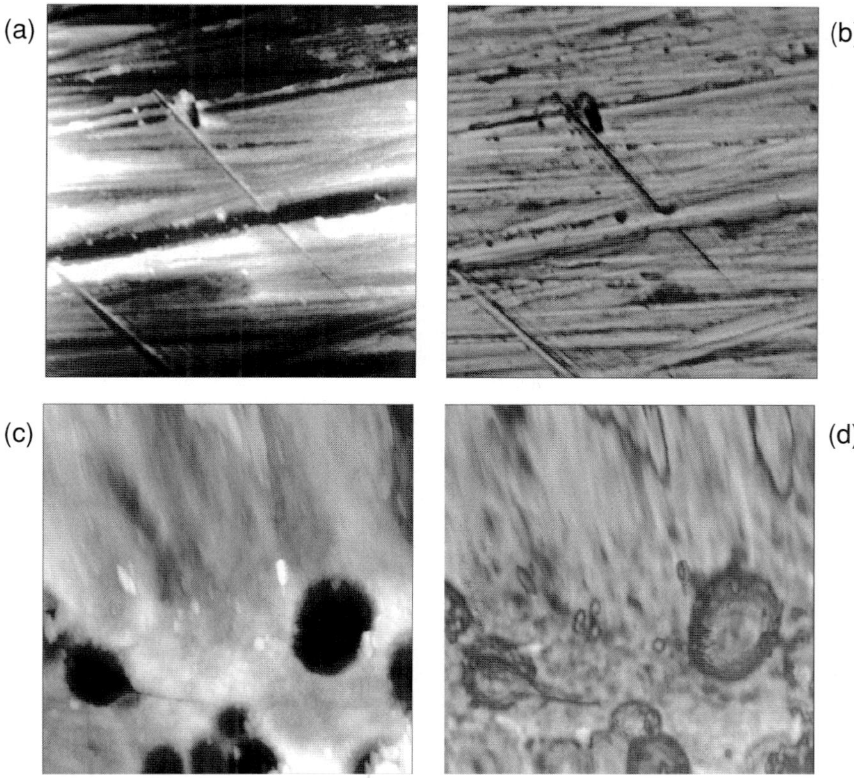

Fig. 2.38. Simultaneous shear-force AFM (**a**) (10 µm, z-range: 300 nm) and (**c**) (z-range: 1.6 µm) and apertureless SNOM (**b**, **d**) of an abraded nickel alloy; (**a**) and (**b**) before, (**c**) and (**d**) after electrochemical load

the topology (c) changes dramatically, but no contrast is discernable at all in the SNOM image (d). This appears to be a very favorable dental alloy as far as the corrosive behavior is concerned. Numerous further alloys have been similarly studied under various conditions of corrosion [61] and classified. Hence, a reliable nanoscopic standard technique is available now that it cannot be neglected. A major feature of the reliable technique (with high enhancement factors in the field of metals) is the possibility to distinguish artifacts from impossible topology and genuine materials contrast in the same image.

The procedure found already due interest in the study of high-grade steel [62]; however, the reported images show typical signs of an artifact.

2.9.2 SNOM on Glazed Paper

Paper has a very rough surface and paper glazing is an important industrial task. Various finely divided minerals are applied in the glazing process.

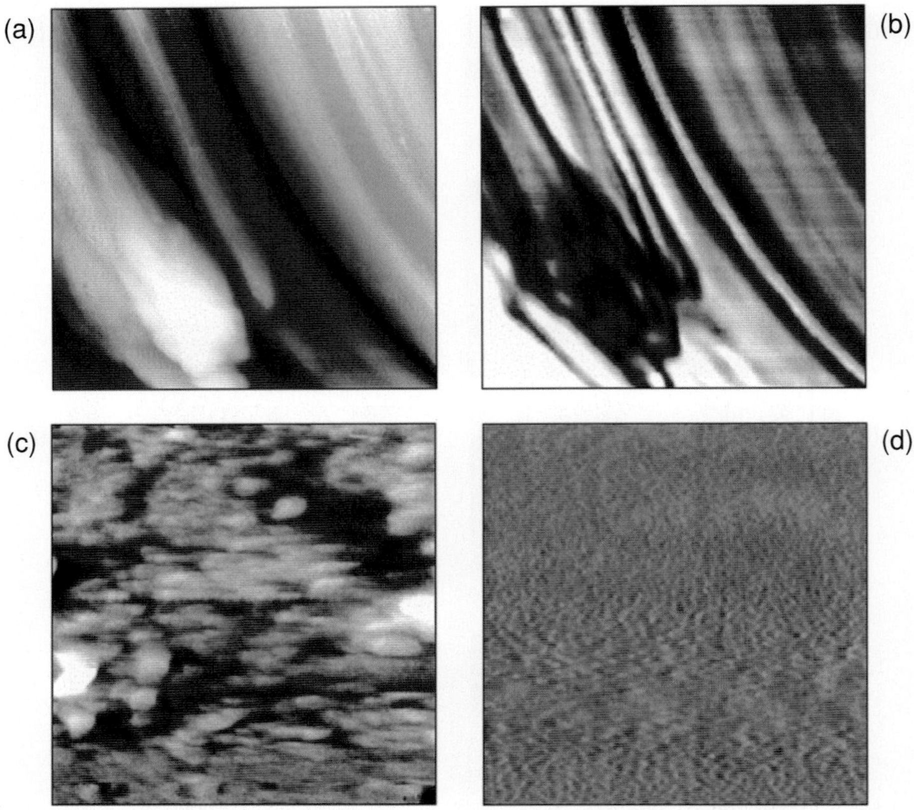

Fig. 2.39. Simultaneous shear-force AFM (**a**, **c**) (10 μm, z-range: 400 nm) and apertureless SNOM (**b**, **d**) of an abraded copper alloy; (**a**) and (**b**) before, (**c**) and (**d**) after extended electrochemical load securing very uniform corrosion as judged by SNOM

The success depends on various parameters. It is of high interest to study the results by monitoring the submicroscopic roughness and the distribution of the materials for optimizing and saving cost of materials. Two samples of different roughness are suitable examples for demonstrating the success of SNOM on rough surfaces for this technical application. It can be seen in the surface plots and in the top views of Figs. 2.40 and 2.41 how different particles with different near-field reflectivity combine to the total surface. The identification of the particles at the various sites can be obtained from local Raman SNOM (Sect. 2.11.1). Any added fluorescent components can be located by local fluorescence SNOM. Clearly, such knowledge helps the designer in the optimization of the glazing procedure. The sample for Fig. 2.40 was glossier and felt smoother than the one of Fig. 2.41.

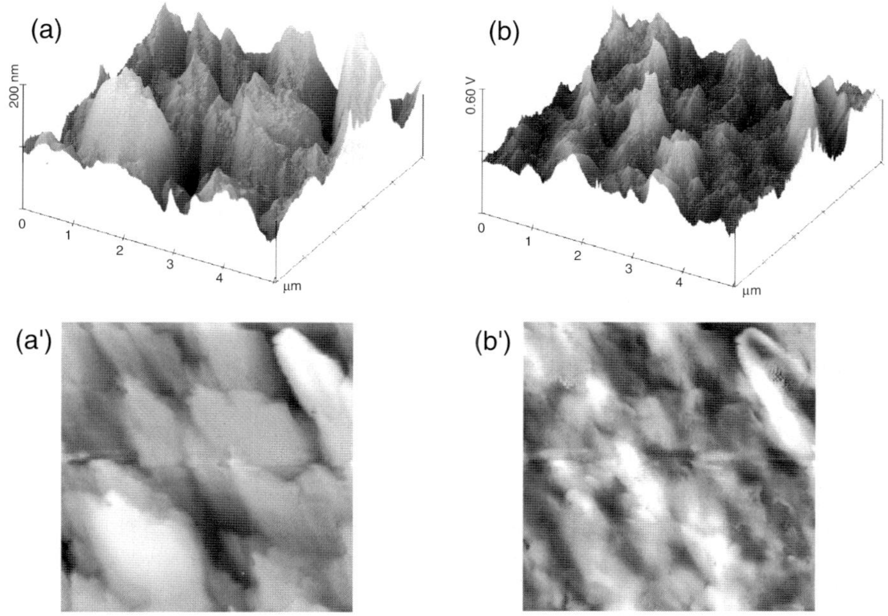

Fig. 2.40. Shear-force AFM (**a**, **a**′) and apertureless SNOM (**b**, **b**′) of a glazed paper showing roughness and distribution of high, medium, and low optical contrast

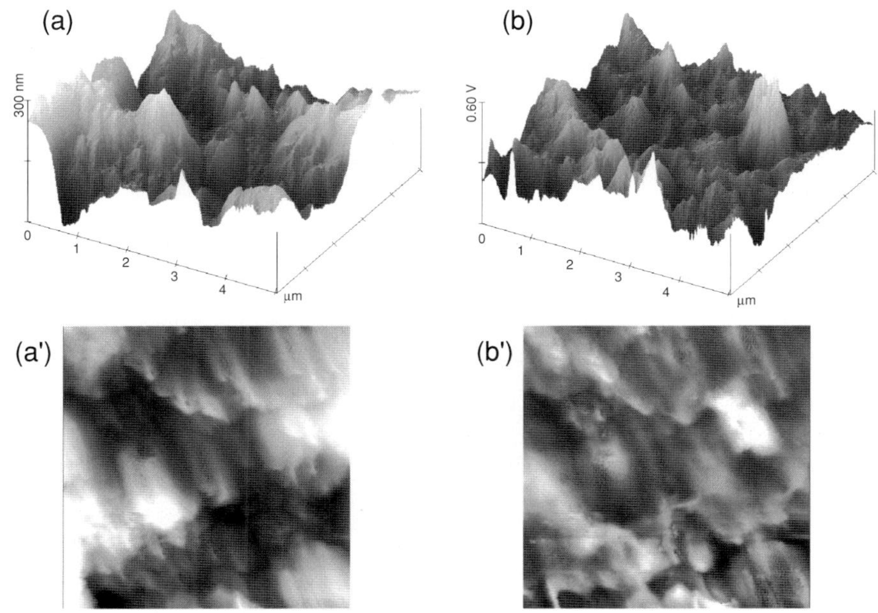

Fig. 2.41. Shear-force AFM (**a**, **a**′) and apertureless SNOM (**b**, **b**′) of a glazed paper showing roughness and distribution of high, medium, and low optical contrast

2.9.3 SNOM on Blood Bags

AFM measurements at flat parts of commercial highly textured fresh blood bags indicate little change of the submicroscopic (100 nm range) surface after treatment with physiological sodium chloride solution. After short- or long-term storage of blood the surface changes considerably [61]. This fact is of importance for the allowable storage time of blood and submicroscopic investigations are badly required for economical reasons. While standard nanoindentation hardness and elasticity modulus of the bag (14 and 60 MPa) and of the deposits (16 and 31 MPa) exhibit large differences, more detailed information yields chemical contrast by SNOM measurements with uncoated tips at constant shear-force distance. A fresh blood bag at flat or purposely rough surface does not exhibit SNOM contrast (Fig. 2.42). This means that any topology found in a blood bag giving SNOM contrast must be a deposit out of the blood.

The measurements with the charged blood bags are shown here for the flattest regions that could be found. Organic deposits give chemical contrast at blood bags that had been filled with blood for less than one week, washed, and dried (Fig. 2.43). It can be clearly seen that deposition is widespread with a brighter optical contrast at the shallow depression. The chemical contrast proves the deposition of blood components to the surface of the blood bags that had not previously been detected because the hardened solid deposits do not exceed 100 nm widths. It is still unclear if such deposition is helpful or detrimental for the storage time of native blood. However, further studies are

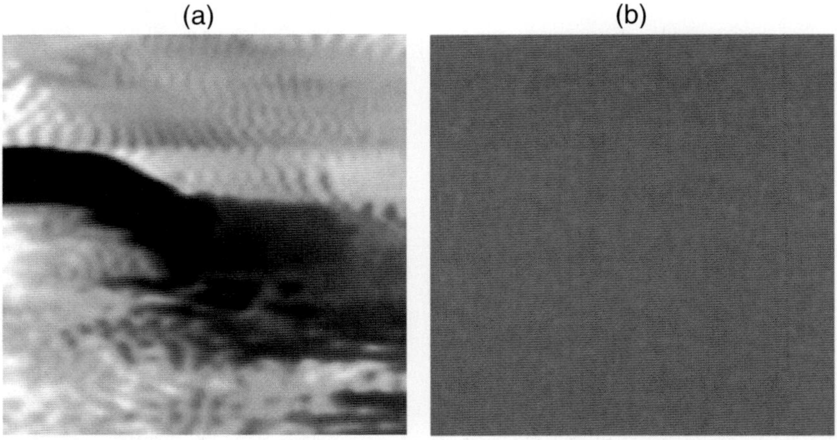

Fig. 2.42. Shear-force AFM (**a**) (670 nm, z-range: 500 nm) and apertureless SNOM (**b**) of a pristine blood bag surface purposely at an unusually rough site (in the flat part) showing no optical contrast despite the topology due to homogeneous composition

essential because all chances for safe extension of blood storage times must be followed.

More distinct and localized is the chemical contrast at expired blood bags (Fig. 2.44). Again it appears that the brightest contrast is obtained at depression sites.

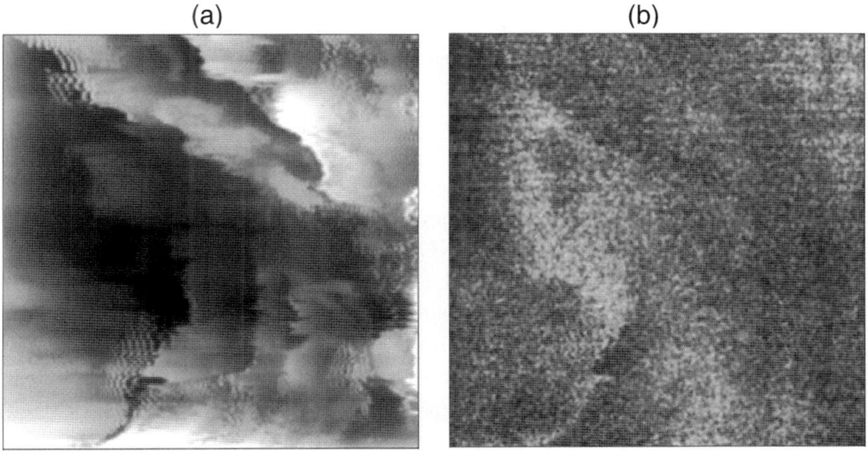

Fig. 2.43. Shear-force AFM (**a**) (800 nm, z-range: 200 nm) and apertureless SNOM (**b**) of a shortly charged blood bag surface at a particularly flat site showing widespread optical contrast

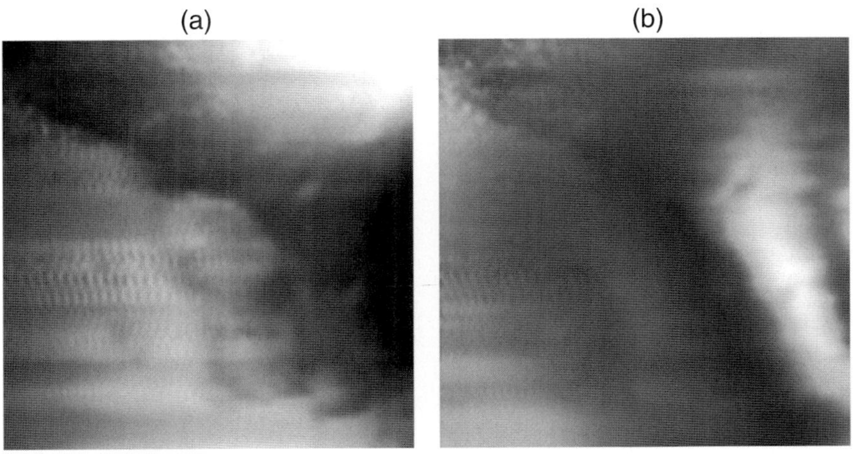

Fig. 2.44. Shear-force AFM (**a**) (500 nm, z-range: 100 nm) and apertureless SNOM (**b**) of an expired blood bag surface at a particularly flat site showing widespread and pronounced localized optical contrast

2.9.4 Quality Assessment of Metal Sol Particles for SERS by SNOM

Suppliers of silver and gold sols for surface enhanced Raman spectroscopy (SERS) must ensure that their preparations are pure and suitable for the desired large enhancement. Apertureless shear-force SNOM with uncoated tips is the proper analysis technique. Figure 2.45 shows silver particles of different size in a common preparation after drying on a microscope slide without further treatment. The topology cannot decide if the larger particles are aggregates or impurities. However, SNOM does immediately prove impurity. It further shows that the small particles are not uniform, some exhibiting bright, others medium optical contrast. This is also a good example for proving the SNOM image being free of topologic artifacts: different optical contrast at positive topologies is featured. The chemical nature of the different particles should be clarified by local Raman SNOM (Sect. 2.11).

The attainable enhancement of SERS experiments relies severely on the quality of the colloidal particles. An actually inactive gold sol with nominal 50-nm particles is analyzed in Fig. 2.46. AFM and SNOM disclose large grains with dark chemical contrast, which must be impurity. Furthermore, the higher resolved image in an impurity free region exhibits 400–500 nm wide particles of different height that are too large. Furthermore, only the higher and regular spheres give the bright chemical contrast that is expected for gold. None of the specifications is met in this preparation. Checking every charge of the rather delicate gold soles prior to sale and purchase with apertureless shear-force SNOM is easy, economic, and necessary. Also the SERS applicant will want to control the stability of the sol prior to the experiment by our reliable technique.

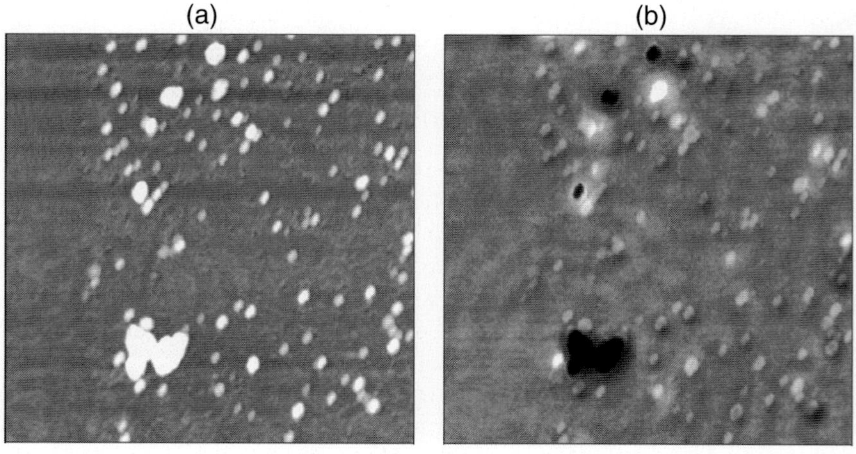

Fig. 2.45. Shear-force AFM (**a**) (10 µm, z-range: 200 nm) and apertureless SNOM (**b**) of dried silver sol on a microscope slide

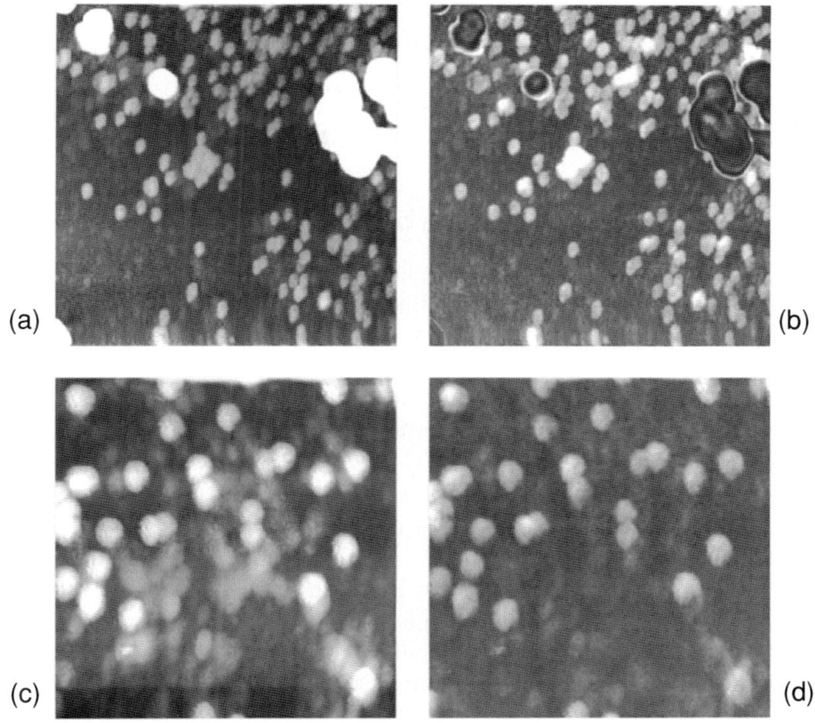

Fig. 2.46. Shear-force 10 μm (**a**) and 5 μm AFM (**c**) (z-range: 300 and 200 nm, respectively) and apertureless SNOM (**b**, **d**) of dried inactive Au sol on a microscope slide

Further industrial applications of SNOM in textile dyeing, including determination of diffusion coefficients, are treated in Sect. 2.11.4.

2.10 Applications of SNOM in Biology and Medicine

Biological and medicine-related surfaces, including microtomes, are generally very rough. They should be studied without further treatment by AFM and SNOM if the highest resolution is desired in order to gain nondestructive subcellular information at the level of organelles. Apertureless shear-force SNOM with enhanced reflection back to the sharp fiber is the only technique that can achieve this goal. Confocal microscopy does not detect chemical contrast, and it does not reach the possibilities of SNOM on rough surfaces even in the 4-π configuration. It is, however, useful for pre-evaluations and positioning. The point-spread-function-engineering could reach comparably high resolution for flat surfaces, but the versatility of the STED-fluorescence microscope is restricted. Molecular biological sample preparations can also be studied with photon scanning tunneling microscopy (PSTM). Apertureless SNOM deals with advancement of biocompatibility, implantology, submicroscopic histology, and pathology, including oncology, blood storage, prehistoric tissues, etc. Of particular importance are also local fluorescence and Raman SNOM on fresh and prepared soft and solid tissue, including life organisms under physiological conditions.

2.10.1 Prehistoric Bacterium, Pyrite Formation upon Petrification

A finely polished dolomite stone (from Dr. T. Schöberl, Erich-Schmid Institut, Leoben, Austria) contains fossil-petrified bacteria with diameters of 50 µm. The AFM shows a sharp edge due to differences in the abrasive resistance. The SNOM shows dark chemical contrast next to the edge contrast at the steep interface and at the five steep protrusions in addition to some bright spots that derive from embedded pyrite particles (Fig. 2.47). The structure in the fossil bacterium part is chemical contrast. Part of it is higher resolved in Fig. 2.48. The pyrite geodes derive from petrification of the organic sulfur content of ancient biological materials in the presence of iron ores. They are identified in Figs. 2.8 and 2.66.

The higher resolved Fig. 2.48 has been taken in the area of the petrified bacterium. It clearly contains some still conserved structural details with brighter contrast at the site of the higher more scratch resistant feature due to some chemical difference. Further analysis will require local Raman SNOM.

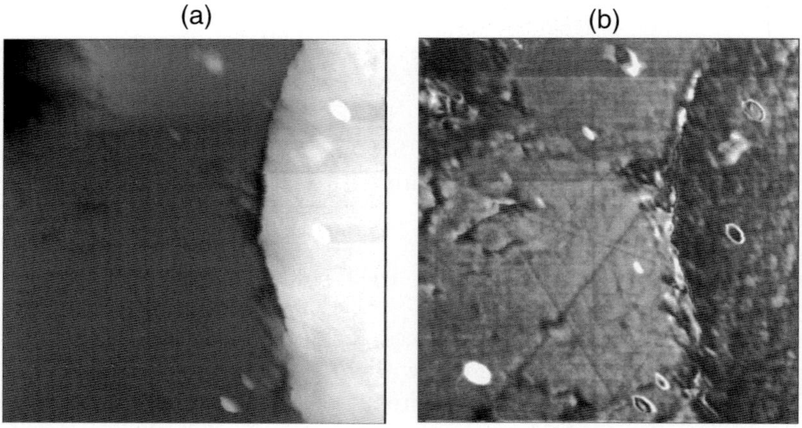

Fig. 2.47. Shear-force AFM (**a**) (20 µm, z-range: 2 µm) and apertureless SNOM (**b**) of a part of a petrified bacterium (dark optical contrast) at a polished dolomite sample with some pyrite geodes under the surface and dust particles on the surface

2.10.2 SNOM on a Human Tooth

A human tooth was cut with a microtome (sample obtained from R.P. Franke, Universität Ulm). It could be used for a study of the tubuli structure.

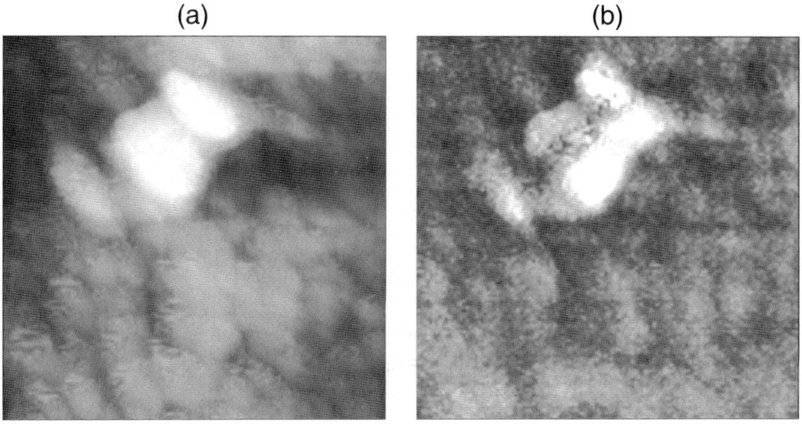

Fig. 2.48. Structure details of a fossil bacterium in a polished dolomite sample; (**a**) shear-force AFM (5 µm, z-range: 300 nm); (**b**) simultaneous apertureless SNOM image indicating the chemical difference at the bright contrast

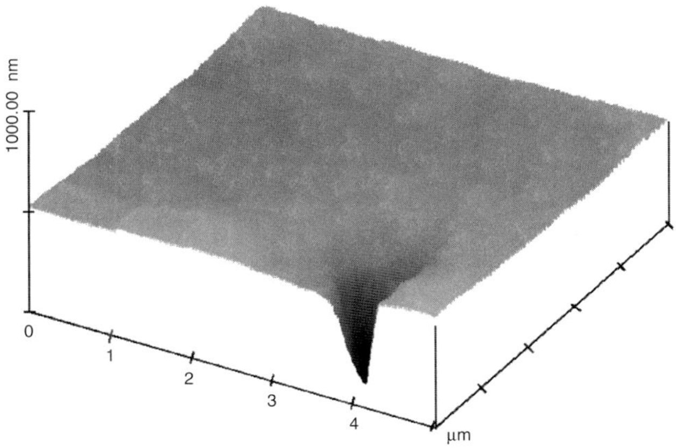

Fig. 2.49. Shear-force AFM topology of a human tooth on a microtome cut orthogonal to the tubuli direction showing deep penetration of the tapered tip

Shear-force AFM with the sharp tapered tip was quite successful. The vertical cylinder was deeply descended without tip breakage (Fig. 2.49).

The simultaneous AFM and SNOM images provide unexpected structural results of the tubulus, which are subject to histological interpretation (Fig. 2.50). The rings at the impossible topology (open cylinder) are stripes artifact (Sect. 2.6.2). The nature of the strongly reflecting material around the triangle and extending into the fissure in (b) must be the subject of further interpretation. It should be noted that the topologic triangle in (a) is probably a rupture result of the microtome cut, which may have changed the stressed environment around the edges by putting pressure and redistributing

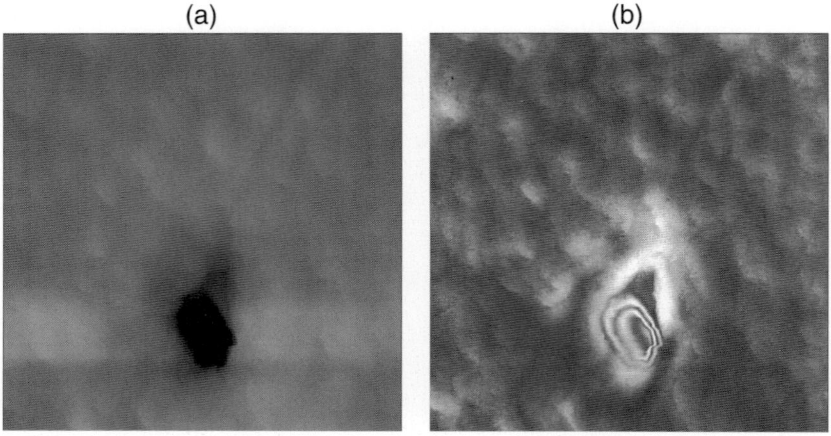

Fig. 2.50. Simultaneous shear-force AFM (**a**) (5 µm, z-range: 500 nm) and apertureless SNOM (**b**); the SNOM image shows both artificial rings at the impossible tube structure and an interesting chemical contrast close to the dark triangle

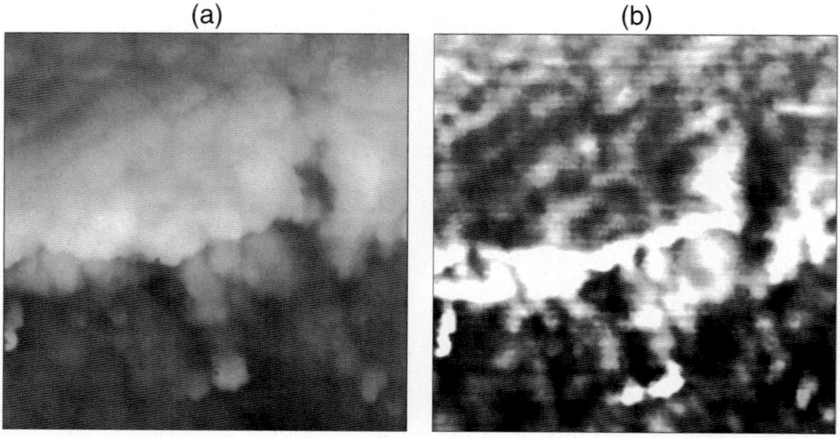

Fig. 2.51. Microtome ruptures at the enamel–dentin boundary as determined by (**a**) shear-force AFM (10 µm, z-range: 1 µm) and (**b**) apertureless SNOM

collagen-free dentin from the inner walls of the tubulus. The constituents of intertubulus dentin are hydroxylapatite, mucopolysaccharides, and collagens.

More pronounced microtome ruptures occur at the boundary of enamel and dentin. This type of preparation artifact is more distinctly seen by SNOM than by AFM, notwithstanding the additional steepness artifact.

Figure 2.51 shows very extended ruptures in both parts (a) and (b). The enamel (upper part) forms high rims and extends far into the dentin, and there are also deep fissures into the enamel. The rims are due to the different elasticity of the materials upon cutting with the microtome. Most prominent in the SNOM image is the edge artifact at the rim sites. It can however be seen that enamel has a brighter contrast than dentin. The rupture is shown by the dark contrast entering the enamel region. However, some of the interpenetration of the two materials could be natural. Therefore, similar SNOM should be measured at the uncut surface of the tooth in order to clarify this point. Topology that may occur at the natural interface will not be impeding to that endeavor.

2.10.3 SNOM on Polymer Beads

Prior to the discussion of soft biological samples the validity of apertureless shear-force SNOM shall be demonstrated with 430-nm polymer beads. The aggregation of polymer beads has important parallels in the aggregation of biological objects. The beads are fractionated copolymers of styrene and acrylamide with uneven distribution of the polar component and therefore polar sides that are of importance in their aggregation from water suspension on mica. Clearly, submonolayer beads have difficult topology due to overhangs below the vertical. Such edge artifact gives rise to highly resolved (<20 nm) stripes, and this can be seen in Fig. 2.52b,c (see also Sect. 2.4). Figure 2.52a images small aggregates and it is clearly seen in the simultaneous SNOM image that a bright contrast occurs at the site of contact of four or five beads that are not elongated when isolated or in a closed monolayer image (Fig. 2.53). Clearly, the more polar and thus more hydrated ends with considerably brighter optical contrast search each other upon aggregation and form the polar center. The contact is maximized by the elongation of the beads.

The arrangement and shape of the beads in closed monolayers is of course different. Figure 2.53b indicates that the optical contrast of some beads is considerably brighter than the contrast of others due to different orientation of their polar ends that accumulate water layers. The interpretations and further use of these results have been detailed in [63, 64].

2.10.4 SNOM on Cryo Microtome Cuts of Rabbit Heart [61]

The state-of-the-art cryo microtome cut of a rabbit heart, unstained but stabilized with glutaraldehyde, was obtained from R.P. Franke, Universität Ulm. The shear-force AFM and apertureless SNOM easily resolves organelles despite the remarkable roughness even in a 2 μm scan. The images are interpretable at this range. Shorter ranges are possible in view of the high-resolution capacity but would require knowledge about the organelle that is searched and hit. Clearly, the AFM surface plot (Fig. 2.54a) shows convex features. One of them is identified as a lysosome (in front) due to its size and dark optical contrast in the SNOM image. The top views (Fig. 2.54b,c) distinguish the location of various organelles topologically (b) and by the chemical contrast (c). At least three types of contrast (dark, medium, and bright) are

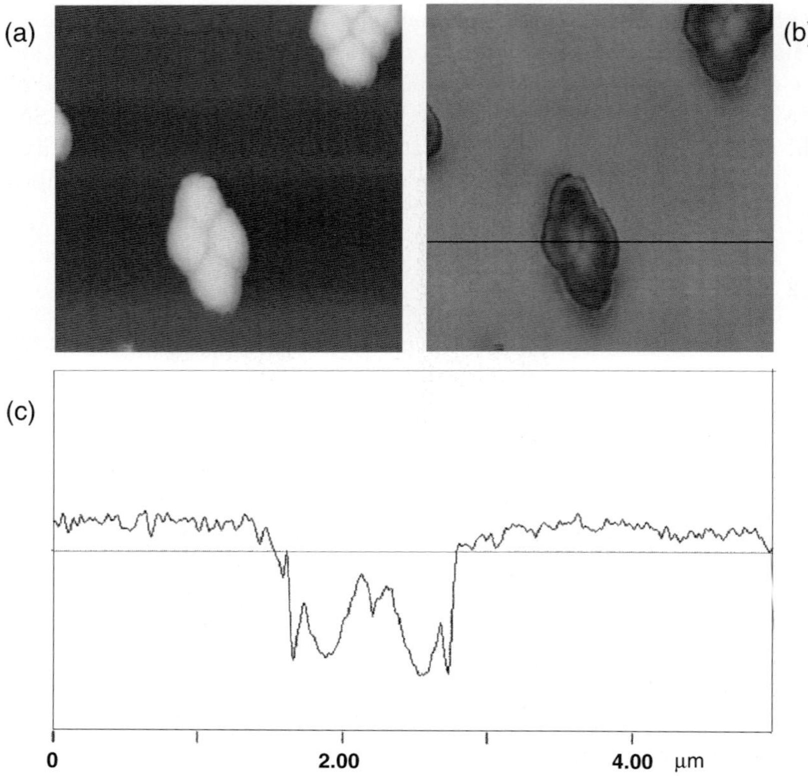

Fig. 2.52. Shear-force AFM (**a**) and apertureless SNOM (**b**) of small aggregates of 430 nm styrene–acrylamide copolymer lattices on mica and cross section (**c**) through the optical contrast showing both stripes contrast and materials contrast

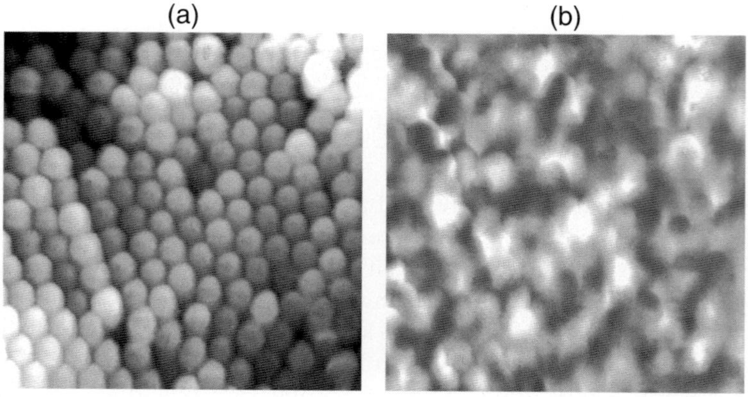

Fig. 2.53. Shear-force AFM (**a**) and apertureless SNOM (**b**) of a monolayer of 430 nm styrene–acrylamide copolymer lattices on mica showing different brightness of the optical contrast due to different orientation

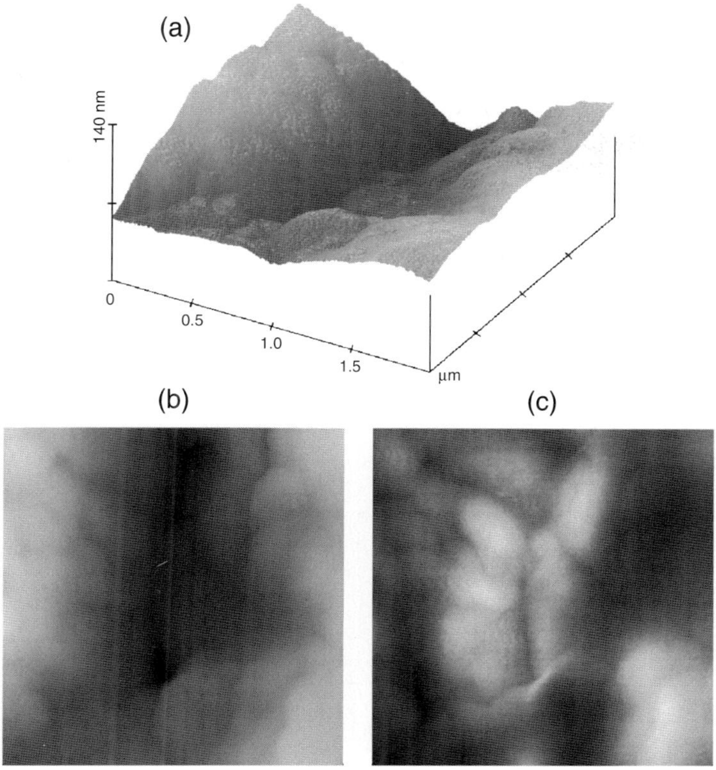

Fig. 2.54. Surface plot (**a**) and *top* view shear-force AFM (**b**) of a cryo microtome cut of a rabbit heart and simultaneous apertureless SNOM (**c**) showing a lysosome (*bottom*, dark optical contrast), and mitochondria (bright optical contrast)

immediately distinguished. The bright contrast belongs to mitochondria. The differences are obtained without staining. Staining would perhaps secure the interpretation in additional apertureless SNOM measurements. Furthermore, higher resolved SNOM would trace the fine structure of the organelles (cf. Fig. 2.60b), the interpretation of which will be subject to histological interpretation that opens new fields in the recognition of biostructures and bioprocesses, but this is not outlined here due to the enormous complexity. The histologist is offered the unadulterated organelle for detailed study with a nondestructive technique. Such information cannot be obtained by aperture SNOM or confocal microscopy. A major advantage of apertureless SNOM in shear-force distance is the possibility for investigation without extensive preparation. This includes measurements at 0°C of fresh cuts and scan areas of organelle size.

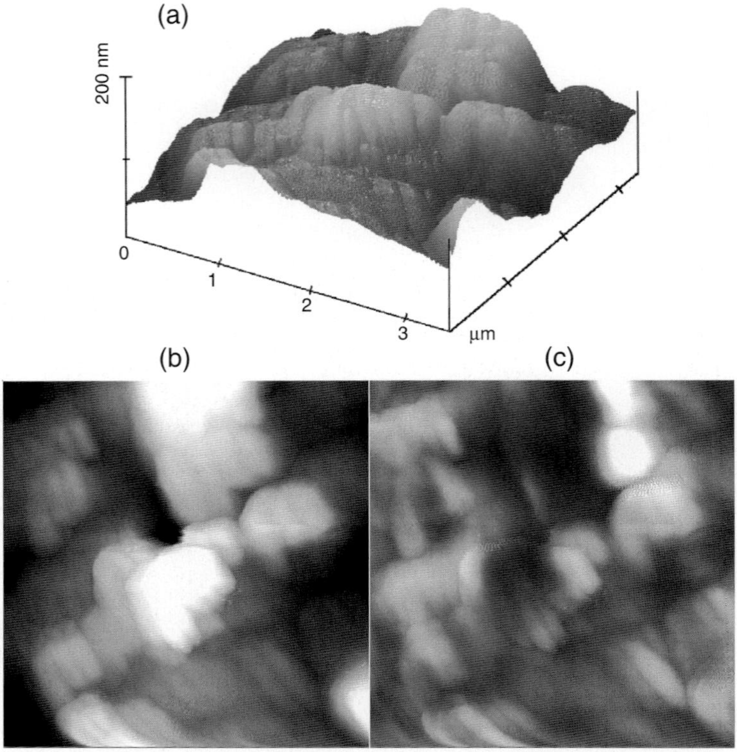

Fig. 2.55. Shear-force AFM (**a,b**) and apertureless SNOM (**c**) of a cryo microtome cut of a rabbit heart showing lamella structures of mitochondria (bright optical contrast), lamella structures with dark contrast (part of the Golgi apparatus) next to two most brightly reflecting Golgi vesicles (upper part), and a central dark reflection (lysosome)

A very different region in heart cells is shown in Fig. 2.55b. It exhibits the mitochondria with their characteristic lamella structure (a,b) and bright reflection (c). The higher central vesicle with dark optical contrast points to a lysosome. The upper high lamellae with dark optical contrast are part of the Golgi apparatus with two distal secretionary Golgi vesicles to the right that give the brightest optical contrast (c). Clearly, the SNOM image is interpretable, as there is no topologic artifact.

A lysosome (right side with bright optical contrast of the membrane) is distinctly seen in Fig. 2.56b. The rich dark contrast below the center in (b) and the structured bright optical contrast left to it indicate large chemical variations in the respective organelles. The features might easily be more highly resolved. The remarkable variations of the optical contrast will certainly find more detailed histological interpretation.

An interesting series of organelles with bright optical contrast is obtained roughly along the ditch of Fig. 2.57a, which seems to occur along a cell membrane. The dark optical contrast at the lower right edge of Fig. 2.57b points to a lysosome at that position. Many further details are subject to histological interpretation.

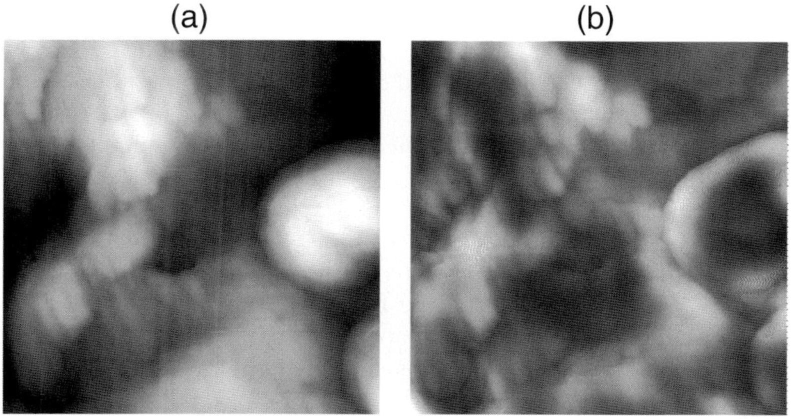

Fig. 2.56. Shear-force AFM (**a**) (3.5 μm, z-range: 100 nm) and apertureless SNOM (**b**) of a cryo microtome cut of a rabbit heart showing lamella structures and a lysosome (bright membrane and dark interior optical contrast). The various organelles show very diversely structured optical contrast in line with highly varied chemical differences at the corresponding sites

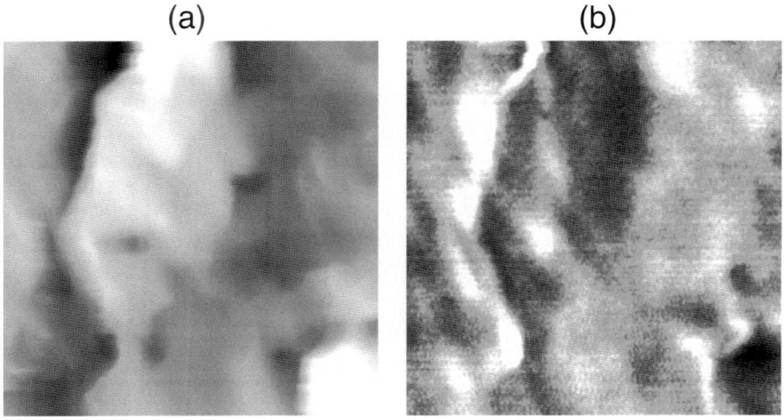

Fig. 2.57. Shear-force AFM (**a**) (3.5 μm, z-range: 150 nm) and apertureless SNOM (**b**) of a cryo microtome cut of a rabbit heart showing a series of high-reflecting and low-reflecting organelles but not always following the topology

2.10.5 SNOM on Stained and Unstained Shrimp Eye Preparations [61]

Shrimp eye preparations (cryo microtome cuts stabilized with glutaraldehyde) were supplied by P. Jaros, Universität Oldenburg. The organelle scheme of nuclei, reticula cell, rhabdoms, microvilli, tapetum cell, basal lamina is known [65]. Again, the chemical contrast serves for the identification of the organelles. That is even possible with rather old preparations. Cell organelles give the same optical contrast with SNOM as in the case of the rabbit heart. Figure 2.58 shows the comparison of a microphoto of the stained sample with the higher resolved shear-force AFM measurement at very high topology. The sides are too steep for artifact-free SNOM. There is the risk of adulteration due to extensive staining.

An unstained preparation of the shrimp eye cryo microtome cut with glutaraldehyde stabilization is measured in Fig. 2.59 at much higher resolution in a flat region. Numerous granulae (containing vision pigments) with dark optical contrast and a lysosome (large ball with structured optical contrast) are resolved and identified by their shape in the surface plot (a) or in the top view (b), as well as by their optical contrast in the SNOM image (c). The bright contrast in the optical image is attributed to the mitochondriae, which are rich in highly reflecting materials. A complicated array of these and other organelles is recognized. The power of apertureless SNOM in shear-force distance is even more clearly evidenced by the higher resolved 2.5 μm scan of the lysosome in Fig. 2.60, which images structure details despite the position at a steep slope (Fig. 2.59a). The membrane is bright and the interior shows much more structural details. The resolution can be further improved in view

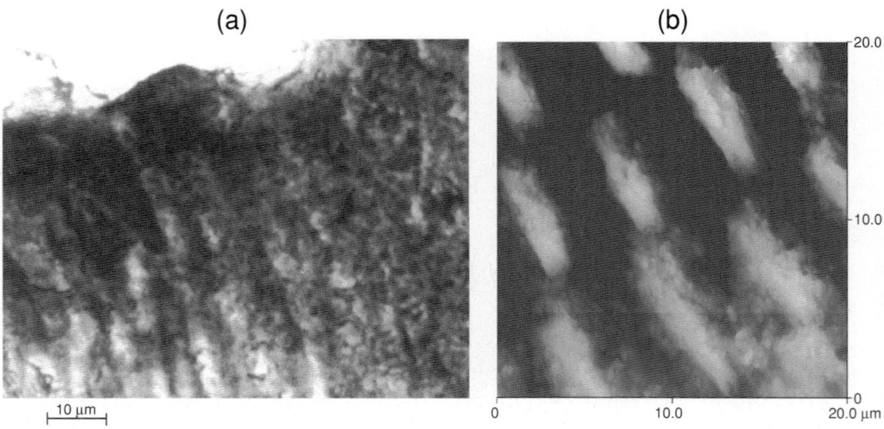

Fig. 2.58. Micro-photo (**a**) and shear-force AFM (**b**) of a stained (toluidine blue and osmium tetroxide) shrimp eye in the peripheral region; the z-range in the AFM is 4 μm

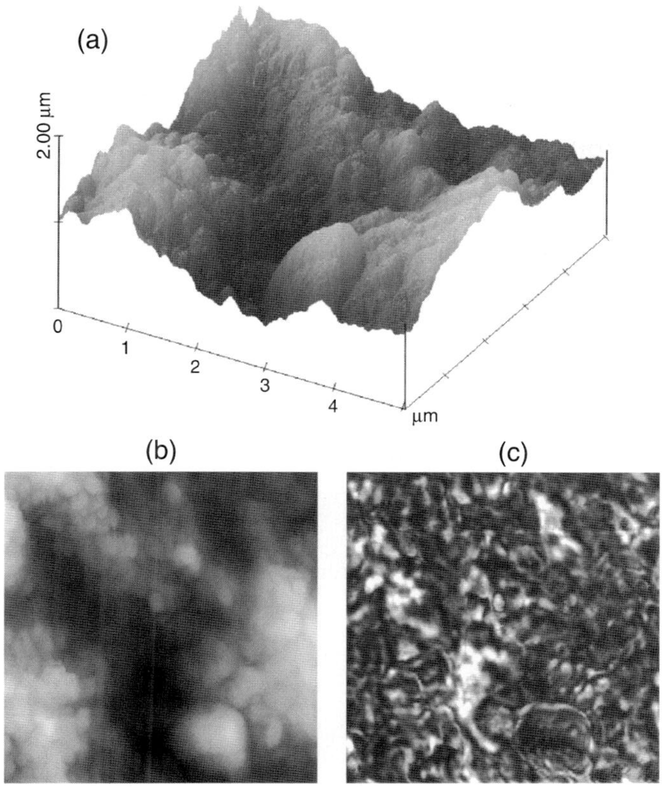

Fig. 2.59. Shear-force AFM (**a**, **b**) and apertureless SNOM (**c**) of an unstained cryo microtome cut of a shrimp eye in the peripheral region resolving various organelles by topology and chemical contrast in the optical image

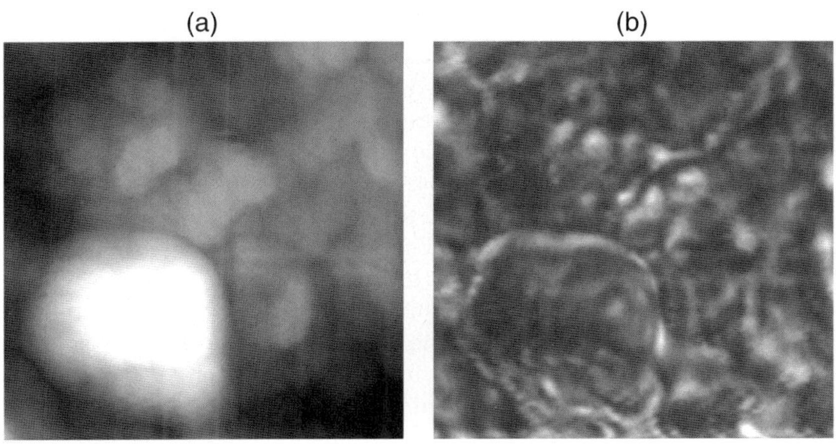

Fig. 2.60. Higher resolved (pixel size 5 nm) AFM (**a**) (2.5 µm) and SNOM image (**b**) of the lysosome and adjacent organelles in an unstained cryo microtome cut of a shrimp eye in the peripheral region

of the high lateral resolution of this SNOM technique (Sect. 2.4). Interestingly, SNOM shows much more details than AFM. Further assignments are the subject of cell biologists. They are offered a robust photophysical tool to study the suborganelle structures with unadulterated preparations.

2.10.6 SNOM in Cancer Research [61]

Subcellular chemical contrast with high and very high local resolution enables screening for distinctive marks of cancerous tumors in the organelles of the pathological cells as compared to the healthy cells. Apertureless constant shear-force SNOM is well suited, while other submicroscopic techniques are unsuitable as there is very high roughness of delicate tissue. The high potential for precancerous diagnosis by chemical contrast derives from Rayleigh and Raman scattering, fluorescence emission, light absorption, and shear-force efficiency by the site-specific mechanisms. Cancer prediagnosis is of eminent medicinal importance. Bladder tissue and bladder cancer tissue was cooled with water to $-10°C$, cut, and dried in air. The parallel cuts serve for selective staining to mark the cancerous regions. The samples were provided by G. Müller, LMTB, Berlin.

The staining of the samples wipes out all pertinent optical information in apertureless SNOM measurements. This is shown in Fig. 2.61 with 25 and 10 μm scans.

No chemical contrast is recognizable after HE staining, whereas the appearance of the AFM topology remains largely unchanged. This is also evident from Fig. 2.62, which compares typical surface topologies of unstained cuts of cancerous and healthy bladder tissue.

The comparison of Fig. 2.62 and Fig. 2.61c reveals that different sites of biological tissue cuts have different topology at the chosen resolution. However, both the unstained and the stained sample exhibit small sphere-type features at this range that seem to be less frequent in the healthy unstained tissue (Fig. 2.62b). This latter difference in the AFM topology may not be a reliable criterion for cancer diagnosis due to the high variability at different sites. More distinct are the differences in the chemical contrast of SNOM. Figure 2.63 shows the topologic variations at three sites for healthy unstained tissue and also the simultaneously recorded SNOM contrast.

It is clearly seen in Fig. 2.63 that the topologic variation is indeed very strong even in the long-range scans. Furthermore, the chemical contrast exhibits dark, medium, and bright spots that are spread over the surfaces and are not very characteristic. Corresponding images of cancerous bladder tissue are shown in Fig. 2.64. The topology of the cancerous tissue at the long range of 20 μm is not very different in character from the healthy tissue. However, there are distinct differences in the chemical contrast of SNOM between Fig. 2.63 and Fig. 2.64, which are very useful for the differentiation. It is clearly seen in Figs. 2.64a'–c' that characteristically distinct chemical contrast occurs in the form of bright extended plate-like zones that have frequently sharp edges and

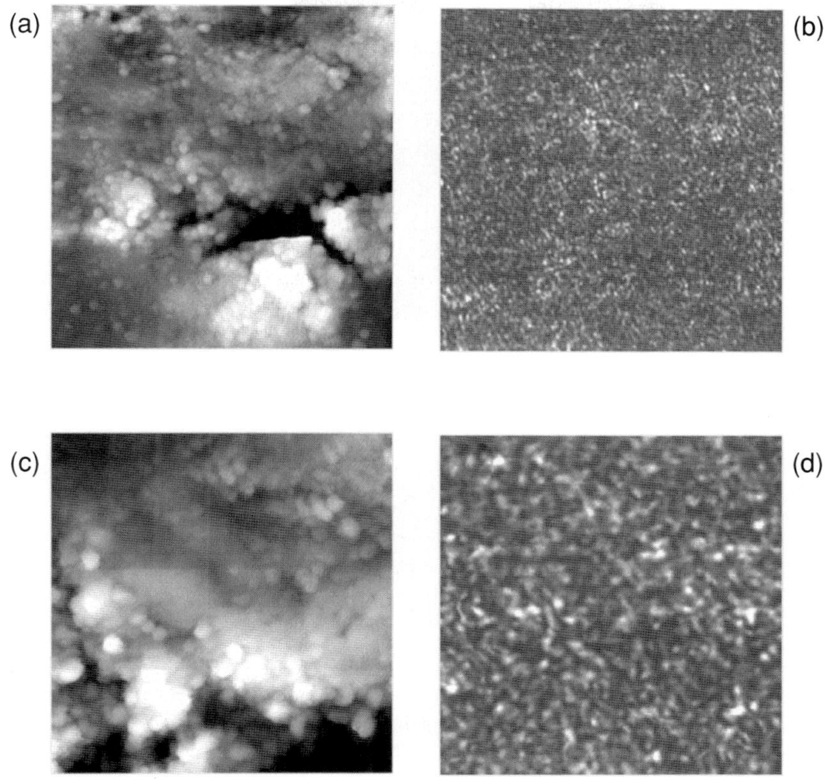

Fig. 2.61. Shear-force AFM (**a**, **c**) and apertureless SNOM (**b**, **d**) of stained microtome cuts of cancerous bladder cells; (**a**), (**b**) 25 µm scan; (**c**), (**d**) 10 µm scan; the z-range is 2 µm in (**a**) and 1 µm in (**c**); any useful SNOM contrast is wiped out by the staining

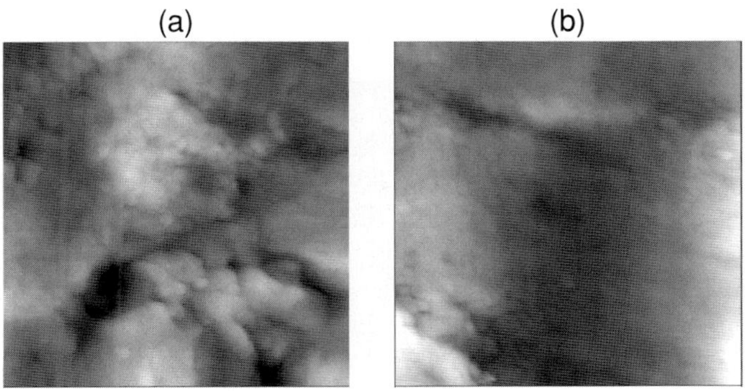

Fig. 2.62. Shear-force AFM (10 µm, z-range: 1 µm) of unstained microtome cuts of (**a**) cancerous bladder tissue and (**b**) healthy bladder tissue

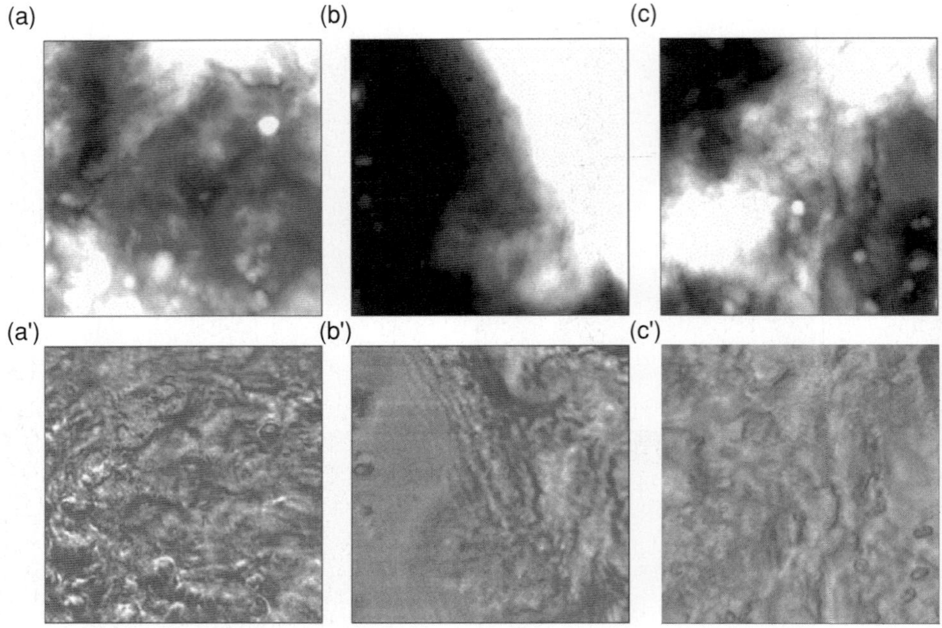

Fig. 2.63. Shear-force AFM topologies (25 μm, z-range: 1 μm) at three different sites (**a**–**c**) and the corresponding optical contrast (**a**′–**c**′) of apertureless SNOM on the rough surfaces of healthy bladder tissue

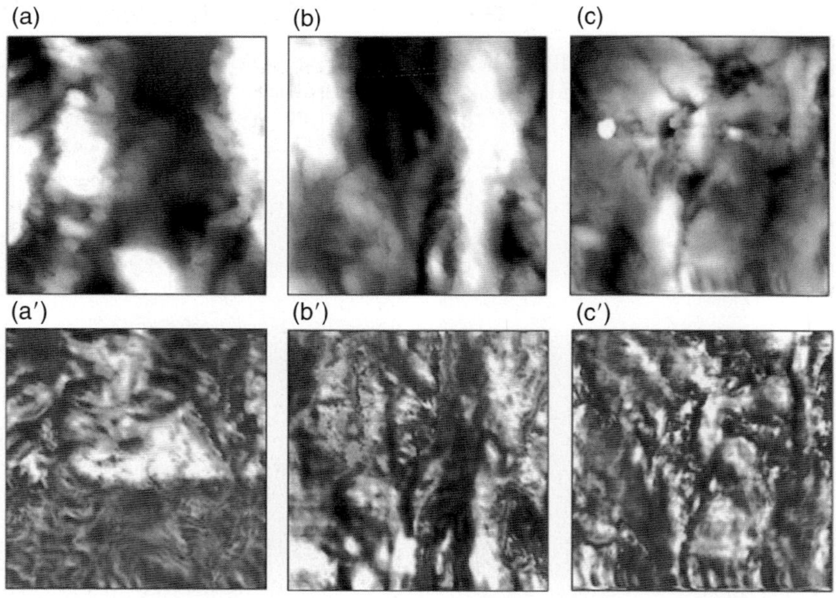

Fig. 2.64. Shear-force AFM topologies (20 μm, z-range: 1 μm) at three different sites (**a**–**c**) and the corresponding optical contrast (**a**′–**c**′) of apertureless SNOM on the rough surfaces of cancerous bladder cells showing characteristic bright plate-like contrasts at all of the different sites

that seem to fit with topologic features. Such correspondence is very important and not artificial as the topologic features are both in sections of depressions and hills in the same image. The latter correspondence points to cells that are highly reflective in the shear-force gap. The brightness is assumed to be due to accumulation of metal ions and mitochondria in the rapidly replicating cells. Such plate-like chemical contrast does not occur with healthy tissue (Fig. 2.63). This observation provides a clue for precancerous diagnosis when the cells start to change their chemistry, and these new chances should not be disregarded in oncology.

The SNOM investigations should be urgently extended to other types of cancer. For example, the breast carcinoma is most easily diagnosed by the microcalcification at a very late stage. Our SNOM should be able to easily detect submicrocalcification in a very early stage, primarily for patients with a genetic risk.

The proper SNOM on rough surfaces is versatile, robust, and economic. The plate-like chemical contrast is not seen in confocal microscopy. The new diagnostic possibilities profit from the fact that apertureless SNOM in constant shear-force distance prefers minimal pretreatment of the samples, which are unstained. The stained samples had lost the optical contrast though (Fig. 2.61).

2.11 Near-Field Spectroscopy

There is presently no alternative to Raman SNOM and fluorescence SNOM if nanoscopic features with chemical contrast have to be locally identified with respect to their chemical structure by vibrational or electronic analysis, as IR SNOM is very complicated and local IR spectroscopy (Sect. 2.11.3) is just emerging with submicron resolution though not yet well established. It is possible to identify geodes, local impurities, or organelle constituents in biological material, etc. by these techniques. The influence of partly aggregating silver or gold colloidal particles for surface enhanced Raman spectroscopy (SERS) is not yet clear, as apertureless SNOM at constant shear-force is itself a surface enhanced technique. It may, however, occur that highly dispersed materials cover the colloidal metal particles and that these are a helpful support inducing further enhancement of the reflected Raman intensity. The detection of Raman SNOM signals requires a highly sensitive optoacoustic spectrometer with waveguide optics and single-electron counting capability. Also edge and notch filters are essential. The spectrometer can also be used for fluorescence SNOM. If spectroscopic analysis of the emitted light is not essential an edge filter in front of the photodetector may block the primary light and pass only fluorescence light.

2.11.1 Direct Local Raman SNOM

The limitations of aperture Raman SNOM are severe, due to very hot and blunt tips and long collection times. Some reports do not look very promising, especially for moderately rough surfaces [66, 67]. Apertureless SNOM by focusing laser light from below through the sample to a noble metal cantilever tip in short distance to the surface was used for enhancing the scattered Raman efficiency by factors up to 40 [68]. However, the Raman signals of C_{60} were shifted to the red from bulk spectra by charge-transfer interaction with the gold tip, while unaltered bands were weak. Plasmon-enhanced Raman contrast in SNOM images with a scanning SERS active metal tip has been put forward [69]. Scattering apertureless Raman SNOM can also be obtained under shear-force control for rather flat surfaces by focusing a laser beam from the side to the very end of a damped vibrating sharp metal tip. Edge and notch filters eliminate primary light for the spectrometric analysis. Shear-force controlled apertureless AFM/SNOM operates at about 5 nm distance to the surface, and there is the enhancement of reflection back into the sharp tapered fiber. It is very likely that the enhancement is also at work for the near-field collection of Raman SNOM and fluorescence SNOM light in the shear-force gap [39]. It was thus possible to detect the local Raman spectrum on the highly reflecting geode at or under the polished dolomite sample in Fig. 2.65 using the illumination (488 nm) through the uncoated sharp tip. The received spectrum after collection of 145 scanlines in the region of the vesicle is depicted in Fig. 2.66. The spectrum exhibits the known lines of pyrite and is a resonant one.

Further direct Raman SNOM spectra of silicon (519.7 cm^{-1}) and its freshly grown (5 nm) silica layer (500 cm^{-1}) as well as of nonresonant gallium nitride

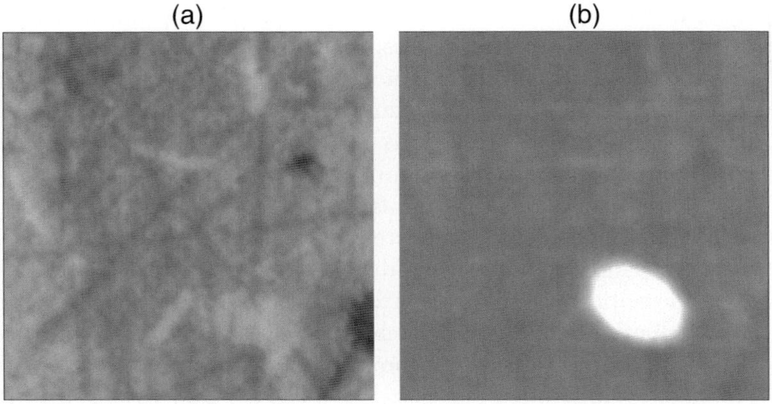

Fig. 2.65. Shear-force AFM (**a**) (5 µm, z-range: 100 nm) and chemical contrast by SNOM (**b**) of a pyrite geode at the surface of polished dolomite that cannot be traced in the 5 µm AFM topology

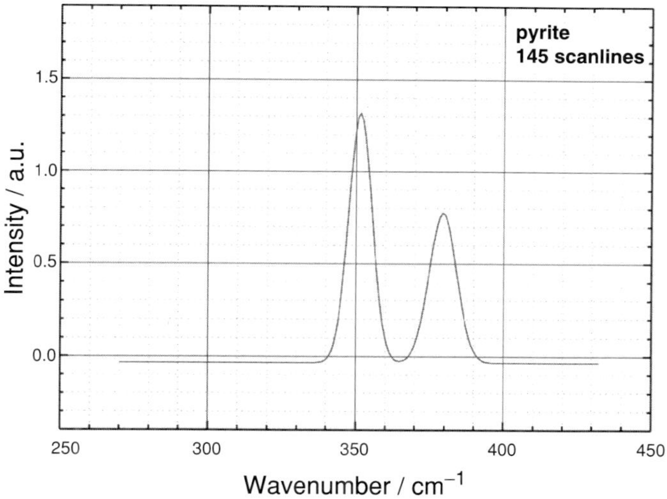

Fig. 2.66. Local resonant Raman SNOM spectrum of the pyrite geode (cf. Figs. 2.8 and 2.47) in dolomite; the spectrum is flattened and a baseline subtraction has been executed

[E_1(TO) and E_2 Raman modes at 560.8 and 570.4 cm^{-1}] on alumina (subtraction of the Raman response of the fiber) with the shear-force apertureless technique (using 488-nm light) at total collection times of 10 min have been reported in [39].

2.11.2 SERS SNOM

Aperture tips have enormous difficulties with topology. Thus, dye-labeled DNA on silver coated Teflon beads gave Raman spectra [70], but these were not correlated with the sites of the beads and do not aim to analyze the DNA structure. Surface enhanced Raman spectroscopy succeeds in the presence of silver or gold colloidal particles [71]. An AFM tip has been coated with gold in order to create a SERS probe with a laser beam directed on the tip or 15–25 μm away for comparison [72]. The Raman signal was locally enhanced underneath the tip. Also coating of an AFM tip with silver increased the Raman signal of both rhodamine 6G and crystal violet (size of the metallic tip <40 nm) [73]. These techniques are not suitable for rough surfaces. A sharp silver tip at a distance of 1 nm in shear-force feedback from the dichroic laser beam focused on the sample gave local variations in the Raman spectrum along a single-walled nanotube. The scattered light was collected with the same objective, transmitted by the beam splitter, filtered by a long pass filter, and detected by a single-photon counting avalanche photodiode after a narrow band pass filter. The Raman image was obtained by raster scanning and continuous recording of the scattered Raman signal [74]. The height of the nanotube was 1.6 nm, and the lateral resolution 25 nm.

SERS can also be used in the comparatively uncomplicated apertureless shear-force SNOM setup. If the standard compound adenine is studied in the presence of silver particles (Fig. 2.45) from dried solutions, it is possible to obtain the characteristic Raman line at $735\,\text{cm}^{-1}$ from a $100 \times 100\,\text{nm}^2$ area. Suitable interference and notch filters are used for the elimination of source irradiation light. Usually sodium chloride is used for the stabilization of the metal colloid. However, under the near-field conditions sodium chloride was avoided. $10\,\mu$l of aqueous solution containing $10^{-3}\gamma$ of adenine was added to the silver sol evaporating to a circle of 2 cm diameter in air. The 20-mW Ar-laser at 488 nm was coupled to the tip (Fig. 2.5), and the collection time was 10 min. The SERS SNOM spectrum obtained is depicted in Fig. 2.67. It was not necessary to fix the dried components to the glass.

The experiments indicate that Raman scattering can be collected by the sharp tip in the shear-force gap and diffracted in an optical spectrometer. Numerous applications can be envisioned, as sufficiently sensitive spectrometers are available also for local nonresonant Raman SNOM. Collection times must be low for the local experiment, as thermal drift cannot be fully impeded.

The sodium chloride content of common SERS applications is troublesome in SERS SNOM not only because of the cubic crystals that form upon evaporation but also because of increased aggregation of the silver particles if the adenine ($10^{-3}\gamma$) is added to two drops of silver sol. This can be seen from shear-force AFM scans in Fig. 2.68 on evaporated samples at initial NaCl

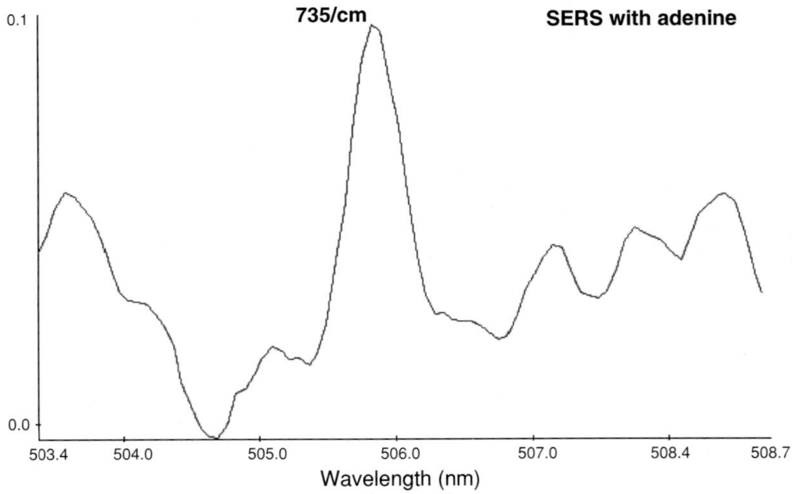

Fig. 2.67. Local SERS SNOM spectrum from a $100 \times 100\,\text{nm}^2$ area of adenine in the presence of Ag sol without NaCl in a dried drop on a microscope slide without flattening or baseline subtraction

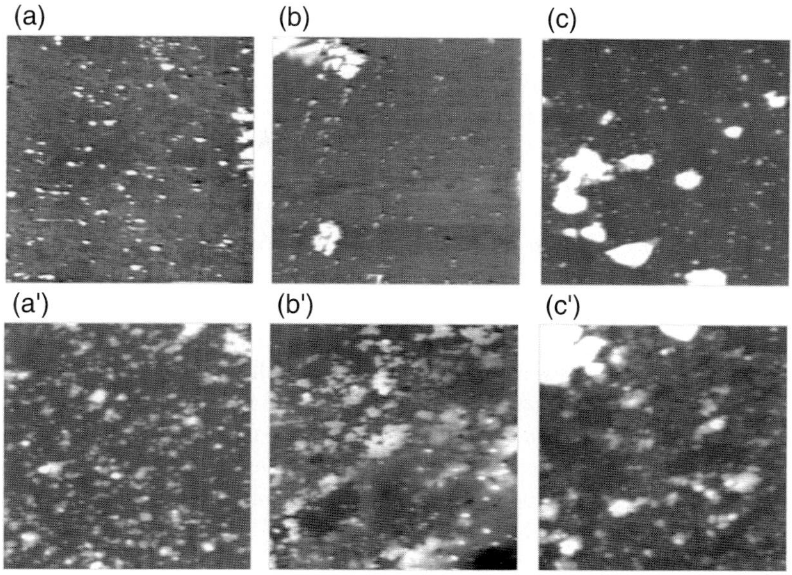

Fig. 2.68. Shear-force AFM measurements (10 µm) of evaporated silver sol on glass with various concentrations of NaCl with (**a**′–**c**′) and without (**a**–**c**) addition of $10^{-3}\gamma$ adenine

concentrations of 0 (a), 0.1 (b), and 0.5 mol l^{-1} (c) without (a–c) and with the addition of adenine a′–c′).

Even at sites where no NaCl crystals are discernible, the sol particles are less uniform the higher the initial salt concentration is. Furthermore, the consistent enlargement of the sol particles by addition of a minute quantity of adenine in water and more so at the higher NaCl concentrations is puzzling. It was also detected by microscopic inspection that the sodium chloride crystals grow smaller in the samples with added adenine [61]. The experimental conclusion is that sodium chloride can and should be avoided in local SERS SNOM. It may be expected that Ag sol when evaporated over biological preparations would facilitate the rapid collection of apertureless shear-force SERS SNOM spectra.

2.11.3 Near-Field Infrared Spectroscopy and Scanning Near-Field Dielectric Microscopy

It would be very profitable to have similarly versatile local IR spectroscopy with its particularly high potential for compound identification. However, such endeavor was only rarely pursued. Early developments of IR SNOM (monochromatic) and local IR spectroscopy have been reviewed [75]. These used apertured tips or metal-scattering tips but suffered from topologic errors so that only flat surfaces could be studied. Also the heat of IR light absorption has been used with thermal probes (thermoelectrical signal) to obtain local spectra similar to regular IR spectra (differences in the relative band intensities and overall band shapes) and the combination with common FT-IR spectrometers appears promising for vibrational and photothermal spectroscopy combined with near-field imaging at submicrometer resolution. More recent improvements have used a broadly tunable IR laser light source producing ultrafast pulses with an FWHM bandwidth of 150 cm^{-1} and a single mode fluoride glass fiber probe to map chemical functional groups by IR absorption (2,000-4,500 cm^{-1}) with subwavelength spatial resolution [76]. Thin metal and polymer films have been studied in transmission with apertured tips. Mesoscale structures can be described. It is suggested that apertureless probes would achieve a higher spatial resolution than 1/8 at 2,900 cm^{-1} [77]. The technique requires broad tunability and bandwidth, parallel spectral detection for high image acquisition rates, and IR-transparent aperture probes. The analysis of the near-field IR spectra indicates that the image contrast in the C–H stretching region is largely due to near-field coupling and/or scattering effects [78].

Furthermore, a scanning near-field microscope was presented operating at microwave frequencies for the characterization of advanced dielectric materials. The key component of the microwave-evanescent-probe microscope is a resonant-cavity/transmission line structure [79]. A broad development of these techniques will be highly rewarding.

2.11.4 Fluorescence SNOM

Fluorescence SNOM is possible even under conditions of topologic artifacts, as the longer wavelength light can be separated from the primary reflectance. Again, aperture tips in constant height provide low local resolution. Scattering apertureless SNOM can be studied with fluorescing test samples [31].

Two-photon near-field fluorescence microscopy has also been used [80]. Furthermore, second harmonics generation [81] at the metal tip as well as a white continuum has been observed [82]. These techniques require pico- and femtosecond laser pulses to the metal tip on the surface. Local fluorescence SNOM is most easily differentiated from topology as the emitted light can be diffracted and the emission spectrum measured. A possible drawback is photobleaching. Virtually all SNOM modes tried fluorescence applications, and some techniques have been used for the detection of single fluorescent molecules at extreme dilution [10]. However, such single molecule detection (but not resolution) is also possible with conventional microscopy. Important fields of research include local information in the chloroplasts of the photosynthesis in green leaves, DNA analysis, organelle structure in biological cells, fluorescing labeled biological samples, fluorescent probes in oncology, living cells in wet environment, quantum wells, optoelectronic semiconductors, etc. The highest optical resolution in macromolecules may be obtained in PSTM experiments.

For "real-world" samples that have strong irregular topologies the technique of choice is the apertureless back to the fiber emission in constant shear-force distance. Again the collection of the emitted light by the illuminating tapered tip in shear-force distance is best suited for technical or biological applications that must avoid topologic artifacts [39]. Furthermore, this technique is by far less complicated and most economic. The technically applied fluorescing disperse dyes **1**–**3** of the merocyanine type (Fig. 2.69) were used in a high-temperature exhaust or heat-transfer dyeing procedure for even or ring coloration of polyester textile fibers. These were embedded into an appropriate resin (Technovit 7100, Heraeus Kuzer Ltd. or Agar 100, Lannet Ltd.), cut with a microtome and subjected to shear-force apertureless SNOM while eliminating the primary light by an edge filter (OG 515, transmitting from 506 nm onward, Schott) in front of the detector for removal of all primary 488-nm light.

Figure 2.70 shows the AFM and fluorescence SNOM of the microtome cut with dye **1**. It is clearly seen that the polyester fiber is several μm down due to different hardness and elasticity of the materials. Nevertheless, the fluorescence is evenly strong over the whole area but only at the area of

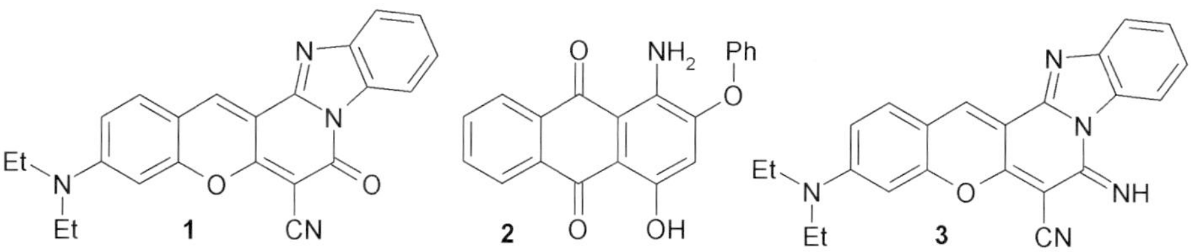

Fig. 2.69. Structural formula of the dyes **1**, **2**, and **3**

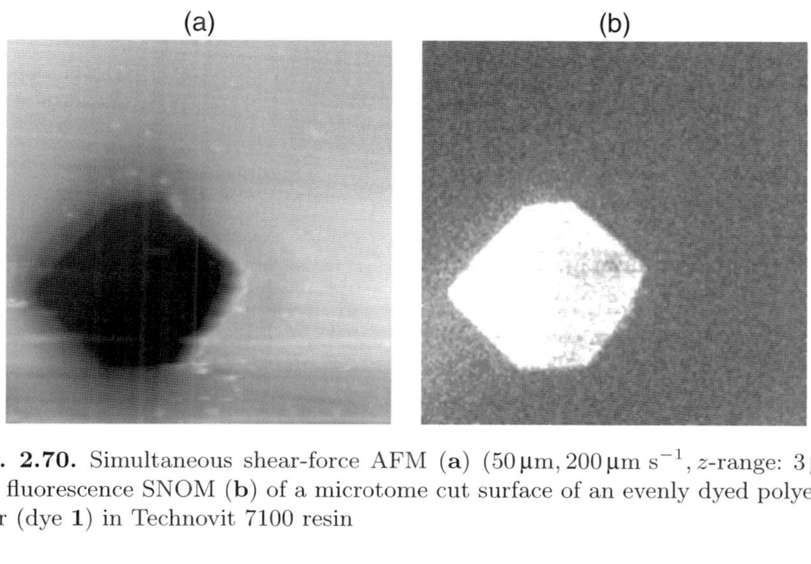

Fig. 2.70. Simultaneous shear-force AFM (**a**) (50 μm, 200 μm s^{-1}, z-range: 3 μm) and fluorescence SNOM (**b**) of a microtome cut surface of an evenly dyed polyester fiber (dye **1**) in Technovit 7100 resin

the polyester fiber and not at the slopes, which consist of uncolored resin. Furthermore, the minor deficiencies of the very deep AFM are not reproduced in the optical contrast (perspective images are in [39]).

The polyester fiber evenly colored with dye **2** and embedded in soft Technovit 7100 resin shows again homogeneously distributed fluorescence (Fig. 2.71b) over the fiber cross section in a very skew microtome cut (Fig. 2.71a). The slight reduction in the deep hole (3 µm wide, 800 nm deep) and the upper right corner appear to relate with the unsuitable topology.

The light fastness of dye **2** has been studied by its fluorescence bleaching in the SNOM experiment. Figure 2.72 shows the effect of 10 min scanning

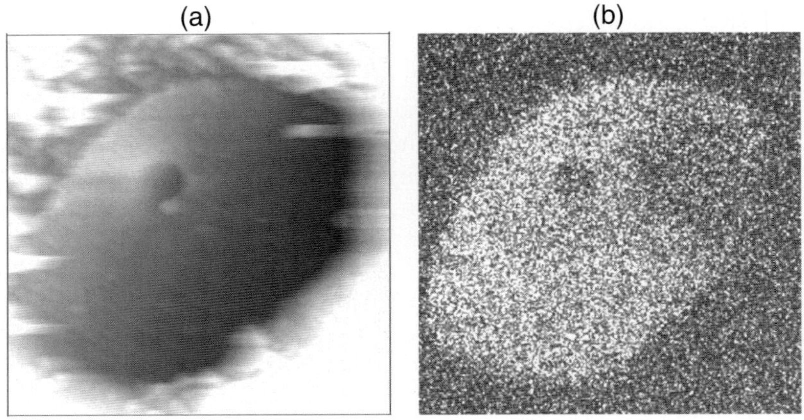

Fig. 2.71. Simultaneous shear-force AFM (**a**) (25 µm, z-range: 5 µm) and fluorescence SNOM (**b**) of a microtome cut surface of an evenly dyed polyester fiber (dye **2**) in Technovit 7100 resin

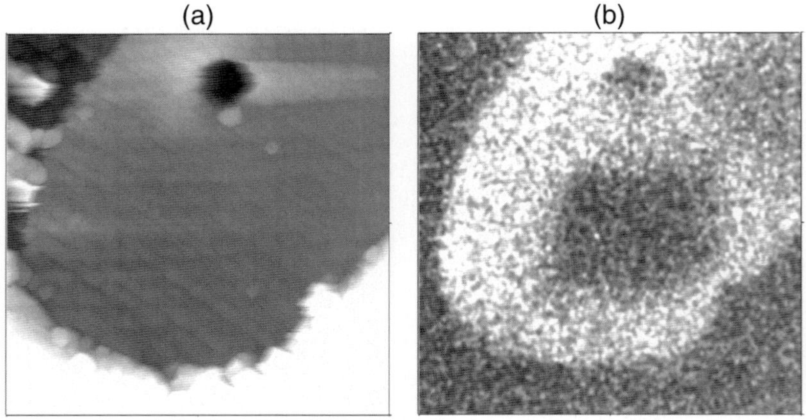

Fig. 2.72. Simultaneous shear-force AFM (**a**) (20 µm, z-range: 5 µm) and fluorescence SNOM (**b**) of a microtome cut surface of an evenly dyed polyester fiber (dye **2**) in Technovit 7100 resin after 10 min illuminated scanning in a $5 \times 5\,\mu m^2$ area of the dyed fiber showing fluorescence bleaching

in a $5 \times 5\,\mu m^2$ area with the uncoated illuminated tip (1.6 mW) giving a $6 \times 6\,\mu m^2$ area of photobleaching in the optical image that cannot be seen in the topology. This is a clear technical application for the determination of relative light fastness of dyes in the actual fabrics of interest by large-scale SNOM that works despite very high topology within the scan. This application of SNOM will help designers to improve their products based on clear-cut data.

By using the resin Agar 100 the ring-dyed polyester fiber with dye **2** comes up in the microtome cut as shown in Fig. 2.73a. The ring dyeing can be seen in

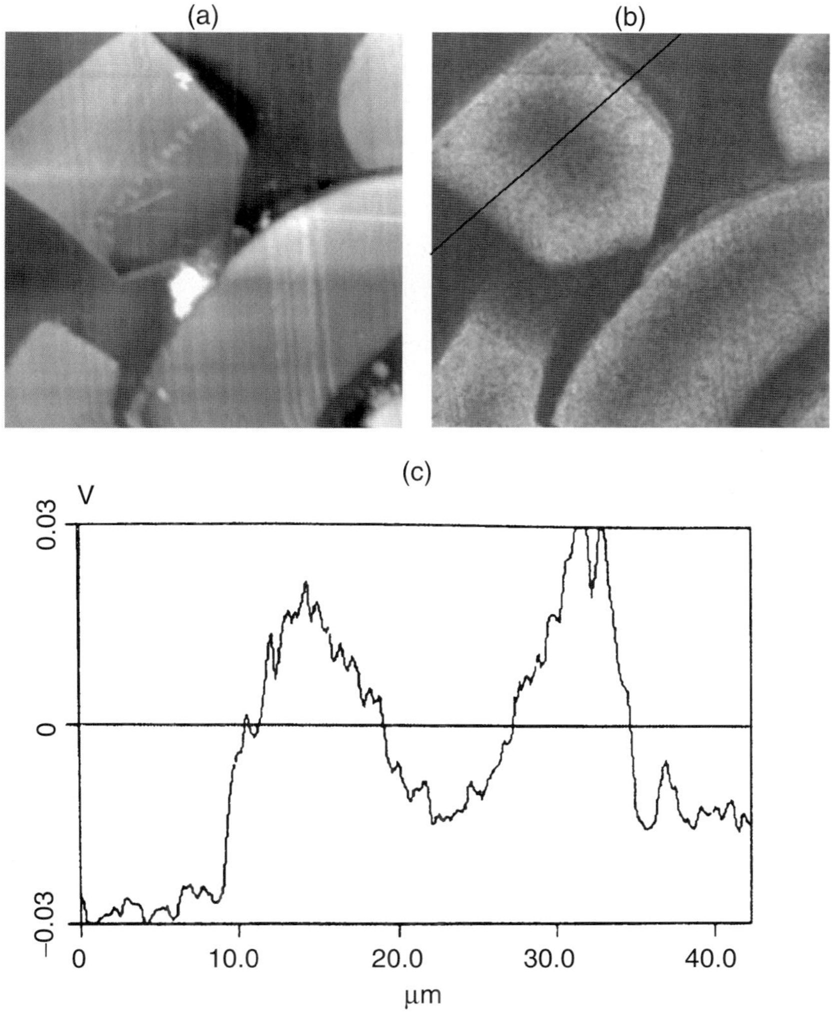

Fig. 2.73. Simultaneous shear-force AFM (**a**) ($50\,\mu m$, $200\,\mu m\ s^{-1}$, z-range: $1\,\mu m$) and fluorescence SNOM (**b**) of a microtome cut surface of a polyester fiber (dye **2**) embedded in Agar 100 resin, showing ring dyeing in the optical contrast and the cross section (**c**) through the fluorescence intensity as indicated in (**b**)

both orientations of the textile fiber. The very high topology of an impurity in the center does not give an artifact in the optical image. The flat lying fiber is flatly cut. The analysis of the near-field fluorescence curve (Fig. 2.73c) offers a very easy and versatile new possibility for the determination of the industrially important diffusion coefficient of the particular dye in the heat-transfer dyeing process at the actual object rather than in tedious determinations at model systems. The profiles of the near-field fluorescence intensity in Fig. 2.73c are fitted to Fick's second law of diffusion, which is numerically solved by using the tabulated values of the "error function" in the usual way. The resulting D-values (diffusion coefficient) are found constant over the complete curve starting at the maximum (right curve chosen). This indicates that the near-field fluorescence intensity is proportional to the local dye concentration. The absolute value of the diffusion coefficient at the dyeing temperature is calculated by using the dyeing time of 30 min and gives the result for the dye **2** of $D_{115^\circ C} = 4.7 \times 10^{-11}\,\mathrm{cm^2 s^{-1}}$. This value is consistent with similar values in the literature [39].

A further ring dyeing with dye **3** gave closely related results. Figure 2.74 has the polyester fiber up in Agar 100 resin (a) and clearly indicates the ring coloration gradient (b). The images (c) and (d) are added for showing that the nonfluorescing topologic impurities do not affect the SNOM contrast in shear-force apertureless SNOM.

The diffusion coefficient of dye **3** into the polyester fiber is determined with the Technovit 7100 resin microtome, which has the fiber down (Fig. 2.75) for demonstration of the versatility of the invaluable new application of artifact-free SNOM in the presence of very high topology. The right profile is selected for the determination applying Fick's second law as above. The D-values are again constant over the complete decay curve and the absolute value calculates to $D_{130^\circ C} = 2.4 \times 10^{-11}\,\mathrm{cm^2 s^{-1}}$. This value is consistent with similar values in the literature [39]. The ease of experiment and setup guarantee widespread industrial use of this unexpected application of SNOM. Heat-transfer techniques in dyeing can be used to control the penetration depths of dyes into the fibers if the diffusion coefficients are known. Such technically important applications of SNOM are only possible with the easy and economic shear-force SNOM with apertureless (uncoated) sharp fiber tips but not with any one of the other approaches that cannot handle (high) topology.

Following these apertureless SNOM measurements [39], confocal laser scanning microscopy was tried to study penetration of fluorescing dyes into human hair at 240 nm resolution [83]. However, no diffusion coefficients could be determined. Even less details are obtained with shear-force aperture SNOM (nominal width 50–100 nm) on the same hair samples giving artificial transmission images, whereas the application of a long-pass filter and a very slow scan speed of 6 s per line and 200 × 200 pixels (the temperature of the tip in shear-force distance was not specified) gave a fluorescence image with a claimed optical resolution of 130 nm [83]. A perfect correspondence with the less resolved confocal fluorescence image is not shown. It would have been

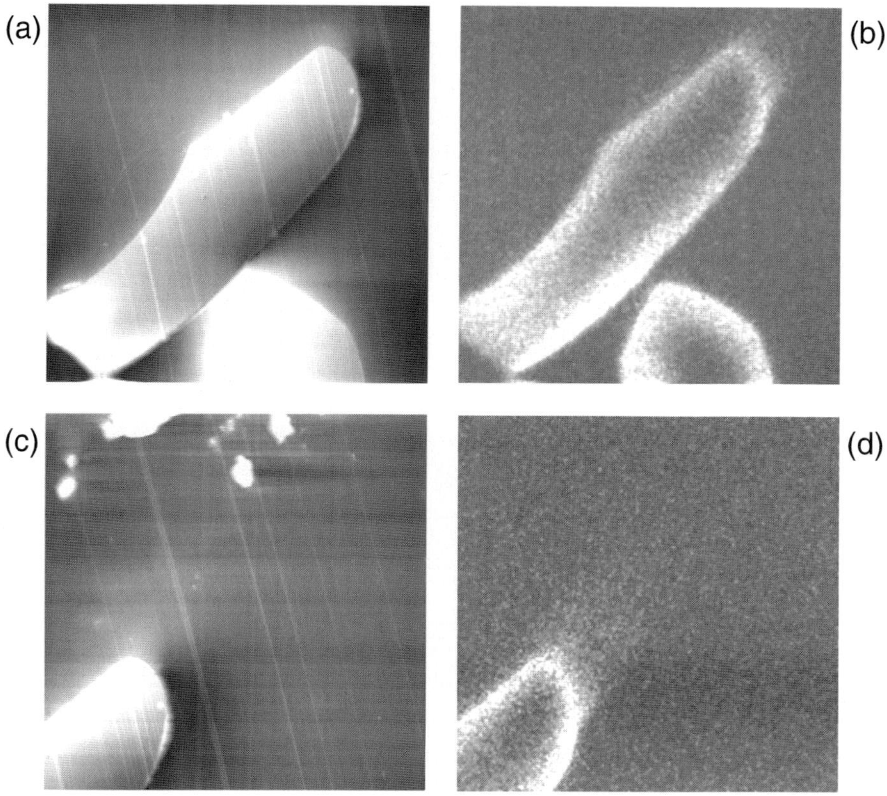

Fig. 2.74. Simultaneous shear-force AFM (**a**) (50 μm, 200 μm s^{-1}, z-range: 1 μm), (**c**) (50 μm, z-range: 560 nm) and fluorescence SNOM (**b**, **d**) of a microtome cut surface of a partly dyed polyester fiber (dye **3**) embedded in Agar 100 resin, showing the ring-dyeing in the optical contrast

easier to measure fluorescence SNOM at the much higher resolution capabilities of 10–20 nm with a cold and sharp uncoated tip without topologic artifact and enhanced light detection in the close near field of the shear-force gap. A wealth of highly reliable and additional information would certainly arise by this highly superior technique.

The present SNOM technique with its lateral resolution power better than 10 nm is well suited for the investigation of nanoparticles such as technical dye pigments that are embedded in resin or varnish. The color shade depends strongly on their possible aggregation that can be determined by the comparison of topology width and SNOM width even if the particles are covered by the embedding material. Degree and type of aggregation can be studied at the surface if part of the particles swim up in the hardening resin/varnish and contribute to the surface roughness. For example, fluorescent technical dye nanoparticles of 100–200 nm diameter in polyvidone resin of 1 μm thickness

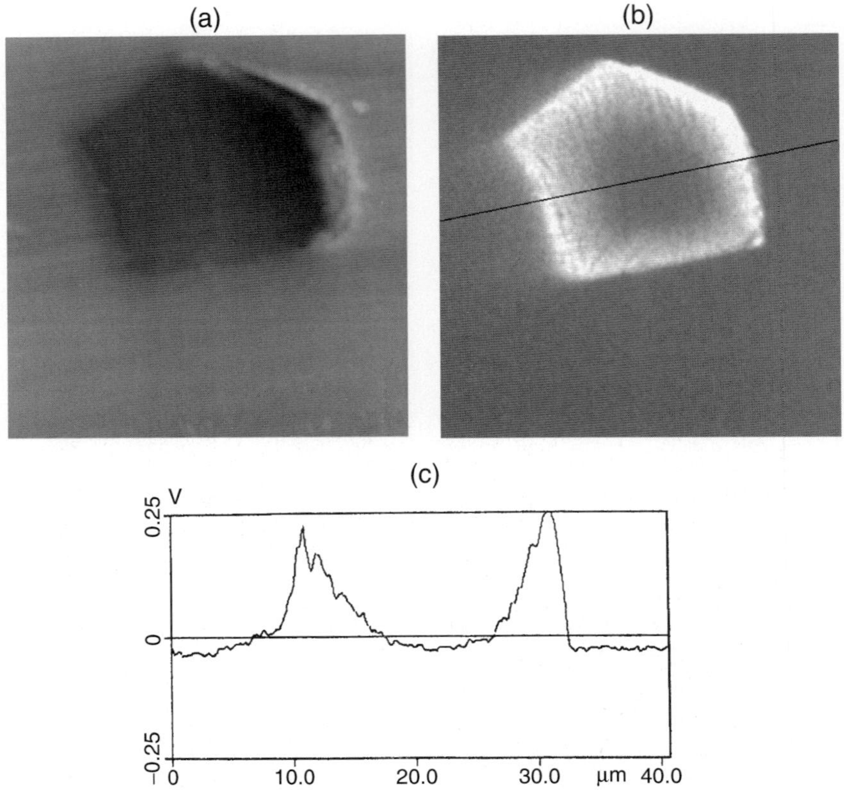

Fig. 2.75. Simultaneous shear-force AFM (**a**) (40 µm, 200 µm s^{-1}, z-range: 2 µm) and fluorescence SNOM (**b**) of a microtome cut surface of a polyester fiber (dye **3**) embedded in Technovit 7100 resin, showing ring dyeing in the optical contrast (**b**) and the cross section (**c**) through the fluorescence intensity as indicated in (**b**)

on glass can be subjected to fluorescence SNOM with uncoated sharp tips in the shear-force gap at 488 nm using the edge filter OG 515 (transmitting from 506 nm onward; Schott) in front of the detector [39]. Figure 2.76 shows the topology (a) on the surface and the local fluorescence (b) at the sites of the protrusions. The sizes of the protrusions indicate both single particles and aggregated ones. The particles emerge over the surface by up to 69 nm (R_{ms} = 15.3 nm). The single particles have widths of 110–210 nm in the AFM and consistently 100–200 nm in the SNOM image, respectively. Only the optical image gives the correct size of the particles. It is thus clear that the polyvidone cover on the surface particles is about 5 nm thick. This important result underlines the high resolving power of the apertureless shear-force SNOM. Due to the resin cover, the nanoscopic fluorescence areas are indeed smaller than those of the topologic features.

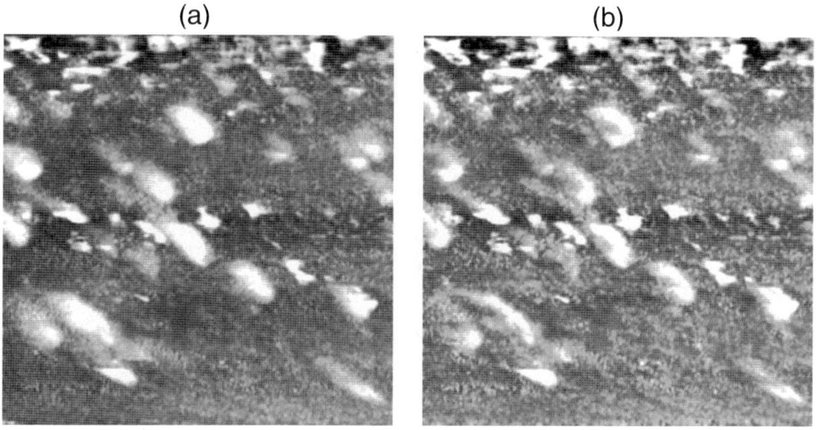

Fig. 2.76. Simultaneous shear-force AFM (**a**) (5 µm, z-range: 200 nm) and apertureless fluorescence SNOM (**b**) of 100–200 nm nanoparticles in polyvidone resin showing both the occurrence of single and aggregated particles

Furthermore, there are some remarkable differences between topological and optical image in the larger features that correspond to aggregates. There are three types of aggregation: linear, angular, and dense. In particular the angular ones require scrutiny at higher magnification (Fig. 2.77). It is clearly seen in the fluorescence images (b, b') that the aggregation is angular (and there is also a single particle). As these nanoparticles form a loop, the resin cover fills the void for closing the topologic object. Only SNOM differentiates between the different chemical species. Such information is of major importance for the industrial application in varnish formulations (surface quality and color shade). It can only be obtained with the most reliable apertureless shear-force SNOM with sharp tips under the conditions of enhancement in the shear-force gap [4–9, 40].

Further results with shear-force apertureless fluorescence SNOM have been reported on biological samples. Thus, the green fluorescent protein GFP in intact *Escherichia coli* bacteria expressing GFT has been locally imaged [84]. A dichroic mirror achieved separation of the fluorescence light from the primary light. The location of the fluorescence in the topology of the bacterium was seen. This was also possible for immunofluorescence, which gave fluorescence sites independent of the topology, which improves the understanding of clustering processes in breast cancer tissue. However, a diagnostic value was not claimed or recognized [85]. Unfortunately, the authors did not study the non-fluorescent SNOM contrast and it is not known if it was enhanced with respect to the far-field background (cf. [4–9, 40]). Only that additional information would have allowed assessment of the size of the fluorescing areas (unknown optical resolution).

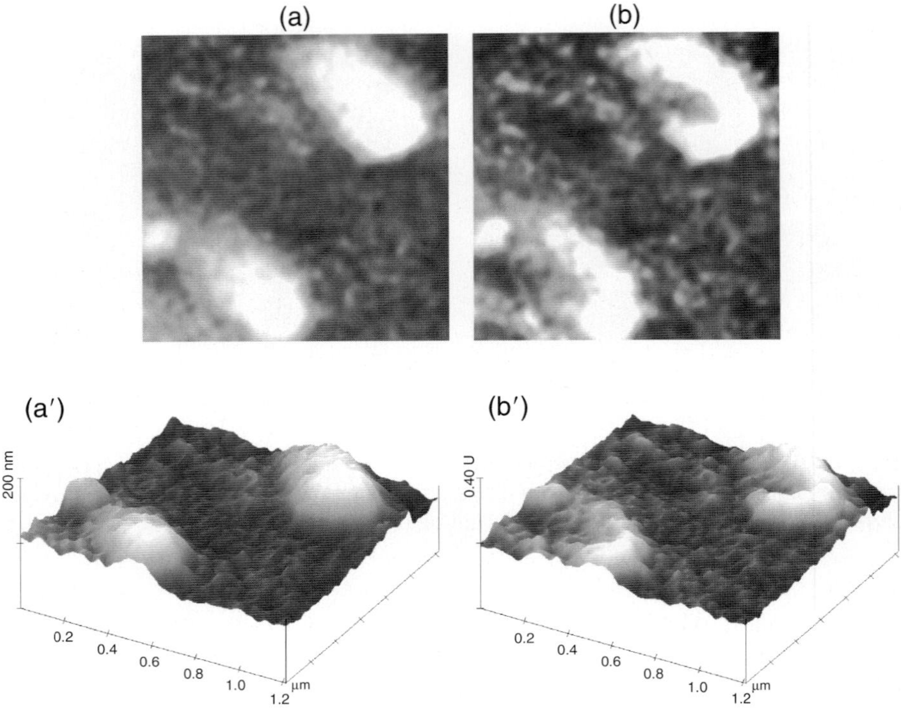

Fig. 2.77. Simultaneous shear-force AFM (**a**, **a′**) and apertureless fluorescence SNOM (**b**, **b′**) of dye nanoparticles in polyvidone resin showing angular aggregation by the difference in shape of topologic and optical image both in the *top-views* (**a**, **b**) and in the *surface-views* (**a′**, **b′**)

Chromosomes and various organelles in cells of biological or medicinal interest have been labeled with fluorescent agents for fluorescence SNOM in the apertureless shear-force mode using multiphoton absorption at considerable topologies [86, 87]. The fluorescence SNOM images are sharper and provide more information than the topologies in extended scan areas. The smallest resolved optical features had a width of 200 nm (∼1/6 of the wavelength). Again the determination (or assessment) of the enhancement of reflected light intensity in the shear-force gap [4–9, 40] is missing. It would be helpful for the assessment of the resolution. Fluorescence resonance energy transfer (FRET) between excited donor and acceptor molecules occurs over a range of 1–10 nm. FRET of different dye molecules embedded in polyvinyl alcohol films and bound to cell surfaces could be detected by selective photobleaching of donor and acceptor fluorophores using SNOM with sharp uncoated tips [88].

2.12 Nanophotolithography

The possibilities of SNOM-based data storage have been analyzed [89]. It would include SNOM-based writing and reading. The reading may require 4,000 photons per bit in order to achieve a readout speed of 10 MHz. This requires high illumination intensities. Also photo- and thermostability of film and SNOM head are important issues. High-speed AFM reading may be another choice. Apertureless SIAM in the patented version has been proposed for CDROM reading at 400 bits μm^{-2} with pit sizes of 50 nm × 50 nm size and 25 nm depths at data rates in the tens of MHz range [90]. Conversely, the pits could not be read by an AFM tip at 10 nm constant height above the pattern. Aperture SNOM photolithography was tried in [91], and a resolution of 30–50 nm was claimed. Apertureless optical near-field fabrication of >100 nm lines to a 110 nm thick positive photoresist by focusing a laser beam through the sample to an AFM tip well below the threshold energy dose for polymerization has been described [92]. The energy required for inducing polymerization decreased at the metallic tip. Apertureless scattering SNOM at metallic tips has been used to form topological nanodots by a photoisomerization [93]. A similar technique was used for two-photon nanolithography under a metal tip using polarized 120 fs laser pulses producing features with 70 nm resolution for the polymerization in a photoresist [94]. This resolution is by a factor of 2 better than techniques based on far-field two-photon lithography. Aperture tips suffer from the propagation cutoff so that only a very small fraction of the light can be transmitted and are thus unsuitable. The scattering apertureless techniques are only applicable to very flat surfaces.

Another approach uses the heat of aperture near-field probes. These can be used for photolithography by taking advantage of their enormous local heat if high melting metals (for example chromium) are applied for the coating instead of the more common lower melting aluminum. Thus, the heat of sharp metal-coated tips has been used for the fabrication of pits on anthracene crystals at a lateral resolution of 70 nm [14] or on polymers [95]. It proved difficult to write deep structures sharply into photoresists, and the practical use of monolayer lithography is doubtful, except for the construction of nanostructures [10].

The cold apertureless shear-force SNOM offers totally different possibilities for thermal lithography, if thermal insulators cannot dissipate locally absorbed light energy. Local heat by radiationless deactivation will accumulate in molecular crystals, if these are lacking hydrogen bond networks, and melt them locally. Unlike the pits formation under the action of a hot metal coated aperture tip, there is protrusion formation under these conditions of retarded local surface heating. The formation of isolated cones by surface heating of a photostable merocyanine dye without hydrogen bonds [2-(*p*-dimethylaminobenzylidene)-3-*oxo*-2,3-dihydrothionaphthene] (S-merocyanine) and having an appropriate melting point (172–173°C) can be demonstrated with surface heating. A medicinal ceramic source IR heater with a reflecting mirror is applied for a predetermined time (e.g., 10 min) from a distance of 20 cm, while the crystal is in contact with a gold plate and preheated to 112°C. The required time for reaching surface melting is detected with a telescope. The heaters are immediately switched off after the first signs of surface change and AFM measurement follows at room temperature. The obtained cones of Fig. 2.78 exhibit heights up to 180 nm, half-widths of 280–500 nm, and base diameters of 450–980 nm. Their frequency is typically 40 cones per 100 μm^2 [13]. The cone formation takes material from the ground and profits from the decreased surface tension of the viscous melt, which crystallizes rapidly upon cooling.

The successful cone formation of Fig. 2.78b can be used for their precise placement using the illuminated sharp tip either far-field apertured or uncoated. The local heating by light absorption and radiationless deactivation is slow enough to form the cones. The 400 nm far-field apertured tip (nominal apex radius: 15 nm) is obtained by melting off the Al coating during adjustment of the light coupler for highest throughput at common laser settings. The result is seen in Fig. 2.79.

It is not clear if the initial tip had an open aperture before illumination. However, the confined light heated it up to more than the melting point of Al (660°C) so that the apex lost its Al coating at a length of 1.2 μm. Thereafter the tip was stable upon further illumination (>3 h). The light throughput of this far-field apertured tip can be adjusted up to 100μW at 488 nm. Figure 2.80a shows a surface of the S-merocyanine dye with some landmarks. By choosing a position of the far-field apertured tip under shear force control with a proper offset device, illumination for 10 s at 1 μW throughput gives the cone in (b). It was lost by repeated AFM scans, but two more cones were placed at a close distance by 10 s spot illumination at shear-force distance [13]. The base of the cones (420–440 nm) is slightly stretched due to the horizon-

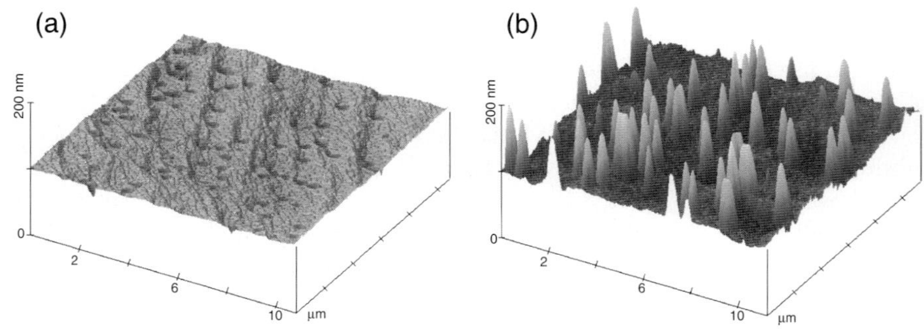

Fig. 2.78. Contact AFM topology of the main surface of the organic S-merocyanine crystal without hydrogen bonds; (**a**) before and (**b**) after short partial melting at its surface by sharply interrupted IR-heater application leading to random cones of considerable height

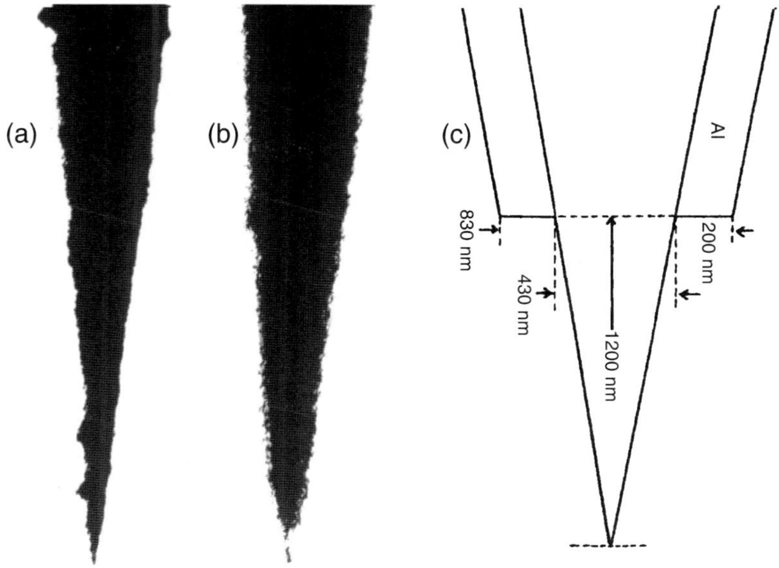

Fig. 2.79. Microimage (400 × microscopic enlargement) of (**a**) an Al-coated tip with apex radius of 15 nm and 200 nm coating, (**b**) the same tip from a different side after losing the apical coating upon adjustment of the light coupler for maximum throughput with common intensity settings for SNOM, and (**c**) a sketch of the final end of the far-field apertured tip according to the photograph

tal tip vibration but corresponds closely to the far-field aperture at 1.2 μm distance. The height of the cones is 60 nm, the half-width 300 nm, and the peak-to-peak distance of the clearly separated cones 600 nm in (c). These values could be easily improved, the organic material stabilized into molds by suitable coatings.

The regular shape of the cones is an indication of a good tip taper. Apart from data storage considerations the technique is a simple means for judging the quality of particular SNOM tip specimens.

In support of the thermal insulation it was shown that no cones were formed within 10 s by the same far-field apertured tip at 0.5 μW light intensity; however, after 45 s the same cones with about the same size were observed as in Fig. 2.80b, c. Furthermore, the analogous photostable NH–merocyanine dye [2-(*p*-dimethylaminobenzylidene)-3-*oxo*-2,3-dihydroindole] with NH instead of S, which can dissipate the heat of radiationless deactivation in its crystals via the hydrogen bonds network, does not give similar cone formation upon surface heating (starting temperature 176°C; melting point 236°C) or with the far-field apertured tip at 488-nm illumination.

If the irradiation time and throughput energy were increased for the S-merocyanine dye to 45 s and 80 μW the resulting protrusion was much more extended in Fig. 2.81. The double nature may have different causes due to the

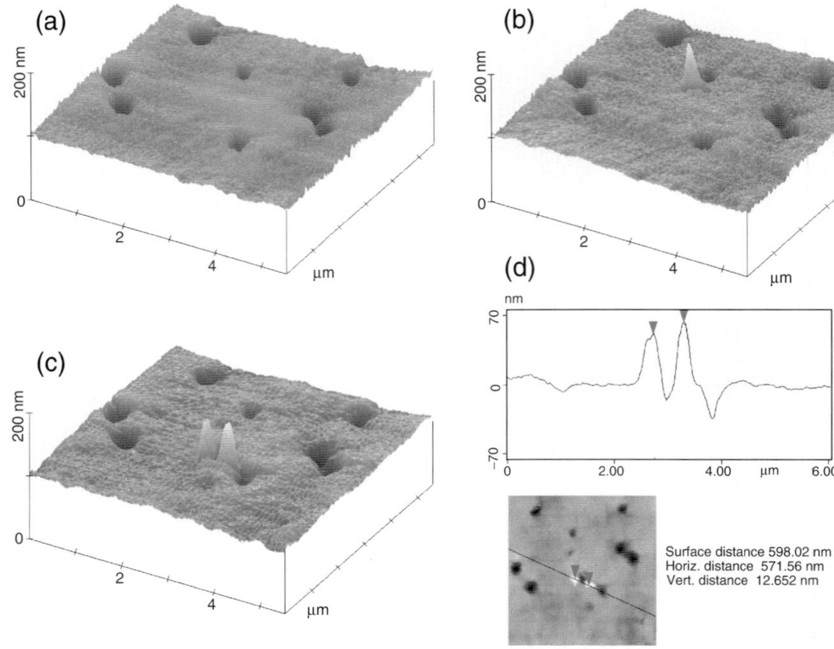

Fig. 2.80. Shear-force AFM on the most prominent face of the S-merocyanine dye without hydrogen bridges; (**a**) fresh: (**b**) after 10 s irradiation at a preselected spot with the far-field apertured tip at 1 µW; (**c**, **d**) after 10 s irradiation at two preselected spots 600 nm apart with 1 µW after loss of the feature in (**b**)

excessive irradiation in shear-force distance if the vibrating tip might have immersed in the melt or if the recrystallization started from two different points, or if there were minor imperfections of the tip, or if there was some thermal drift. The cone structure is retained and the feature starts sharply from the ground (1.2 µm wide and 200 nm high). The sharpness of the cone even at high overirradiation makes this system particularly versatile.

The SNOM image in Fig. 2.81c does not exhibit any optical contrast in accordance with a suitable tip (a blunt or broken tip would create topologic artifacts). This confirms again that the features derive from local melting but not from photochemistry [13]. Also the features in Figs. 2.78 and 2.80 gave no SNOM contrast.

The versatility of the new technique of submicroscopic tip evaluation has been demonstrated with further suitable systems. For example, 9-chloroanthracene (platelets) is known to give floe-like surface features when irradiated on its main surface by dimerization upon low-intensity photolysis [96]. The 9-chloroanthracene crystals absorb at 488 nm. A single crystal gave the features in Fig. 2.82b with an apparently deficient apertureless tip from 10 nm distance at a high local light intensity of 1.2 mW.

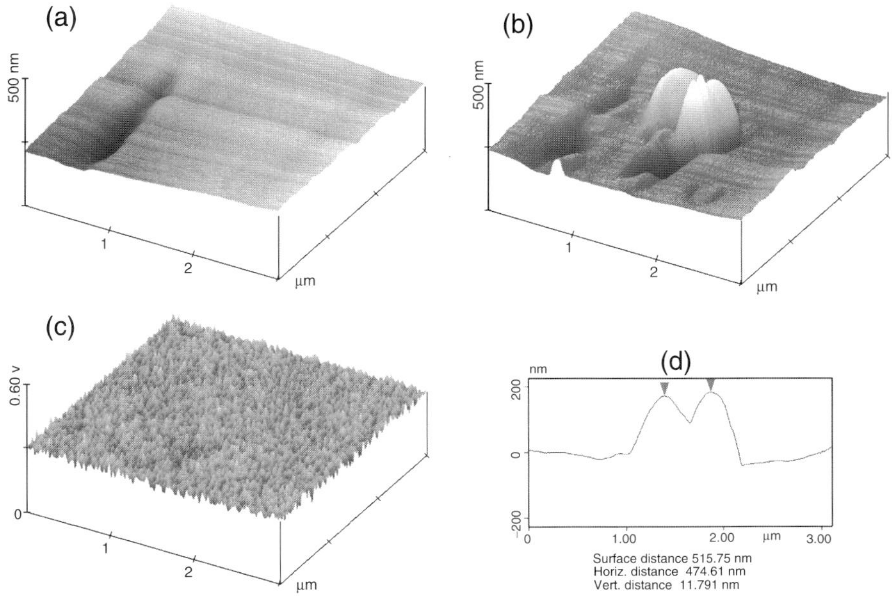

Fig. 2.81. Simultaneous shear-force AFM (**a**, **b**) and apertureless SNOM (**c**) on the most prominent face of the S-merocyanine dye: (**a**) fresh; (**b, c, d**) after 45 s irradiation at a preselected spot with the far-field apertured tip at 80 µW light intensity

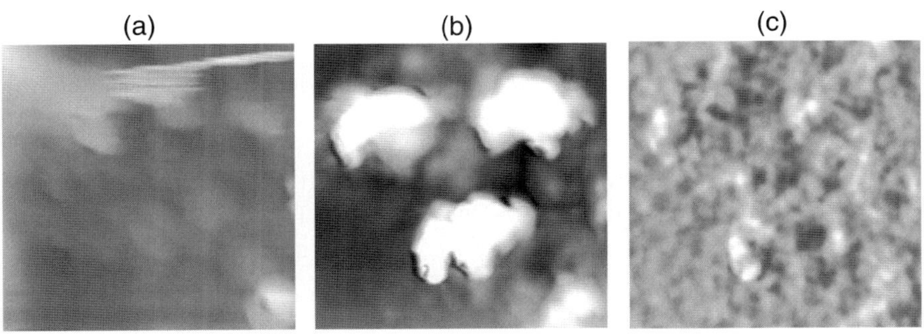

Fig. 2.82. Photolithography with a deficient uncoated tip (nominal radius: 20 nm) illuminated with 488-nm light from 10 nm distance to the main face of 9-chloroanthracene at three different locations for 20 min each; (**a**, **b**) shear-force AFM analysis (6 µm, z-range: 500 nm); (**c**) with (**b**) simultaneous apertureless SNOM exhibiting only minor artificial contrast; (**a**) before, (**b**) and (**c**) after the lithography

The features are up to 2.5 µm wide and of irregular shape but with sharp limits (typical slopes 40°), indicating a non-conospheric shape of the applied tip. The question arises if the features are the result of photodimerization [54] or of local melting, which would prevent solid-state photodimerization. Again SNOM gives a clear-cut answer in terms of melting (Fig. 2.82c). The SNOM image with a good tip exhibits only very minor artificial contrast at the steep slopes (50–65°) and considerable heights, which is not materials contrast. The surface stays chemically uniform despite the topology. The features in (b) are thus not the result of photodimerization but of melting under the influence of the absorbed light. Radiationless decay produces the heat that cannot be dissipated by the insulator (no hydrogen bonds are available). The viscous melt piles up and crystallizes upon cooling forming also the depressions next to them. The irregular shape of the islands in (b) reflects deficiencies of the illuminating tip that may have been broken before the experiment, while leaving a sharp apex. Every lithographic approach (the sequence is top right, top left, bottom) was followed by an AFM measurement with the same tip, and this did not alter the shapes of the islands that are firmly anchored in the bulk due to deep light penetration into the crystal.

Also 2,5-dibenzylidenecyclopentanone crystals on (001) have been used for such tip testing and the regular shape produced by a good uncoated tip can be seen in [5].

2.13 Digital Microscopy for Rough Surfaces

While light microscopy cannot reach subwavelength resolution there is some useful overlap with the SNOM capabilities. Digital microscopes with large working distance lenses are now able to provide very high depth of focus. They image 3D topologies of surfaces also on nontransparent samples. The optical enlargement reaches the value of 1,000 and local colors are differentiated in the VHX Digital Microscope of Keyence Inc. The depth of focus is achieved by a collection of several images at various focal points and electronic buildup of the sharp image either in 2D projection or as a 3D topography.

The power of the system may be demonstrated with particularly rough cracked shells of a Macadamia ternifolia nut from Hawaii. The $30 \times 20\,\mu\text{m}^2$ image of the fracture area in Fig. 2.83 covers a remarkable height range of $20\,\mu\text{m}$. All parts are sharp and well resolved. Clearly, SNOM imaging on rough surfaces (it is also possible to add the optical information of SNOM to the 3D topology of the simultaneous AFM) is nicely continued by new digital light microscopy. Different colors are seen in the original image, which are very useful for the botanical analysis.

The enormous power of 3D resolution of such digital microscopy can be demonstrated with a fraction edge of the same Macadamia nutshell at 100 fold optical magnification (Fig. 2.84). The topology composition out of 12 images at $50\,\mu\text{m}$ focal distances does not exhibit breaks in the vertical part structures. The observed depth of focus of $600\,\mu\text{m}$ is quite remarkable for an optical microscopy technique.

Still another approach tries to break the refraction limit in light microscopy with materials of negative refraction index. Whether highest resolution light

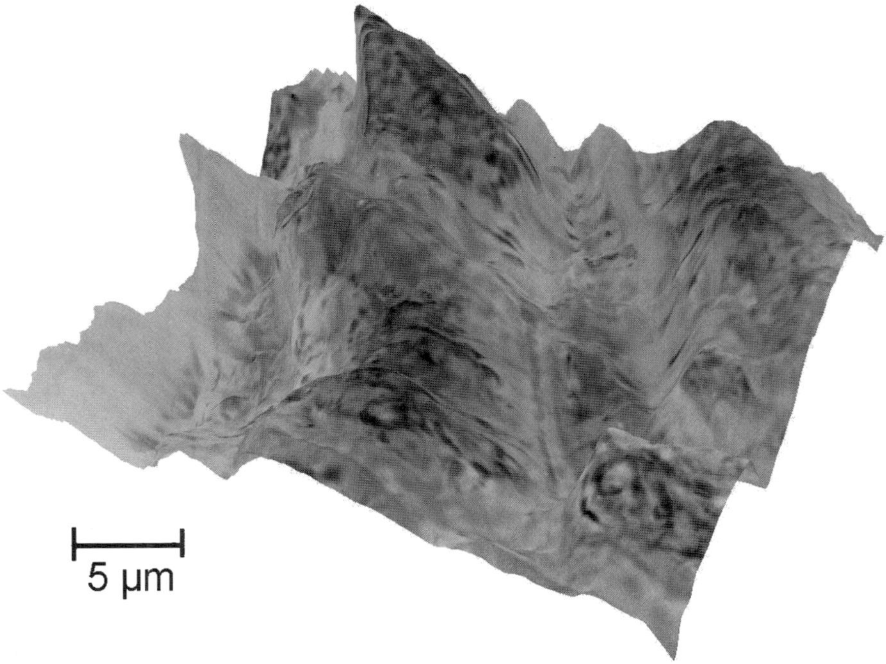

Fig. 2.83. Microscopic 3D plot at 1,000-fold optical enlargement of a region in the fracture zone of a Macadamia nutshell; $x = 30\,\mu\text{m}$, $y = 20\,\mu\text{m}$ and maximum height $20\,\mu\text{m}$, as composed from 20 images at $1\,\mu\text{m}$ focal distance [97]

microscopy with negative refraction index materials will be practically possible is however a subject of intensive research (e.g. [98]).

2.14 Conclusions

Scanning near-field optical microscopy SNOM provides nanoscopic contrast or spectroscopy (absorption, emission, Rayleigh, Raman) and a new light source for direct local photochemistry (chemical reaction, radiationless deactivation). Only apertureless shear-force SNOM with very sharp tapered tips is suitable for (very) rough delicate surfaces and provides chemical (materials) contrast. The strong enhancement of reflection in the shear-force gap is a new unexpected near-field effect that is absolutely essential for the success. It depends on the preset damping and works only for very sharp illuminating and collecting tips with end radii of $<20\,\text{nm}$ that must not break or abrade. A fast, light based control circuit avoids tip breakage at fast scanning (up to $200\,\mu\text{m s}^{-1}$). Such fast scanning is essential for minimizing thermal effects that occur if the illuminating light is absorbed by the sample. The wavelength of the light may be freely chosen from lasers with good beam quality. Multimode fibers with

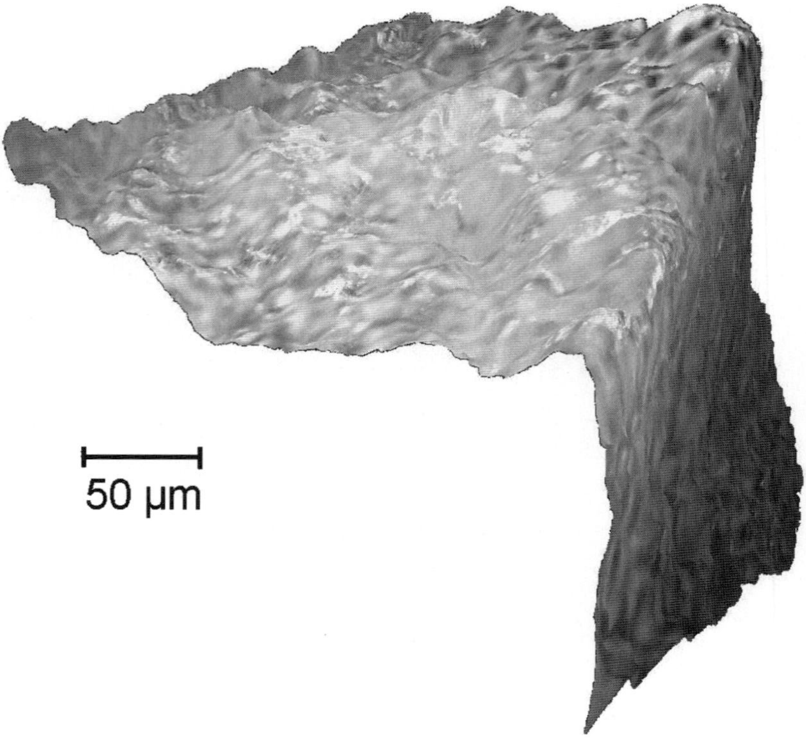

Fig. 2.84. Microscopic 3D plot at 100-fold optical enlargement of an outer fracture edge of a Macadamia nutshell with a height range of 600 µm as composed from 12 images in 50 µm focal distance [97]

loss of polarization are the best choice. Local fluorescence and Raman SNOM can be easily executed. Absorbed light may give rise to photobleaching and local heat production with nanopit formation, the latter particularly elegant with far-field apertured tips. The versatile technique is robust and economic also due to the uncomplicated pulling of the very sharp probes at low cost. High resolution at 10–20 nm is routinely obtained even on corrugated surfaces. Topologic artifacts do not occur due to the new physical effect of very strong enhancement of reflectivity in the shear-force gap of typically 5 nm. Assessment of such enhancement by an energy meter during every single measurement is indispensable, as blunt tips produce artifacts. These have been imaged and classified according to their varied appearance. They occur also with unsuitable SNOM measurements in all of the other SNOM techniques (there even on very flat surfaces). Realworld samples without further pretreatment are successfully and artifact-free measured for the first time under ambient conditions if a useful shear-force mechanism exists such as a water layer or dangling bonds. There seem to be no restrictions for the application of shear-force apertureless SNOM to nonsticky surfaces (except for the rare lack of a useful shear-force mechanism), and virtually all types of materials have been successfully studied. Scientific, (nano)technical, industrial, and medicinal applications, including cancer prediagnosis, are legion. The still not widely known and used apertureless shear-force SNOM does not have the problems of all other SNOM techniques with topology and also not the heat problems of aperture SNOM in illumination mode. It will therefore enjoy a great future for practical use.

While SNOM is yet the only practical method of breaking the refraction limit of microscopy, it should be noted here that recently a new way of breaking the refraction limit is sought by use of metamaterials with negative refraction index. It is, however, still a subject of intensive research (e.g. [98]) and we are looking forward to successful applications of metamaterials for optical components.

References

1. D. Courjon, J.M. Vigoureux, M. Spajer, K. Sarayeddine, S. Leblanc : External and internal reflection near field microscopy: experiments and results. Appl. Opt. **29**, 3734–3740 (1990)
2. M. Spajer, D. Courjon, K. Sarayeddine, A. Jalocha, J.M. Vigoureux: Microscopie en champ proche par reflexion. J. Phys. (Paris) Ser. III **1**, 1–12 (1991)
3. V. Sandoghdar, S. Wegscheider, G. Krausch, J. Mlynek: Reflection scanning near-field microscopy with uncoated fiber tips: How good is the resolution really? J. Appl. Phys. **81**, 2499–2503 (1997)
4. G. Kaupp: Supermicroscopy in Supramolecular Chemistry: AFM, SNOM, and SXM, in *Comprehensive Supramolecular Chemistry*, (eds. J.E.D. Davies, J.A.Ripmeester),Vol. **8**, 381–423 + 21 color plates, Elsevier, Oxford (1996)
5. G. Kaupp, A. Herrmann, M. Haak: Near-Field Optical Microscopy with Uncoated Tips: Calibration, Chemical Contrast on Organic Crystals, and Photolithography. J. Vac. Sci.Technol. B **15**, 1521–1526 (1997)
6. G. Kaupp, A. Herrmann: Chemical contrast in scanning near-field optical microscopy. J. Phys. Org. Chem. **10**, 675–679 (1997)
7. G. Kaupp, A. Herrmann: SNOM by near-field reflectance enhancement: a versatile and valid technique. J. Phys. Org. Chem. **11**, 141–143 (1998)
8. G. Kaupp, A. Herrmann, M. Haak: SNOM (scanning near-field optical microscopy), a genuine photochemical technique. Internet Photochem. Photobiol., Available from http://www.photobiology.com/IUPAC98/page2.htm (1998)
9. G. Kaupp, A. Herrmann, M. Haak: Artifacts in topography and scanning near-field optical microscopy (SNOM) due to deficient tips. J. Phys. Org. Chem. **12**, 797–807 (1999)
10. U.C. Fischer: Scanning near-field optical microscopy, in *Scanning Probe Microscopy*, (ed. R. Wiesendanger), Springer, Berlin Heidelberg, New York, 161–210 (1998)
11. A.H. La Rosa, B.I. Yakobson, H.D. Hallen: Origins and effects of thermal processes on near-field optical probes. Appl. Phys. Lett. **67**, 2597–2599 (1995)
12. M. Staehelin, M.A. Bopp, G. Tarrach, A.J. Meixner, I. Zschokke-Graenacher: Temperature profile of fiber tips used in scanning near-field optical microscopy. Appl. Phys. Lett. **68**, 2603–2605 (1996)

13. G. Kaupp, A. Herrmann: Positive submicron lithography using uncoated or far-field apertured SNOM tips on organic crystals. Ultramicroscopy **71**, 383–388 (1998)
14. D. Zeisel, S. Nettesheim, B. Dutoit, R. Zenobi: Pulsed laser-induced desorption and optical imaging on a nanometer scale with scanning near-field microscopy using chemically etched fiber tips. Appl. Phys. Lett. **68**, 2491–2492 (1996)
15. A. Naber, H.-J. Maas, K. Razavi, U.C. Fischer: Dynamic force distance control suited to various probes for scanning near-field optical microscopy. Rev. Sci. Instrum. **70**, 3955–3961 (1999)
16. S. Kühn, C. Hettich, C. Schmitt, J.P. Poizat, V. Sandoghdar: Diamond colour centres as a nanoscopic light source for scanning near-field optical microscopy. J. Microsc. **202**, 2–6 (2001)
17. N. Chevalier, M.J. Nasse, J.C. Woehl, P. Reiss, J. Bleuse, F. Chandezon, S. Huant : CdSe single-nanoparticle based active tips for near-field optical microscopy. Nanotechnology **16**, 613–618 (2005)
18. M. Labardi, P.G. Gucciardi, M. Allegroni: Near-field optical microscopy. Riv. Nuov. Cim. **23/24** 1–35 (2000)
19. R.C. Reddick, R.J. Warmack, T.L. Ferrel: New form of scanning optical microscopy. Phys. Rev. B **39**, 767–770 (1989)
20. S. Aubert, A. Bruyant, S. Blaize, R. Bachelot, G. Lerondel, S. Hudlet, P. Royer: Analysis of the interferometric effect of the background light in apertureless scanning near-field optical microscopy. J. Opt. Soc. Am. B **20**, 2117–2124 (2003)
21. H. Brückl, H. Pagnia, N. Sotnik: Optical near-field characterization of submicron structured silicon films. Scanning **17**, 24–27 (1995)
22. F. Zenhausen, M.P. O'Boyle, H.K. Wickramasinghe: Apertureless near-field optical microscope. Appl. Phys. Lett. **65**, 1623–1625 (1994)
23. F. Zenhausen, Y. Martin, H.K. Wickramasinghe: Scanning interferometric apertureless microscopy: optical imaging at 10 Angstrom resolution. Science **269**, 1083–1085 (1995)
24. B. Knoll, F. Keilmann: Enhanced dielectric contrast in scattering-type scanning near-field optical microscopy. Opt. Commun. **182**, 321–328 (2000)
25. F. Keilmann, R. Hillenbrand. Near-field microscopy by elastic light scattering from a tip. Philos. Trans. R. Soc. London, Ser. A **362**, 787–805 (2004)
26. A. Kramer, W. Trabesinger, B. Hecht, U.P. Wild: Optical near-field enhancement at a metal tip probed by a single fluorophore. Appl. Phys. Lett. **80**, 1652–1654 (2002)
27. S.A. Vickery, R.C. Dunn: Scanning near-field fluorescence resonance energy transfer microscopy. Biophys. J. **76**, 1812–1818 (1999)
28. G. Carbone, B. Zappone, R. Barberi, R. Bartolino: Apertureless SNOM microscopy on a commercial AFM. Mol. Cryst. Liq. Cryst. **372**, 373–382 (2001)
29. R. Hillenbrand, F. Keilmann, P. Hanarp, D.S. Sutherland, J. Aizpurua: Coherent imaging of nanoscale plasmon patterns with a carbon nanotube optical probe. Appl. Phys. Lett. **83**, 368–370 (2003)
30. H. T. Chen, S. Kraatz, G.C. Cho, R. Kersting: Identification of a resonant imaging process in apertureless near-field microscopy. Phys. Rev. Lett. **93**, 267401–1–267401–4 (2004)
31. S. Patane, P.G. Gucciardi, M. Labardi, M. Allegroni: Apertureless near-field optical microscopy. Riv. Nuo. Cim. **27**, 1–46 (2004)

32. W.-X. Sun, Z.-X. Shen: Optimizing the near-field around silver tips. J. Opt. Soc. Am. A **20**, 2254–2259 (2003)
33. M. Schüttler, M. Leuschner, M. Lippitz, W. Rohle, H. Giessen: Towards the origin of the shear force in near-field microscopy. Jpn. J. Appl. Phys. **40**, 813–818 (2001)
34. C. Adelmann, J. Hetzler, G. Scheiber, T. Schimmel, M. Wegener, H.B. Weber, H. v. Löhneysen: Experiments on the depolarization near-field scanning optical microscope. Appl. Phys. Lett. **74**, 179–181 (1999)
35. S. Madsen, T. Olesen, J.H. Hvam: Surface modifications via photo-chemistry in a reflection scanning near-field optical microscope. Optics at the nanometer scale: imaging and storing with photonic near fields, NATO ASI Series E, Appl. Sci. **319**, 263–275 (1996)
36. G. v. Freymann, T. Schimmel, M. Wegener: Computer simulations: subwavelength resolution with an apertureless SNOM: Appl. Phys. A **66**, S939–S942 (1998)
37. B. Hecht, H. Bielefeldt, Y. Inouye, L. Novotny, D. W. Pohl: Facts and artifacts in near-field optical microscopy. J. Appl. Phys. **81**, 2492–2498 (1997)
38. S. Madsen, S. I. Bozhevolnyi, J. M. Hvam: Sub-wavelength imaging by depolarization in a reflection near-field optical microscope using an uncoated fiber probe. Opt. Commun. **146**, 277–284 (1998)
39. G. Kaupp, A. Herrmann, G. Wagenblast: Scanning near-field optical microscopy (SNOM) with uncoated tips. Proc. SPIE-Int. Soc. Opt. Eng. **3607**, 16–25 (1999); Available from http://kaupp.chemie.uni-oldenburg.de/spie/
40. G. Kaupp: Optische Nahfeldmikroskopie auf rauhen Oberflächen mit Glasfaserspitzen in Kreuzpolarisation. *Forschungsbericht des Bundesministeriums für Bildung und Forschung (BMBF)*, FKZ 13N7640/3, 25.3.2003, pp. 1–24 (2003)
41. J. Koglin, U.C. Fischer, H. Fuchs: Material contrast in scanning near-field optical microscopy at 1–10 nm resolution. Phys. Rev. B **55**, 7977–7984 (1997)
42. U.C. Fischer, J. Heimel, H.J. Maas, M. Hartig, S. Höppener, H. Fuchs: Latex bead projection nanopatterns. Surf. Interface Anal. **33**, 75–80 (2002)
43. A. Jalocha, N. F. van Hulst: Dielectric and fluorescent samples imaged by scanning near-field optical microscopy in reflection. Opt. Commun. **119**, 17–22 (1995)
44. N. F. Van Hulst, M. H. P. Moers, B. Bölger: Near-field optical microscopy in transmission and reflection modes in combination with force microscopy. J. Microsc. **171**, 95–105 (1993)
45. S.I. Bozhevolnyi, I.I. Smolyaninov, O. Keller: Correlation between optical and topographical images from an external reflection near-field microscope with shear force feedback: Appl. Opt. **34**, 3793–3799 (1995); etched fibers were used in that report
46. R.L. Williamson, L.J. Brereton, M. Antognozzi, M.J. Miles: Are artifacts in scanning near-field optical microscopy related to misuse of shear force? Ultramicroscopy **71**, 165–175 (1998)
47. T. Hartmann, G. Gatz, W. Wiegräbe, A. Kramer, A. Hillebrand, K. Lieberman, W. Baumeister, R. Guckenberger, in *Near-Field Optics*, (eds. D.W. Pohl, D. Courjon). Kluwer, Netherlands, p. 33 (1993)
48. Y. Martin, F. Zenhausern, H.K. Wickramasinghe: Scattering spectroscopy of molecules at nanometer resolution. Appl. Phys. Lett. **68**, 2475–2477 (1996)
49. S. I. Bozhevolnyi, B. Vohnsen: Near-field optics with uncoated fiber tips: light confinement and spatial resolution. J. Opt. Soc. Am. B **14**, 1656–1663 (1997)

50. S. I. Bozhevolnyi: Topographical artifacts and optical resolution in near-field optical microscopy. J. Opt. Am. B **14**, 2254–2259 (1997)
51. G. Kaupp, M. Haak: Phase rebuilding of α-cinnamic acid in tail irradiations. Mol. Cryst. Liq. Cryst. **313**, 193–198 (1998)
52. G. Kaupp, M. Haak, A. Herrmann: Chemical SNOM-contrast on organic solids using cold fiber tips. Lecture at the 4th Near-Field Optics, Conference, Jerusalem (1997); Book of Abstracts, p. 92 (1997)
53. G. Kaupp, A. Herrmann: Waste-free quantitative gas/solid diazotization using nitrogen dioxide and triazene synthesis, both avoiding liquid phases. J. Prakt. Chem./Chem.-Ztg. **339**, 256–260 (1997)
54. G. Kaupp, J. Schmeyers, M. Haak, T. Marquardt, A. Herrmann: AFM in solid state reactions. Mol. Cryst. Liq. Cryst. **276**, 315–337 (1996)
55. G. Kaupp: Supermikroskopie bei organischen Kristallreaktionen. Chemie in unserer Zeit **31**, 129–139 (1997); Engl. translation in http://kaupp.chemie.uni-oldenburg.de
56. G. Kaupp, A. Herrmann: Chemical contrast in scanning near-field optical microscopy. J. Phys. Org. Chem. **10**, 657–659 (1997)
57. G. Kaupp: Organic solid-state reactions with 100% yield, Top. Curr. Chem. **254**, 95–183 (2005)
58. G.M.J. Schmidt: Photodimerization in the solid state. Pure Appl. Chem. **27**, 647–678 (1971)
59. G. Kaupp, A Herrmann, J. Schmeyers: Waste-free chemistry of diazonium salts and benign separation of coupling products in solid salt reactions. Chem. Eur. J. **8**, 1395–1406 (2002)
60. G. Kaupp, A. Herrmann, J. Schmeyers, J. Boy: SNOM a new photophysical tool. J. Photochem. Photobiol. A: Chemistry **139**, 93–96 (2001)
61. G. Kaupp, A. Herrmann, M.R. Naimi-Jamal, S. Stepanenko, V. Stepanenko: Nahfeldmikroskopie für organische und biologische Oberflächen, *Forschungsbericht des Bundesministeriums für Bildung und Forschung (BMBF)*, FKZ 13N7519, 30.6.2003, pp. 1–33 (2003)
62. L.F. Garfias, D.J. Siconolfi: In situ high-resolution microscopy on duplex stainless steels. Proc. Electrochem. Soc. **99-28**, 89–103 (2000)
63. E. Teixeira-Neto, G. Kaupp, F. Galembeck: Latex particle heterogeneity and clustering films. J. Phys. Chem. B **107**, 14255–14260 (2003)
64. E. Teixeira-Neto, G. Kaupp, F. Galembeck: Spatial distribution of serum solutes on dry latex sub-monolayers determined by SEPM, SNOM, SC microscopy. Coll. Surfaces, A: Physicochem. Eng. Aspects **243**, 79–87 (2004)
65. W. Hevers, H. Stieve: Ultrastructural changes of the microvillar cytoskeleton in the photoreceptor of crayfish *Orconectes limosus* related to different adaptation conditions. Tissue Cell **27**, 405–419 (1995)
66. D.A. Smith, S. Webster, M. Ayad, S.D. Evans, D. Fogherty, D. Batchelder: Development of a scanning near-field optical probe for localized Raman spectroscopy. Ultramicroscopy **61**, 247–252 (1995)
67. S. Webster, D.N. Batchelder, D.A. Smith: Submicron resolution measurement of stress in silicon by near-field Raman spectroscopy. Appl. Phys. Lett. **72**, 1478–1480 (1998)
68. R.M. Stöckle, Y.D. Suh, V. Deckert, R. Zenobi: Nanoscale chemical analysis by tip-enhanced Raman spectroscopy. Chem. Phys. Lett. **318**, 131–136 (2000)
69. R.M. Stöckle, V. Deckert, C. Fokas, D. Zeisel, R. Zenobi: Sub-wavelength Raman spectroscopy on isolated silver islands. Vibr. Spectrosc. **22**, 39–48 (2000)

70. V. Deckert, D. Zeisel, R. Zenobi: Near-field surface-enhanced Raman imaging of dye-labeled DNA with 100-nm resolution. Anal. Chem. **70**, 2646–2650 (1998)
71. C.A. Michaels, C.E.J. Dentinger, L.J. Richter, D.B. Chase, R.R. Cavanagh, S.J. Stranick: Assessment of sensitivity advances in near-field Raman spectroscopy. Proc. SPIE **4098**, 152–159 (2000)
72. M.S. Anderson: Locally enhanced Raman spectroscopy with an atomic force microscope. Appl. Phys. Lett. **76**, 3130–3132 (2000)
73. N. Hayazawa, Y. Inouye, Z. Sekkat, S. Kawata: Near-field Raman imaging of organic molecules by an apertureless metallic probe scanning optical microscope. J. Chem. Phys. **117**, 1296–1301 (2002)
74. A. Hartschuh, E.J. Sanchez, X.S. Xie, L. Novotny: High-resolution near-field Raman microscopy of single-walled carbon nanotubes. Phys. Rev. Lett. **90**, 095503–1–095503–4 (2003)
75. H.M. Poolock, D.A. Smith: The use of near-field probes for vibrational spectroscopy and photothermal imaging. In *Handbook of Vibrational Spectroscopy*, Vol. 2 (eds. J.M. Chalmers, P.R. Griffiths) Wiley, Chichester, 1472–1491 (2002)
76. C.A. Michaels, L.J. Richter, R.R. Cavanagh, S.J. Stranick: Chemical imaging with scanning near-field infrared microscopy and spectroscopy. Proc. SPIE-Int. Soc. Opt. Eng. **4089**, 102–109 (2000)
77. S.J. Stranick, D.B. Chase, C.A. Michaels: Near-field microscopies for chemical and material analysis on the nanoscale. Polym. Mater. Sci. Eng. **87**, 172 (2002)
78. C.A. Michaels, X. Gu, D.B. Chase, S.J. Stranick: Near-field infrared imaging and spectroscopy of a thin film polystyrene/poly(ethyl acrylate) blend. Appl. Spectrosc. **58**, 257–263 (2004)
79. S.J. Stranick, S.A. Buntin, C.A. Michaels: Scanning near-field dielectric microscopy at microwave frequencies for materials characterization. Polym. Mater. Sci. Eng. **88**, 142 (2003)
80. E.J. Sanchez, L. Novotny, X.S. Xie: Near-field fluorescence microscopy based on two-photon excitation with metal tips. Phys. Rev. Lett. **82**, 4014–4017 (1999)
81. M. Labardi, M. Allegrini, M. Zavelani.Rossi, D. Polli, G. Cerullo, S. DeSilvestri, O. Svelto: Highly efficient second-harmonic nanosource for near-field optics and microscopy. Opt. Lett. **29**, 62–64 (2004)
82. A. Bouhelier, M.R. Beversluis, L. Novotny: Application of field-enhanced near-field optical microscopy. Ultramicroscopy **100**, 413–419 (2004)
83. A. Kelch, S. Wessel, T. Will, U. Hintze, R. Werpf, R. Wiesendanger: Penetration pathways of fluorescent dyes in human hair fibres investigated by scanning near-field optical microscopy. *J. Microsc.* **200**, 179–186 (2001)
84. V. Subramaniam, A.K. Kirsch, R.V. Rivera-Pomar, T.M. Jovin: Scanning near-field optical microscopy and microspectroscopy of green fluorescent protein in intact *Escherichia coli* bacteria: J. Fluorescence **7**, 381–385 (1997)
85. P. Nagy, A. Jenei, A.K. Kirsch, J. Szöllösi, S. Damjanovich: Activation-dependent clustering of the erB2 receptor tyrosine kinase detected by scanning near-field optical microscopy. J. Cell Sci. **112**, 1733–1741 (1999)
86. A.K. Kirsch, V. Subramaniam, G. Striker, C. Schnetter, D. Arndt-Jovin, T.M. Jovin: Continuous wave two-photon scanning near-field optical microscopy. Biophys. J. **75**, 1513–1521 (1998)
87. A. Jenei, A.K. Kirsch, V. Subramaniam, D. Arndt-Jovin, T.M. Jovin: Picosecond multiphoton scanning near-field optical microscopy. Biophys. J. **76**, 1092–1100 (1999)

88. A.K. Kirsch, V. Subramaniam, A. Jenei, T.M. Jovin: Fluorescence resonance energy transfer detected by scanning near-field optical microscopy. J. Microsc. **194**, 448–454 (1999)
89. S. Hosaka: Possibility of SNOM-based storage and J. Tominaga: Super density optical data storage by near-field optics, Chapters 13.5 and 14 in *Applied Scanning Probe Methods* (ed. B. Bushan, H. Fuchs, S. Hosaka), Springer, Berlin Heidelberg, New York, pp. 414ff and 429ff (2004)
90. Y. Martin, S. Rishton, Hand .K. Wickramasinghe: Optical data storage read out at 256 Gbits/in^2. Appl. Phys. Lett. **71**, 1–3 (1997)
91. E. Betzig, J.K. Trautman, R. Wolfe, E.M. Gyorgy, P.L. Finn, M.H. Kryder, C.H. Chang: Near-field magnetooptics and high density data storage. Appl. Phys. Lett. **61**, 142–144 (1992)
92. A. Tarun, M.R.H. Daza, N. Hayazawa, Y. Inouye, S. Kawata: Apertureless optical near-field fabrication using an atomic force microscope on photoresists: Appl. Phys. Lett. **80**, 3400–3402 (2002)
93. P. Royer, D. Barchiesi, G. Lerondel, R. Bachelot: Near-field optical patterning and structuring based on local-field enhancement at the extremity of a metal tip. One contribution of 13 to a Theme 'Nano-optics and near-field microscopy'. Philos. Trans. Math., Phys. Eng. Sci. **362**, 821–842 (2004)
94. X. Yin, N. Fang, X. Zhang, I.B. Martini, B.J. Schwartz: Near-field two-photon nanolithography using an apertureless optical probe. Appl. Phys. Lett. **81**, 3663–3665 (2002)
95. H.J. Mamin, D. Rugar: Thermal writing using a heated atomic force microscope tip. Appl. Phys. Lett. **69**, 433 (1996)
96. G. Kaupp: Photodimerization of Anthracenes in the solid state: New results from atomic force microscopy. Angew. Chem. Int. Ed. Engl. **31**, 595–598 (1992)
97. G. Kaupp, M. Kaupp: Advanced fullerene-type texture and further features of the macadamia nutshell as revealed by optical 3D microscopy. J. Microsc., submitted
98. V.M. Shalaev, W. Cai, U.K. Chettiar, H.K. Yuan, A.K. Sarychev, V.P. Drachev, A.V. Kildishev: Negative index of refraction in optical metamaterials. Optics Lett. **30**, 3356–3358 (2005)

3

Nanoindentation

3.1 Introduction

Hardness and elasticity measurements of solid materials have a long history. Macro- and microhardness have been of major importance for industrial purposes. However, indentation hardness is of particular interest for nanotechnology and nanoscience as the materials damage is much more restricted. Therefore, the underlying effects can be explored and numerous materials properties are revealed and provide new unprecedented knowledge in various branches of science including daily life. While the theory of microindentation appears well developed and the physics of the process seems to be sufficiently described for isotropic materials [1] the reasons for the differences between nano- and microindentation were not known and could therefore not be treated in [1]. Furthermore, various effects in terms of chemical or crystallographic phenomena, including anisotropy, are less understood not to speak of the nanoscopic or molecular basis. As basic differences between nano- and microindentation have not recently been acknowledged or included in the physical treatment it is important to treat the topic in this book also on a more empirical and quantitative basis. Therefore not only selected model systems but as much chemical diversity in the materials as possible are treated in order to find more general relationships.

3.2 Equipment

The distinction between nano- and microindentation is made at a penetration depth of about 200 nm. There are marked differences in these techniques, and nanoindention instruments provide linear load facilities in the range of 20 µN up to 10 mN (usually two regions of calibration), whereas the microindentation instruments span the mN range up to about 4 N. Commercial load-controlled nanoindentation instruments are available with height resolution of less than a tenth of a nm and force resolution of less than a nN. A detailed description of useful specifications of nanoindentation instruments is given in [1]. Some popular commercial brands include the Nano Test®, TriboIndenter®, Nano Indenter®, Nano-Hardness Tester®, and UMIS®. Acoustic detectors may be added for the detection of cracking, fissuring, phase transformation, and slippage beneath the indenter. Depth control instruments at ultralow load for atomistic investigations are not yet of commercial standard [2] (see Sect. 3.9). Modern commercial nanoindenters such as for example the TriboScope® with TriboIndenter® of Hysitron, Inc. apply a constant loading rate through a capacitive force and displacement transducer to a more or less sharp indenter tip and avoid the so-called zero error by a preload procedure according to ISO 14577. Preferably, the indented surface is firmly mounted on an AFM stage (for example of a Nanoscope III), which allows for direct measurement of the initial surface (roughness) and of the final impression obtained using the same indenter tip (though with some tip sample convolution). Separate AFM measurements with sharper AFM tips are bothersome, as it is hard to localize nanoscopic features and as there is a high risk for filling the impressions or ditches with dust from the indentations or scratches of brittle materials. The first AFM after mechanical action is least affected, but repeated AFM scanning may continuously decrease the depths. The problems of sagging are only present for highly compliant samples (for example viscoelastic polymers) and can be handled by an appropriate feedback algorithm [3]. The reason for sagging is the composition of the applied force as the sum of contact force and restoring spring force (of the spring supporting the indenter shaft), which depends on the spring deflection. The spring constant can be kept small enough so that the indentation of stiff samples results in nearly perfect contact force control. Two-dimensional transducers are able to indent and move the indenter horizontally at constant normal load and constant speed when scratching the surface (Chap. 4). The average tilt of the surface even if it may exhibit roughness should be better than 1°. It can be detected by on-site AFM measurements without using the plane-fit facility. A tilting device with rotation facility has been constructed and is available for the TriboScope®. It allows for leveling of the sample normal to the load and to rotate the sample if anisotropic phenomena are to be explored. A force versus displacement curve provides the data necessary to evaluate the plastic and elastic behavior in the test region. Calibration details, analysis software, and full imaging capabilities are provided by the instrument suppliers and exhaustively described in appropriate manuals. This deals with tip radius, compliance, adjusting standard settings, scaling, drift correction, and linearity in open or closed loop feedback control. Complete software for data acquisition and data treatment for example with standard Microsoft EXCEL® and AFM imaging software is required.

3.3 Foundations of the Nanoindentation Technique

3.3.1 General Remarks

For micro- up to macroindentations the normal force – penetration depth relation for the loading was described as an exponential function with varying exponent [4]. On the other hand, a square relation has been theoretically deduced for conical or pyramidal indenters [5, 6] and microindentations up to several hundred mN or the range of several N forces seem to follow that prediction. However these include rather drastic damage of the materials with a number of effects that are not at work in nanoindentations where a 3/2 exponential relation has been consistently found recently [7] so that differentiation is essential. An improved understanding can only be expected from the latter technique. The basic nanomechanical test tries to decrease mechanic damage as much as possible by using low load in the μN-range up to some mN. The higher loads are used in order to connect to the micro range. Solids are probed mostly with sharp diamond indenters (Vickers or Berkovich according to ISO 14577, Knoop, Cube Corner, 60° three-sided pyramid, or sharp conical indenters; spheric indenters with very large end radii produce shallow indents though with nm depth) while constantly increasing the force in the range of 20 μN up to 5 or 10 mN, depending on the hardness, elasticity, and the desired depth. This induces plastic and elastic responses of the material, and upon unloading there will be a residual impression usually accompanied by some pile-up, sinking-in, pop-ins, cracks or fissures, creep, and in the case of highly compliant samples sagging. In the case of time-dependent plasticity effects repeated loading/unloading combined with waiting periods at maximal load are necessary for more reliable hardness and elasticity parameters. Both commercial and home-built nanoindenter instruments record continuous loading and unloading curves with respect to the displacement. Instrument compliance is eliminated by some calibration procedures, which help in obtaining mechanical properties that are, however, not completely independent of the instrumentation.

3.3.2 Load–Displacement Curves

Figure 3.1 shows some typical force-displacement curves of standard materials. Aluminum with a Berkovich indenter (having a total included angle of 142.3°) gives almost pure plastic response (very little elastic recovery). The influence of the tip angle is shown with the 60°, 90° (Cube Corner), and 142.3° (Berkovich) indenter in fused quartz. It is seen that the elastic part increases considerably if the included total angle increases and that much more normal force is required at the larger angles for obtaining the same depth, clearly a consequence of increased pressure with the steeper tips. The loading and unloading curves are smooth, but there is a marked kink in the unloading curve of the 60° tip. Only ideal results without the so-called "pop-ins" (crack/fissure

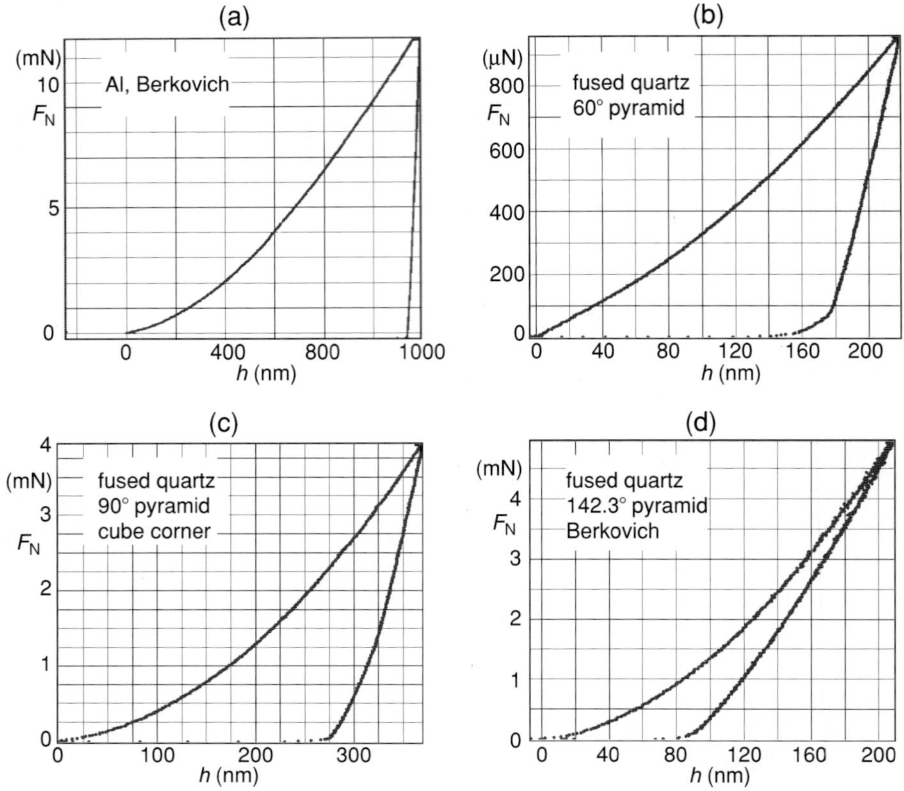

Fig. 3.1. Typical indents of aluminum and fused quartz with various indenters indicating variations in plastic and elastic response of (**a**) aluminum with Berkovich, and (**b**)–(**d**) of fused quartz with 60°, 90°, and 142.3° three-sided pyramids as indenters

formation leading to increase in depth while the load does not increase) were selected for Fig. 3.1. Importantly, the occurrence of pop-ins is more frequently observed with the steeper tip and in the higher load range. It should also be stated that release of strain by pop-ins can be induced by minor shock or noise from the environment. The shape of the unloading curve (and the quantities calculated there from) does not seem to be severely influenced by pop-ins if the curve comes more or less back to its regular continuation. Indentation time is usually chosen in the range of 10–60 s.

There may be time-dependent features such as creep under load. This is determined by hold periods before unloading or preferably by repeated unloading/reloading as in Fig. 3.2. The hardness (H) and reduced elasticity modulus (E_r) are usually determined from iterations of the elastic unloading curves (see below). It is therefore essential that the influence of creep (further penetration without increasing the load) be eliminated. The most secure way

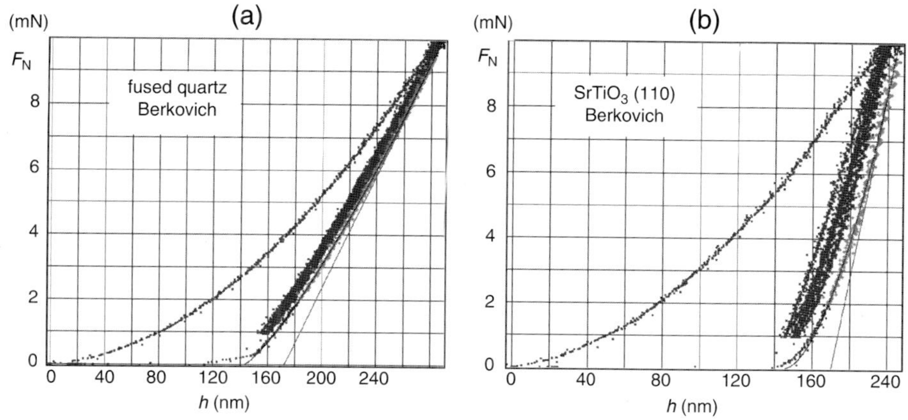

Fig. 3.2. Load and multiple unloads/reloads to/from 10% of the maximal load on (**a**) fused quartz and (**b**) crystalline strontium titanate on its (110)-face

to achieve that requirement is the execution of multiple unloading/reloading until full elasticity is reached. The values in Table 3.1 show the variations of the data when approaching the final values and also some of the important iterated parameters for the calculation of H and E_r of fused quartz and strontium titanate on (110) (the meaning of the parameters is defined in (3.1)–(3.6)). The variations between the consecutive unloadings are remarkable.

The dependence of the various mechanical parameters that are iteratively calculated by reference to "approved values" for fused quartz from the number of reloads is demonstrated in Table 3.1. The variations are importantly high.

The situation is even worse in the case of soft viscoelastic materials such as the polymers PMMA (polymethyl methacrylate), CR39 [polymer from

Table 3.1. Iterative analysis of the Berkovich multiple unloads/reloads of Fig. 3.2 according to ISO 14577

	fused quartz				$SrTiO_3$			
unload #	1	2	3	4	1	2	3	4
E_r (GPa)	67.1	66.0	65.9	66.5	254.5	263.0	251.4	242.0
H (GPa)	7.99	7.83	7.81	7.47	10.36	9.99	10.02	9.37
S (N μm^{-1})	84.4	83.8	83.8	86.5	281.4	296.0	282.7	281.3
A (μm^2)	1.242	1.267	1.270	1.329	0.9595	0.9945	0.922	1.061
B (constant)[a]	13.190	18.696	18.339	18.677	0.794	1.060	5.861	10.114
h_f (nm)	127.97	136.55	136.47	142.66	119.14	126.03	135.76	145.59
m (exponent)[a]	1.314	1.255	1.258	1.261	2.173	2.140	1.794	1.686

[a] The exponent m and the constant h_f and B of (3.3) are iterated for the 20–95% F_N range

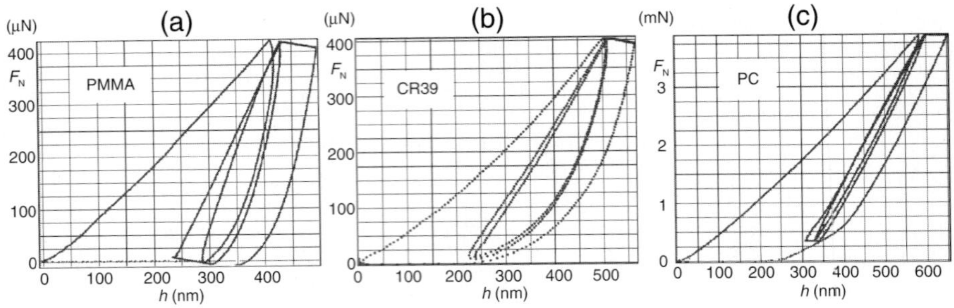

Fig. 3.3. Load and multiple unloads/reloads to/from 10% of the maximal load on the linear (PMMA and PC) or cross-linked (CR39) glassy polymers in $F_N - h$ diagrams; load and unload times are 10 s, waiting time after second reload is 30 s

diethyleneglycol bis(allylcarbonate)], and PC (polycarbonate). This is depicted in Fig. 3.3. Increased amounts of creep (comprehensive creep treatment in [8]), sagging (decrease of nominal load upon holding), and adhesion (unloading curve goes under zero line) occur. The analysis of these unloading curves with marked hysteresis is more troublesome, but the sagging may be removed by an involved feedback system. The unloading and reloading curves are dissimilar in PMMA, PC, and CR39. In PMMA some restoring of the sagging upon a hold after (the first) unloading is observed. With PC creep is less severe [9].

The unloading–reloading hysteresis (Fig. 3.3a–c) derives from the well-known production of macroradicals from main chain rupture by the mechanical stress [10]. The radicals disproportionate (to give shortened chains) until these are too short for further mechanical chain breaking, which leads to the final loss of hysteresis.

Fortunately, the reliability of the loading curves is secured if several indents at different load are compared as for example in Fig. 3.4. This technique will automatically exclude any single experiment that does not fit to the rest. All curves fall on the one with the highest load.

3.3.3 Anisotropy and Far-Reaching Response

Theoretical treatment and simulation techniques assume a circle of contact, which shall include the plastic zone where "geometrically necessary dislocations" occur as a consequence of the loaded indenter (conical or pyramidal). The center of such circle is at the point where the indenter started penetration and the radius is half of the projected contact area diameter [1]. While this is a concept for isotropic materials it may be doubted if the material is only changed in that "core area." Material may be transported along cracks or fissures, pileup may go far beyond the "core area," and the far-reaching effects are more easily shown with anisotropic crystals such as $SrTiO_3$. Figure 3.5 depicts AFM images of some indents. The standard sample fused quartz provides symmetric indents that image the shape of the tip without considerable

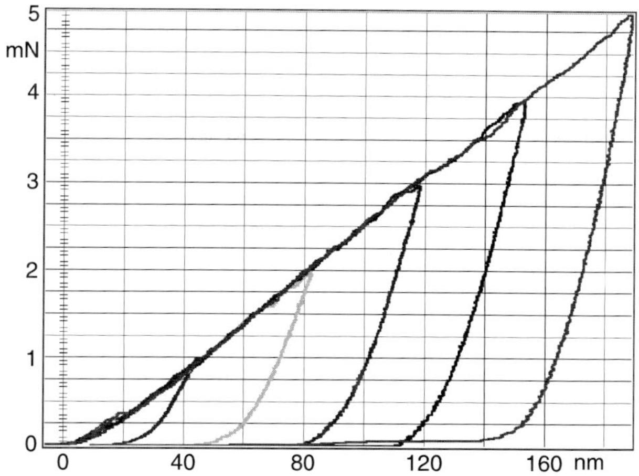

Fig. 3.4. A series of indentations at various final loads to the surface of $SrTiO_3$ on its (100) face, indicating reliability of the data

pileup and apparently well-located depression (there is also a phase transition involved, see Sect. 3.7). Conversely, the cubic strontium titanate shows very pronounced pileup and far-reaching oriented depressions that depend on the crystallographic face and are modeled by the tip. There is much anisotropic pileup on the (100) face but not at one of the edges and shallow depressions form at 0° and 90°. These change direction if the crystal is rotated by 45° (the tip orientation is fixed). The two identical crystal axes on (100) are at 90° to each other. The orthogonal axes are nonidentical on (110) where the pileup is less localized at the expense of long-range depressions primarily at the 90° direction but also at the diagonal axes. On (111) a three-arm depression rosette over very long distances occurs next to pile-up close to the impression. The directions represent the trigonal axes on that face: the rosette depressions align along the crystallographic axes and start preferably at the corners of the indents. The angle mismatch on the (111) face is clearly seen as the heights of the projected impression triangle are not in the 60° axes of the crystal and the rotation of the crystal is 45°. Of course, the rosettes change their orientations if the rotation of the crystal is continued [9]. A similar image on the (111) face of $SrTiO_3$ has been reported in [11] and "dislocation migration" was claimed. It is, however, more likely that the known [7] pressure-induced phase transformation aligns along the highly occupied lattice axes. These observations indicate that some scrutiny may be necessary toward the foundations of theoretical deductions and simulations.

Cracks at the edges of the indenter that are typical in microindentation and the detection of their length for the detection of the fracture toughness [1] are not typical in nanoindentation. Figure 3.5 clearly shows that anisotropic effects are more important at the much lower forces.

Fig. 3.5. AFM images of cube corner indents of (**a**) fused quartz and (**b**)–(**f**) $SrTiO_3$ on (100) at 0° and 45°, on (110) at 0°, and on (111) at 45° and 0° scratch direction (edges orientation of square samples) with a cube corner at 5 mN load, respectively; the x, y-scales are 5 μm

3.4 Elastic and Plastic Parameters

The mathematical definition of the important parameters as well as their description is preceded to their experimental determination. Equations (3.1)–(3.6) are the basis for all interrelations between the parameters and their determination.

$$\textit{Hardness}: \quad H = \frac{P}{A} \quad (\mathrm{N\ \mu m^{-2}}), \tag{3.1}$$

$$\textit{reduced elastic modulus}: E_{\mathrm{r}} = \frac{\sqrt{\pi}}{2}\frac{S}{\sqrt{A_{hc}}} \quad (\mathrm{N\ \mu m^{-2}}), \tag{3.2}$$

$$\begin{aligned} \textit{stiffness}: \quad S &= \left(\frac{\mathrm{d}P}{\mathrm{d}h}\right)_{\max} = \frac{2}{\sqrt{\pi}} E_{\mathrm{r}} \sqrt{A_{\mathrm{hc}}} \\ &= Bm\,(h - h_{\mathrm{f}})^{m-1} \quad (\mathrm{N\ \mu m^{-1}}), \end{aligned} \tag{3.3}$$

$$\textit{compliance}: \quad C = \frac{1}{S} \quad (\mathrm{\mu m\ N^{-1}}), \tag{3.4}$$

$$\textit{projected area}: \quad A_{\mathrm{hc}} = \pi/(4\ E_{\mathrm{r}}^2\ C^2), \tag{3.5}$$

$$\textit{contact depth}: \quad h_{\mathrm{c}} = h - 0.75\frac{P_{\max}}{S}. \tag{3.6}$$

3.4.1 Nanohardness

There are numerous definitions of hardness with very different meaning. The qualitative Mohs scale with 10 standard minerals is a valuable approach for minerals but it does not yield relative values. Rosiwal grinding hardness, drilling hardness, or pendulum hardness that are determined according to standard techniques and use very different materials responses obtain quantitative values. Sclerometers determine either scratch hardness at the critical load or by the size of scratches, and indentation hardness with various tools for specific tasks is determined at rather high load. The macroscopic Brinell (N mm^{-2}), Vickers (N mm^{-2}), or Knoop hardness (N mm^{-2}) uses loads (up to 3,000 kp) to stiff balls with radii of up to 5 mm or to four-sided pyramids with equal or with different semi-angles, respectively. The penetration resistance is determined as the relation of the force and the surface of the impression, or in the Rockwell hardness (cone or balls) by the remaining depth of impression, which is converted to a dimensionless figure in various clearly defined "scales" (1 point of Rockwell hardness corresponds to 0.002 mm depth). Anisotropy of the microhardness is detected with the Knoop indenter [12]. The load can be a static weight, a ball may fall on the surface from a known height, or the rebound height of a drop hammer may be detected with a scleroscope in the Shore rebound hardness (dimensionless) which, however, is more of an elasticity test. For elastomers, Shore-A or -C and -D hardness (dimensionless) is detected as resistance toward the penetration of truncated cones or rounded cone by pressing with a spring. Conversely, micro- and nanohardness H (GPa) refer to the projected area of the impression with spherical, conical, or pyramidal indenter tips, and they signify the resistance to permanent deformation. It is not easy to relate these hardness values H with the macroscopic (including additional more specialized techniques) hardness values even though the projected areas A_{hc} of a Vickers and a three-sided Berkovich indenter are identical and Knoop indenters are also used in microhardness determinations. Despite almost equal A_{hc} values of Vickers or Berkovich indenters the H values of micro- and nanohardness are different. Also, nanohardness values depend on the indenter type, the load, and the data evaluation technique. They have been extracted from the loading curve for Martens universal hardness or from the unloading curve for the eight-parameter procedure of Oliver Pharr (adapted by ISO14577), for the two-parameter procedure of Thurn and Cook, or by the work of indentation procedure ((3.22), and (3.24)). All the different kinds of macro-, micro- and nanohardness have to be specified according to type and load or further experimental conditions. ISO standards exist for several of these, but it would appear that the variation of nanohardness by variation of the tip geometry is still a matter of concern in research and nanotechnology.

The nanohardness H is defined as the normal load F_N over the projected contact area A (not over the impression surface under load) (3.1). It is calculated in units of GPa and MPa and is a quantitative measure of deformation resistance. Hardness values suffer from a sometimes very pronounced indentation size effect (ISE) [13], which limits their practical use. For example, the H values of high-purity aluminum single crystals as measured with a Berkovich or a Vickers indenter vary from 0.8 GPa at low contact height ($h_c = 30$ nm) to 0.2 GPa at large contact height ($h_c = 8\,\mu$m) or the hardness of gold (Berkovich) from 1.4 GPa ($h_c = 130$ nm) to 0.43 GPa ($h_c = 1.2\,\mu$m) [14]. The reasons are not elucidated, yet, but the Meyer power law for the description of ISE in Vickers microindentations does not seem to have a physical meaning but is a mere mathematical artifact to take into account the ISE effect in microhardness tests [15]. Additionally, hardness values depend on the indenter types [16].

Furthermore, and not surprisingly, there is considerable scatter in the single measurements of H, which does not guarantee a high precision even if a multitude of measurements is averaged (typically 10 measurements averaged, three of them with higher deviations discarded). Even worse, there may be a risk of varying response of certain materials in single indents in the same specimen under exactly the same experimental conditions so that the values cluster in two different regions. These should then not be averaged but distinguished and studied more closely. An example is sapphire on (001) when probed with a Vickers indenter of large radius ($R = 400$ nm) at 54 nm contact depth. The H-values clustered at 54–60 and 32–38 GPa [14]. Such behavior is to be expected if the ratio of elastic/plastic deformation varies by pop-in events in the loading curve and as the definition of the contact depth does not appear very reliable due to difficulties with the actual shape of the indent at the periphery of indenter contact under load ($h_c = h - h_s$), where h_s is the surface profile height at the perimeter of contact due to sinking-in. There are certain assumptions to the definition of that quantity and it is well known that any creep (that is frequently occurring) and pile-up will produce errors that cannot easily be accounted for. Still another correction factor ε for h_s is necessary to account for the tip geometry [17]. The value of ε, which influences the hardness calculation, is a source of uncertainty. Usually $\varepsilon = 0.75$ is used for pyramidal indenters, but there is no sufficient foundation for that usage.

This type of hardness is also termed *Meyer hardness*. Different definitions of hardness as derived from the work of indentation are discussed in Sect. 3.8.

3.4.2 Reduced Elastic Modulus

Equation (3.2) defines the reduced elastic modulus E_r that is related to Young's modulus E from dilatation measurements by (3.7).

$$1/E_r = (1 - \nu^2)/E + (1 - \nu_i^2)/E_i. \tag{3.7}$$

The measured quantity can be transformed to Young's modulus by applying Poisson's ratio ν, which takes into account the elasticity parameters of the indenter (usually diamond). The difference between E_r and $E/(1 - \nu^2)$ increases as the elastic modulus of the indented material approaches that of the indenter. Elastic moduli are frequently assumed not to experience strong indentation size effects but to be nearly independent of the contact depth [1] (cf., however, Fig. 3.26). They indicate the elastic materials properties.

3.4.3 Contact Area and Contact Height

The projected contact area A (3.5) is the basic quantity that must be determined for the calculation of hardness H and elastic modulus E_r. The theoretical contact area is $A_{theor} = 24.56\ h_c^2$ for a Berkovich, $24.56\,h_c^2$ for a Vickers, $2.598\,h_c^2$ for a cube corner, $0.6495\,h_c^2$ for a 60° pyramid, and $\pi^2\ R\ h$ for a spheroconical indenter. However, experimental values are far off those values. An area function for A is required due to tip radius of curvature at the end. For example a $\sqrt{A}$ versus h_c plot of a typical Berkovich indenter starts to become linear at about 20 nm contact depth h_c. Precise contact areas are not easily obtained. The meaning of contact area and contact height is sketched in Fig. 3.6, which also defines the various other height designations. The contact area is located at the height $h_c = h_{max} - h_s$ as the area of impression. After unloading and removal of the tip the area of impression is to be searched at the original surface height. Only the latter area can be directly measured by SEM and only if the impressions are deep, but it is hard to find in the case of nanoindents. AFM measurements would not be precise enough due to tip sample convolution. The contact area A under load is therefore more frequently calculated by an iteration procedure after calibration of the instrument with a standard of "known" H and E_r, usually fused quartz. It is assumed that A remains constant upon unloading which, however, is an approximation. Unfortunately, other materials than the standards have different indentation geometries, a fact that is not taken into account when comparison to the standard is made. Rather all relevant quantities that are necessary for the computation of H and E_r are obtained by multiparameter iterations with respect to the unrelated standard. This certainly detracts from the physical meaning of such values obtained in accordance to ISO 14577.

Figure 3.6 depicts the usually assumed situation of sinking-in of the surface outside the contact so that the contact height is $h_{max} - h_s$. Such "sinking-in" is assumed for the data analysis system and (3.6) for h_c is deduced on that basis. It is, however, well known that there may be also piling up reaching actual $h_c > h_{max}$ values, because the material is coming above the initial surface (not depicted). Such behavior does not conform to the data treatment and errors in the calculated contact area A and thus H and E_r may be very large because of that (sometimes >80%). A correction procedure has been given that accounts for these effects based on the contact stiffness and SEM pictures of the residual impression from large indentations [18]. Furthermore, the ratio of h_f and h_{max} was used for predicting the pile-up for a given strain-hardening exponent. The power law exponents for the loading curve were taken to be 2.0 for the loading and 1.35 ($h_f/h_{max} > 0.4$) for the unloading curve in that procedure [19].

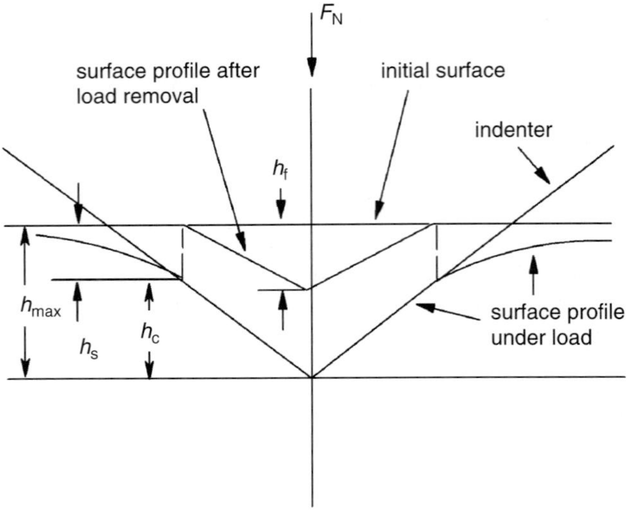

Fig. 3.6. The definition of contact upon loading and unloading of the standard fused quartz

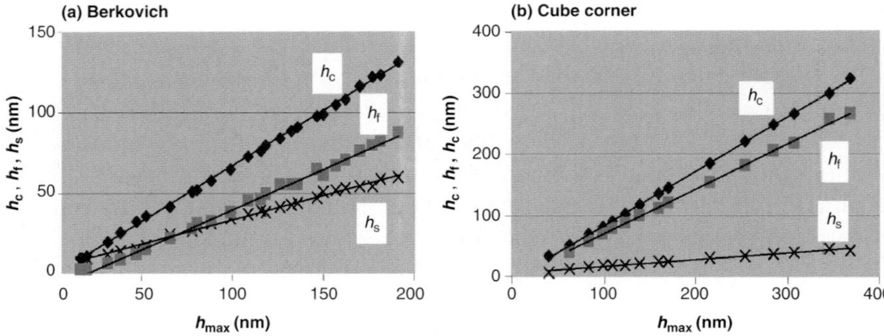

Fig. 3.7. Linearity between h_{max} and h_c, h_f, and h_s from the iterative analysis of indents on fused quartz with (**a**) a Berkovich and (**b**) a cube corner indenter

The various heights are fitted during the same iteration procedure as for A_{hc} (3.9). They are then all linearly related to h_{max} as plotted in Fig. 3.7 for two different indenter tips, but the slopes are different and depend strongly on the indenter [9]. These linearities are taken as the basis for using the directly measured h_{max} instead of h_c in the analysis of the loading curves below. Multicycling tests in fused silica had a kink in the $h_c - h_{max}$ plot. It was, however, deduced that the ratio of h_c/h_{max} relates to H/E_r and that "continuous" hardness measurements with multicycling experiments (at known E_r) can be performed on that basis [11].

3.4.4 Stiffness and Compliance

The unloading stiffness S is the resistance to elastic deformation. It is related to the square root of the projected contact area A at h_c as in (3.3). It is usually determined from the initial slope of the unloading curve, but it can also be continuously measured by some modulation of the load. Its value is required for the calculation of the contact area and thus hardness and elastic modulus. The precision of that value is increased by an iteration procedure as well. The inverse of stiffness is the compliance (3.4). It is needed for enabling a correction of the instrument compliance by the separately determined load frame compliance as part of the instrument calibration.

3.4.5 The Unloading Iteration Process

By measuring the initial unloading stiffness the elastic modulus can be obtained if A_{hc}, the projected area of the elastic contact, is determined. Numerous precautions have to be followed at its determination. These include calculation of the contact height h_c (smaller than the total displacement h_{max}; (3.6) at peak load, minimizing the stiffness errors by consideration of the exponential fit of the unloading curve according to (3.8), where B, h_f, and m

$$F_N = B(h - h_f)^m \tag{3.8}$$

are all determined by a least squares' fitting procedure. The exponent m was deduced to be 2 [20] or according to the model of the virtual sphere indenters 1.5 (in that case h_f was iterated) [17]. However, the exponent m varies between 1.1 and 1.8 depending on the specimen and on the indenter tip and is therefore termed as a "materials constant" and so is B. Furthermore, the iterated exponent m varies in wide ranges depending on the range of the unloading curve that is fitted. The recommendation of ISO14577 to choose the range from 100% to 20% of the load cannot be applied in all cases. Rather the form of the curves (and their deviation down to zero load) points frequently to two or more exponential functions that are involved. Their form is frequently influenced by phase transformations during unloading. For example, some clear examples are recognized in Fig. 3.1 and 3.3, and phase transformations under load occur probably much more frequently than previously envisaged (Sect. 3.7). It is possible to choose the fitting range for all occurring unloading curves starting at 95% (excluding initial noise) so far down that the exponent 1.5 is obtained with corresponding adjustments of h_f and more seriously B. This may be demonstrated with an example of similar elastic and plastic response in Fig. 3.8 and Table 3.2.

The E_r values in Table 3.2 vary by 30%, those of H by 12%, and those of S by 34% due to the choice in the first column. While ISO14577 fixes the iteration to be taken between 20% and 100% of the maximal load and is cautious with unloadings that cannot be used "down to at least 50%," there

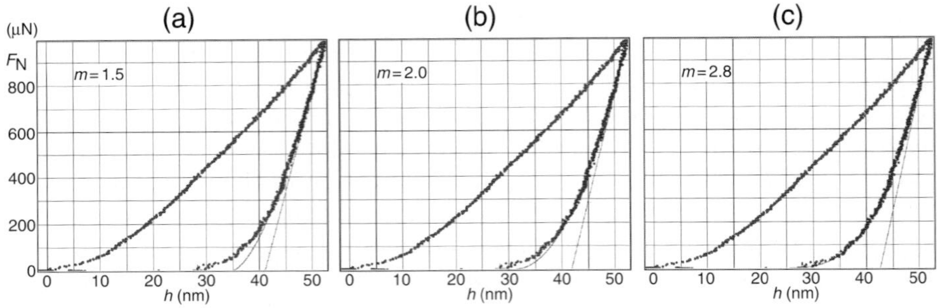

Fig. 3.8. Choosing different exponents m for the unloading curve of $SrTiO_3$ (100) after load with a cube corner for the "correct" tangent at F_{max} in the unloading stiffness determination [9]

Table 3.2. The variation of the parameters B, h_f, and m of (3.8) upon different choices of the unloading range and their influence to the elastic and plastic properties for $SrTiO_3$ (100) [9]

lower end(%)[a]	E_r (GPa)	H (GPa)	S (μN nm^{-1})	h_c (nm)	A (nm^2)	B	h_f (nm)	m
5	327.0	12.9	102.4	45.0	76964.6	0.1	25.5	2.8
10	319.2	12.9	99.7	44.9	76569.4	0.3	27.4	2.5
15	305.4	13.1	94.9	44.6	75793.2	1.4	30.3	2.1
17	300.3	13.1	93.1	44.5	75468.5	2.3	31.4	2.0
20	293.6	13.2	90.8	44.3	75018.1	4.3	32.7	1.8
30	271.6	13.5	83.0	43.7	73251.8	22.8	36.5	1.4
40	283.4	13.3	87.2	44.1	74300.1	7.5	33.7	1.7
50	280.3	13.4	86.1	44.0	74036.3	10.9	34.7	1.6
52	279.4	13.4	85.8	44.0	73947.9	12.9	35.1	1.5
60	277.5	13.4	85.1	43.9	73759.1	18.4	36.1	1.4
70	285.4	13.3	87.9	44.1	74457.4	8.5	34.2	1.6
80	266.9	13.6	81.2	43.5	72646.9	70.9	40.0	1.0
90	230.8	14.6	67.9	41.9	67914.9	39.1	36.1	1.1

[a] Upper end: 95%, $F_N = 989.3\,\mu N$, $h_{max} = 52.7\,nm$

may be concern if that compromise obtained for favorable standard samples is well founded. One may argue that the full curve should be iterated (in Fig. 3.8 $m = 2.8$), that the same exponent as in the loading curve ($m = 1.5$; (3.18)), or a theoretically deduced but in nanoindentations not experimentally supported exponent ($m = 2$) should be fixed rather than the free iteration. Remarkably, the 20/95 iterated exponent 1.26 for the standard fused quartz with Berkovich indenter is very far away from 2. Figure 3.8 shows that the unloading/reloading curves for the determination of reduced modulus and hardness can be fitted to monoexponential functions but with adjusted exponents and further adjusted parameters [21]. The free iteration leads to jumping of the exponent as is evident from the last column in Table 3.2. Instead of the theoretically expected exponent 2 [22] exponents of 1.1–1.8 are actually obtained (depending on the substrate and the indenter tip) in the 20/95 iteration, or from 1 to 3 if the iteration is extended to the full experimental curve. If unloading/reloading curve coincides with the pristine loading curve in the case of fully elastic loading, an exponent of 1.5 is obtained in nanoindentation (an example is depicted in Fig. 3.12). The unloading/reloading curves have nothing in common with the pristine loading curve if there is a residual impression. The material has changed by breaking bonds or intermolecular bonds, phase transitions, "indentation hardening," far-reaching changes, etc. and the geometry of the reloading process is totally different. It would be well possible to fix an exponent of 3/2 for the initial part of the elastic unloading/reloading curve as verified in Fig. 3.12, where the geometry of the process has not changed. The result for the values of the stiffness, hardness, modulus would be slightly different from the ones with exponents of 1.1–1.8 at the 20/95-iteration. However, the uncertainties with the unloading curves (and the contact geometry) rather point to use the loading curves for the mechanical characterization of materials (Sect. 3.6).

The next step in the process is the differentiation of the iterated exponential function (3.8). The derivative at the peak load provides the stiffness (the initial unloading slope) or compliance after corrections for the load frame compliance and geometric indenter deficiencies (with respect to reference materials). The A_{hc} value is then obtained by iterating the area function of the indenter with the exponential series (3.9) using $\mathrm{h_c}$ and eight parameters $C_1 - C_8$, where C_0 is the geometric factor for the ideal indenter, until convergence

$$A_{\mathrm{hc}} = \frac{\pi}{4} E_{\mathrm{r}}^{-2} (C - C_{\mathrm{f}})^{-2} = C_0 h_{\mathrm{c}}^2 + C_1 h_{\mathrm{c}} + C_2 h_{\mathrm{c}}^{1/2} + C_3 h_{\mathrm{c}}^{1/4} + \ldots. + C_8 h_{\mathrm{c}}^{1/128} \quad (\mu\mathrm{m}^2) \tag{3.9}$$

is achieved. The procedure assumes that E_{r} is constant, independent of the load. Calibration standards are fused silica, aluminum (100), and/or tungsten (100). The cut of the stiffness line at zero is h_{c}, which requires correction due to indenter geometry. Therefore it is

$$h_{\mathrm{c}} = h_{\mathrm{max}} - \varepsilon\, F_{\mathrm{N\,max}}/S \tag{3.10}$$

usually calculated by (3.10), where h_{max} is taken from the iterated unloading curve (or calculated as h_{eff}) and ε is taken to be 0.75 instead of 0.7268 as was theoretically deduced for conical and pyramidal indenters. The quantity of h_{f} is obtained from the cut of the calculated exponential unloading curve. This technique of Oliver and Pharr [20] is most frequently used for the detection of H and E_{r} from the unloading curves and adapted in ISO 14577.

It appears that there are difficulties with the assumptions made in the determination of h_{c} if samples other than the standards are chosen. These assumptions are depicted in Fig. 3.6. Sinking-in and pile-up (current techniques for analyzing make no provisions for extra contact area produced by pile-up) cannot be foreseen and they influence the result. The fate of the material under the loaded tip remains largely undetermined. It is termed "strain-hardening of the material." Upward flow is assumed but not always seen (Fig. 3.5) and not included in the basic assumptions for the contact area (Fig. 3.6). For empirical treatments constraint factors (C) and strain-hardening exponents are defined that take care of the stress (σ) and strain (ε) under the indenter [1]. But these techniques are restricted to special types of materials and indenter geometries and lack generality. For more detailed treatment far-reaching effects and phase transitions would have to be additionally considered.

The fate of the displaced material (pile-up, sinking-in, creep (plastic flow), phase transformation, "migration of dislocations," etc.) is disregarded in such mechanical testing and is deemed irrelevant as long as it does not influence the contact area, which is certainly hard to judge. It appears difficult to believe that such omissions are indeed valid in the general case beyond the few well-studied standard materials if all kinds of different materials are included. These may be soft or brittle, anisotropic, inorganic, organic, and biological materials. Anisotropic deformations are known in the case of inorganic and polymer materials and have been interpreted as "plastic deformation," "flow," "some kind of sliding," "cracks along shearing/cleaving planes," "local morphological changes," "migration of dislocations," and the like [23–25], but real materials balances were not provided.

An alternative determination of the contact area uses a two-parameter area function. It was developed and used with equal success as (3.9) in the fitting of unloading curves [26]. This approach uses parameters that correspond to an "effective tip radius" and an "effective cone angle" of the pyramidal tip. The projected area under load is then expressed as in (3.11):

$$A = \pi\, h_c^2/\cot^2\alpha + 4\,R\,\pi\,h_c + 4\,R^2\,\pi\,\cot^2\alpha. \tag{3.11}$$

This method does not rely on the constant modulus assumption, as does the eight-parameter (3.9) (ISO 14577), which is more generally used in the software of the instrument builders. Nevertheless, the difficulties with the area function are severe indeed. There are further problems with creep (errors on H and E_r may be larger than 20% and 50%; approaches for remedy have been discussed in detail [8]) and adhesion and there is the well-known "indentation size effect" ISE for the hardness values, i.e., its change with the maximal load applied. The reasons for ISE remained largely unexplored. It can, however, be expected that such changes arise also from pressure-induced phase transitions (Sect. 3.7). There is a possibility to extract hardness values from the loading curves (3.12), however, usual Martens universal hardness (HM_s) determinations, using the slope n of the $h - \sqrt{F_N}$ curve preferably at 50–90% of $F_{N\,max}$, assume a load–height relation that is not valuable in nanoindentation and should not be used there (see (3.18)). Furthermore, the HM_s technique does not correct for the actual geometry of Vickers or Berkovich indenter, neither does it settle possible problems with the time dependent plasticity.

$$\mathrm{HMs} = n^{-2}\ A_{\mathrm{s}}(h)\ h^{-2} \tag{3.12}$$

$$(A_{\mathrm{s}}(h)/h^2 = 26.43 \text{ for Vickers, } 26.44 \text{ for Berkovich}).$$

There are (draft) standards available such as ISO14577-1 to 3 or ISO/CD 14577-4 (for coatings) and at the European level TC 184/WG5, WI 121–132 (by the committee CEN) for the measurement of hardness and Young's modulus and materials parameters by nanoscopic depth sensing. An account for progress in the determination of the area function of indenters for nanoindentation has appeared [27]. The properties of test materials have been discussed [16].

The special needs and particularities of coatings and films for hardness and elasticity measurements have been thoroughly and exhaustively covered by the European INDICOAT project [28]. Generally, no differences with bulk materials are expected as long as the depth of the layers is at least 10 times the indentation depth. But important results are also obtained if multilayers are totally penetrated down into the support and there exist proper techniques for ultra low penetration in nanoindentation. It should also be noted here that pile-up errors might be very large if the coating is detached. For example, soft Al films on glass have been judged to produce errors that reach 80% for A and H and 35% for E_{r}, as the extra contact area produced by the pile-up is not accounted for by the unloading data [29]. Nanomechanical properties of thin films and multilayers have also been discussed in [30].

The severe difficulties with the area function of the nanoindenters and the uncertainties when going from standard materials to varied materials of practical interest make it necessary to look for further analytical treatments of the nanoindentation process and for further techniques. It should be again stressed that the loading of a pristine surface is different in nature from the unloading/reloading. In successive loadings there is no longer plastic deformation and the indenter experiences only the elastic response at the preformed relaxed impression close to the shape of the indenter. The exponent m for the exponential unloading/reversible reloading curves (iteration from 20% to 95% $F_{\mathrm{N\,max}}$) varies from 1.1 to 1.8 and depends both on the specimen and on the indenter. But a new experimentally secured universal power law for the initial loading curve of nanoindentations with exponent 3/2 will be presented for the nanoindentation to flat surfaces in Sect. 3.5 and 3.6. It indicates a profound difference between nanoindentation and microindentation.

3.5 Improved Indentation Parameters

In view of the difficulties for obtaining reliable values of hardness and elasticity modulus more reliable mechanical properties should be used. With the assumption that the loading F_{N} versus h^2 plots should be linear the parameter $F_{\mathrm{N}}\ h^{-2}$ ($\mathrm{N\,\mu m^{-2}}$) was proposed [31,32]; however, the tangents to $F_{\mathrm{N}} - h^2$ plots do not hit at zero in nanoindentation [7,9], evidently, because the assumption is not correct (experimentally disproved) for nanoindentations.

In another approach, combining (3.1) with (3.4) and (3.5) we obtain (3.13) with a constant parameter $F_N \cdot S^{-2}$(μm^2 N^{-1}) that may be useful for both nano- and microindentation as long as H and E_r are constants.

$$S^2 \cdot F_N^{-1} = 4\pi^{-1} (E_r)^2 H^{-1} \tag{3.13}$$

Recent experiments used this constant parameter $S^2 \cdot F_N^{-1}$ in microindentations (loads up to 700 mN) as a new materials characteristic by performing continuous stiffness measurements along the loading curve that is independent of tip shape (presumably with a Berkovich indenter) [33]. It means that hardness values can be obtained without any knowledge of the detailed indenter shape if E_r is known. Twelve inorganic materials (H: 1.03–39.24 GPa; E_r: 70–450 GPa) gave the linear relation between F_N and S^2 with a Berkovich indenter in microindentations up to large forces (up to 500 mN; e.g., 160 mN in the case of fused quartz). However, there is a dependence of H on the indenter tip, which does not show up in (3.13), and the indenter tip should therefore be indicated [9]. Furthermore, it has been shown [9] that similar plots of S^2 (from the initial slope of the unloading curves) versus F_N for fused quartz at nanoindentation (F_N up to 1, 4, and 5 mN) are also linear, as expected. This has been tested with the 60°, 90°, and 142.3° three-sided pyramids. These plots showed a dependence of the elasticity parameters on the indenter tip whatever the reason may be even if their contact areas are not involved (Fig. 3.9). Interestingly, the micro-scale value obtained using continuous stiffness measurement (up to 160 mN load) on fused quartz ($F_N \cdot S^{-2} = 1.020\,\mu\text{m}^2\,\text{N}^{-1}$) [31] did not agree with the nanoscale plot (1–5 mN load) based on the extrapolated slopes of the unloading curves giving values on fused quartz of $F_N \cdot S^{-2} = 1.361$ for the Berkovich indenter, 0.901 for the cube corner, and 1.622 [μm^2 N^{-1}] for the 60° pyramid [9]. It may reflect the uncertainties in the determination of S and/or the principal differences between nano- and microindentation techniques. No area function whatsoever has been used, and the indentation size effect on the hardness of fused quartz is known to be not very pronounced. But the fused quartz system is complicated due to the kink in the nanoindentation $F_N - h^{3/2}$ plot ((3.8) and phase transition in Sect. 3.7), which is totally obscured in microindents at very high load (e.g., 160 mN). Figure 3.9 shows the marked influence of the indenter shape. The slope change was drastically demonstrated with a blunt ($R = 250$ nm) cube corner, which gave a vastly different parameter of 1.622 (μm^2 N^{-1}). The missing correlation of the $F_N \cdot S^{-2}$ values with the totally included angle of the tip is certainly due to an interplay of phase transition and effective cone angle as well as tip radius or other tip deficiencies.

A different approach is the more detailed analysis of the loading curves, but these are at variance. Hertzian theory of elasticity [1] deduces the load–displacement relationships for different punches as exponential functions

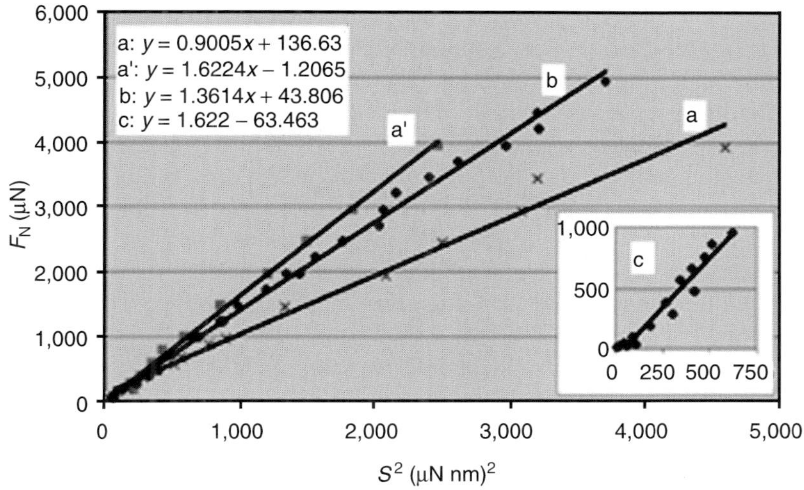

Fig. 3.9. Nanoscopic F_N–S^2 plots for indents on fused silica with (**a**) cube corner, (**a′**) blunt cube corner, (**b**) Berkovich, (**c**) 60° pyramidal indenter tip; 95–20% of the unloading curves were iterated

(3.14) with $m = 1$ for flat cylinders, $m = 1.5$ for spheres at small displacements, $m = 1.5$ for paraboloids, and $m = 2$ for cones.

$$F_N = \text{const } h^m. \tag{3.14}$$

For the loading curves the flat cylinder model ($m = 1$) was generalized for covering all indenter geometries [22]. However, the loading curves are far from being straight lines. Another approach by Loubet et al. [34] and Hainsworth et al. [31,32] claims linear relationships between load and displacement squared. For a general conical indenter of half-angle θ and pyramidal indenters, which are attributed "effective cone angles" α, a complicated $F_N \sim h^2$ relationship has been formulated [1] (3.15), but the assumptions for its deduction use data from the unrelated unloading curve

$$F_N = h^2\ E_r\ [\sqrt{E_r}/\sqrt{H}\ (3\sqrt{3}\tan^2\theta)^{1/2} + (\pi - 2)\sqrt{H}/\sqrt{\pi}\sqrt{E_r}]^{-2}. \tag{3.15}$$

Refs. [17,33] put forward the square relation. However, the tangents to the curves of such plots do not hit at the origin in nanoindentations [7,9]. Further expressions for the constant in (3.15) have been deduced for various situations of elasticity and plasticity [35]. However, the square relation is only verified for microindentations (above about 10 mN load) and deviations of the square relation for nanoindentation at lower load have been located. More specifically, it was claimed that H and E_r determinations should be performed by going beyond a transition region "roughly at 30 mN load," where the basic physical derivations become valid with the recommendation to measure H and E_r in the microindentation region (e.g., at 40 or 50 mN load) by reasoning that "initial contact" and/or the imperfectness of indenter tip may be responsible for the transition region [35]. Nanoindentation experiments clearly show that there is no experimental verification of (3.15) by experimental loading curves in the load region below 10 mN where a new universal 3/2 exponent holds (cf. (3.18)) [7, 9]. Clearly, there are basic differences between nano- and microindentation, and the present physical treatment is only valid for micro- and presumably macro-indentations [7, 9]. Furthermore, the well-known "indentation size effect" was invoked for explaining deviations from the exponent 2 in the case of ceramics [36], but such claim does not provide reason. It should be kept in mind that the generally assumed square dependence of the loading curve (which is not valid for nanoindentation) was used to establish a formula for the relation of elastic/plastic parameters E/σ_y (σ_y = yield stress), h_f/h_{max}, or loading slope/elastic stiffness S_l/S_u (3.16), or for the prediction

$$S_l/S_u = 1.48 \cdot (1 - h_f/h_{max}), \text{ for } h_f/h_{max} > 0.4. \qquad (3.16)$$

of constraint factor (H/σ; σ = flow stress at 10% strain), or for the definition of a "true hardness" (considerably larger than the generally used projected area hardness from the slope of $F_N - h^2$ plots, E_r and geometric factors) [37] (a further definition of "true hardness" or "work of indentation hardness" is described in Sect. 3.8, (3.22)), or for judging the pileup during indentation in bulk materials if the work-hardening characteristics of a material are known [38].

Another description of the loading curve did not acknowledge the differences of nano- and microindentation but used two parameters and two different exponents for the load [39] (3.17):

$$h = C_1 F_N^{1/3} + C_2 F_N^{1/2} \qquad (3.17)$$

Such attempt to separate plastic and elastic penetration can better approach the shape of the loading curve even at lower load than the quadratic fit. But the author soon returned to the quadratic hypothesis for the loading curve using a model of "the virtual sphere indenter" [17].

In that situation an experimental study aimed at finding a reliable and general relation for the loading curve of nanoindentations in metals, salts, oxides, silicon, polymers, and organic crystals was performed. It was consistently found that the exponent 3/2 holds in the F_N versus h relation (for a 60° triangular pyramid [9], cube corner [7,9], Berkovich [7,9], and spherical [7] indenter). The experimental relation is given in (3.18).

$$F_N = k\, h^{3/2} \text{ or } F_N^{2/3} = k^{2/3} h. \qquad (3.18)$$

$F_N - h^{3/2}$ plots reveal straight lines that hit the origin (in the absence of surface layers), unlike the linear or the square approach. In other words, the new general power law for the nanoscopic loading curve suggests that the indentation depth h is proportional to $F_N^{2/3}$ but not proportional to F_N and $\sqrt{F_N}$, irrespective of the chosen indenter tip, elastic/plastic ratio, and chemical bonds in the materials. Therefore, the differences of nano- and microindentation are secured, and nanoindentation provides new effects and new insights over microindentation. For example, the pressures of loaded indenters are high enough to induce phase transitions that are obscured in microindentation curves but can be located as kinks in the nanoscopic loading curve (Sect. 3.7).

It appears that the new nanoindentation parameter, i.e., the slope k of the $F_N - h^{3/2}$ plot with the dimension $\mu N\,nm^{-3/2}$ is a more direct and more reliable new parameter than the $F_N \cdot S^{-2}$ parameter, as the former is available without any assumptions or iterations at low loads [7, 9]. The latter is charged with additional problems of adhesion at various classes of solids and increased cracking/fissuring probability if load oscillation is applied. Importantly, if $F_N - h^{3/2}$ plots reveal kinks this is diagnostic for surface layers or phase transitions under pressure (such kinks cannot be detected in false square plots!). The 3/2 power relation holds for almost purely elastic, almost purely plastic, and elastic–plastic materials, and probably for all indenter types (though with different slopes, of course) except the flat punch as long as pop-ins or macroscopic cracks and fissures are avoided. Such analysis differentiates between loading to a flat surface and reloading to an impression.

Unloading and reloading curves depend on the specimen and indenter geometry (Figs. 3.1–3.3). Various power laws have been proposed varying from $F_N \sim (h - h_f)^2$ [37] to $F_N \sim (h - h_f)^{1.35}$ $(h_f/h_{max} > 0.4)$ [19]. However, the iteration of such exponential relation in the determination of unload stiffness (20–95% range) adjusts varying exponents ranging from 1.1 to 1.8, and severe further variations depend on the range of the unloading curve according to such iteration as is demonstrated in Table 3.2.

A more elaborate treatment of microindentation used the limiting cases that are completely elastic with $F_N \sim (h - h_f)^2$ and completely plastic with $F_N \sim (h - h_f)$. In between a linear combination bounded by the elastic solution and the plastic solution was assumed, the weight being governed by the strain hardening parameter [35]. Evidently this differs from the result of nanounloading where the fully elastic extreme obeys the relation $F_N \sim (h - h_f)^{3/2}$ (Fig. 3.12) and the fully plastic extreme approaches a vertical line (nearly so with Al in Fig. 3.1). In between a very large portion of the unloading curves follows the 3/2 exponential relation (e.g., in Figs. 3.12, and 3.13), which is also characteristic for fully elastic unloading ending with a flat surface. The lower part of the nanounloading curves may represent the reversal of phase transitions in some cases. The kink in Fig. 3.1 (SiO_2, 60° pyramid) may be interpreted that way. In the interpretation of the power laws, the different starting points of loading (to flat surface) and reloading (to impressed surface) should be reminded. The pressure distribution is different, except in the case of totally elastic pristine indentation, where no impression remains upon unloading. It has, however, been theoretically concluded that for spherical indenters the load is proportional to $h^{3/2}$ both in the loading and unloading/reloading curves. This conclusion was analogously applied to extend the hypothesis of a square relation (h^2) for loading also for the reloading with the (pseudo)conical indenter [22]. However, it cannot be experimentally verified for nanoindentations in both cases. The new experimental findings are described now in more detail in Sect. 3.6.

3.6 Linear Plots for the Loading Curves – the New Universal Exponent 3/2

In nanoindentations the experimental analysis of loading curves provides linear plots with a general $F_N \sim h^{3/2}$ relation (3.18). This result has been obtained with a wide variety of material types and indenters (from fully elastic to almost fully plastic behavior) that revealed the generality. Nanoindents with 60° triangular pyramid, 90° cube corner, 142.3° Berkovich or cono-spherical indenter ($R = 1\,\mu m$) indicate the power law of (normal force) – (normal displacement)$^{3/2}$ up to considerable forces in the nanoindentation range (3.18) and with all the various appearance of the indents (Fig. 3.5), which may exhibit sinking-in, pileup, long-range depressions, and even molecular migrations (Sect. 3.10). The experimental parameter k with dimension ($N\,\mu m^{-3/2}$) is a new indentation parameter that is more directly available than elastic modulus and depth depending hardness from unloading curves. These k-values depend also on the indenter tip that must be specified.

Equation (3.18) differs from the varying Meyer exponents ($n = 1.65$–1.9) reported for microindents at forces higher by two to three magnitudes under Vickers indenters [4]. Phase transitions under pressure may change the slope of the $F_N - h^{3/2}$ plot, which will exhibit kinks in those cases. Similar kinks will occur at very low loads if surface layers (e.g., surface hydrates or oxides with water layers) are present, as these are different material. These experimental observations provide new insights into the mechanical properties of a material that cannot be obtained by microindentation unloading curves, which monitor the phase transformed state in such cases and the parameters are, of course, measured for that state. The 3/2 exponential relationship in nanoindentations is valid for amorphous (fused quartz or organic polymers) and crystalline materials such as aluminum, silica, strontium titanate, or crystalline α-quartz, polycrystals such as gold, and organic crystals of various types. The validity of the experimental 3/2-power law is most easily demonstrated by comparison with plots of the previously claimed linear and quadratic relationships. The most prominent standard (fused quartz) is used even though there is a complication due to its amorphous-amorphous phase transition (Fig. 3.10). The slopes of the $F_N - h^{3/2}$ plots are the new indentation parameters k with dimension ($\mu N\,nm^{-3/2}$). The physical meaning of the slope k ($\mu N\,nm^{-3/2}$) reveals from a dimensional analysis: it is the geometric mean of the work per contact area formed ($\mu N\,nm\,nm^{-2}$) or ($\mu N\,nm^{-1}$) and the contact pressure

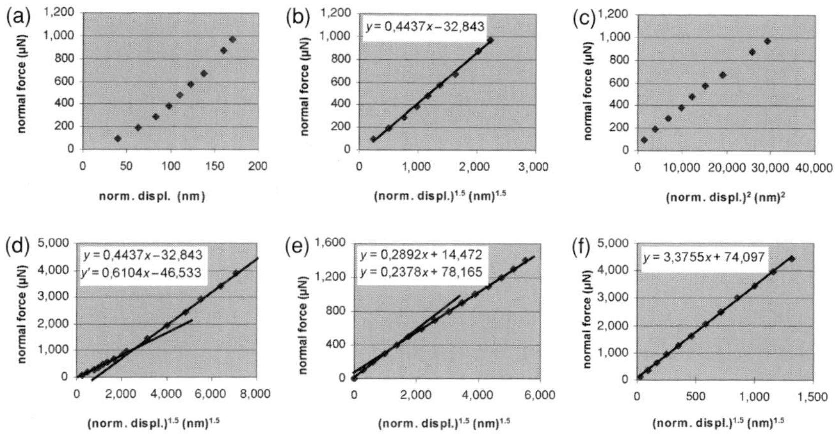

Fig. 3.10. Nanoindents of fused quartz (SiO_2) under various conditions of experimentation and analysis: (**a**–**d**) sharp cube corner, (**e**) sharp 60° pyramid, (**f**) conosphere ($R = 1\,\mu m$); the minor axial intercept in (**b**) and (**e**) reflects the hydrated surface water layer; the intercepts of the second slope lines in (**d**) and (**e**) (the first points excluded from the regression) are caused by the pressure-induced phase transitions, further deviations at very low load are due to the surface roughness; images (**a**) and (**c**) are shown in order to exclude validity of such types of plot

imposed ($\mu N\,nm^{-2}$). Figure 3.10 shows a series of linear plots for fused quartz indentations under various conditions.

The usual shape of the normal force versus normal displacement plots (every point is an average from various indents at the respective load) is nonlinear. An initial part of the fused quartz indent (cube corner) is shown in Fig. 3.10a. Neither a linear (Fig. 3.10a) nor a quadratic (Fig. 3.10c) force law holds, as the trial regressions miss the origin. However, the plot of (normal force) versus (normal displacement)$^{3/2}$ gives a straight line through the origin (Fig. 3.10b,f). The slight deviations at the beginning and a very minor deviation from zero of the cut at the F_N -axis are caused by the hydration layer on SiO_2 in ambient atmosphere. Figure 3.10d depicts a larger force range with the cube corner, and it is seen that a kink is present in the 3/2 exponential plot. Its first linear part is extended in Fig. 3.10b, and the second linear part with a slope that is decreased by 25% with respect to the initial part (regression without the points of the first part) has a finite cut at the F_N-axis. This behavior points to the well-known amorphous to amorphous transition from the "floppy" to the "rigid" region of fused quartz under pressure [40], which explains the change in response. Such kink is not present in the straight $F_N - h^{3/2}$ plot that is obtained by indentation with a conospherical indenter ($R = 1\,\mu m$), which exerts less local pressure on the sample at the same load (Fig. 3.10f). However, the very sharp triangular diamond pyramid with effec-

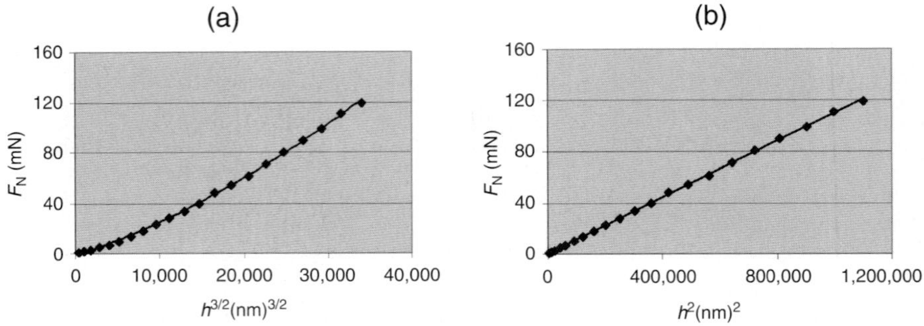

Fig. 3.11. Analysis of load displacement data of a microindentation on fused quartz [20] by trial plotting. The $F_N - h^2$ plot (**b**) is linear, whereas the $F_N - h^{3/2}$ plot (**a**) fails above 10 or 20 mN load; the hardly seen deviation at the beginning of the squared plot points to the different law of nanoindentation where (3.18) is obeyed

tive cone angle of 24.45° has its kink at 170 µN load and 98 nm penetration depth as compared to the cube corner with the values 1,000 µN and 77 nm (Fig. 3.10e). The projected area of the 60° pyramid is four-fold lower at the same penetration.

Interestingly, the new nanoindentation parameters (k (µN nm$^{-3/2}$)) are very sensitive to the indenter shape. Thus, a monotonic increase is found on fused quartz (after the kink) if the pyramidal tip becomes wider and penetrates less deeply at the same load (60° pyramid: k = 0.2378; 90° pyramid: k = 0.6104; 142.3° pyramid (Fig. 3.12): k = 1.6317 µN nm$^{3/2}$).

This is the place to compare the new results from nanoindentation with those of previous microindentations on fused quartz. For microindentations the Hertzian theory deduced an $F_N - h^2$ relation that seems to be obeyed. This may be exemplified in Fig. 3.11 for a Berkovich tip that had been loaded up to 160 mN [20]. The plot of F_N against h^2 is linear (with hardly seen deviations at the very beginning), but not the one against $h^{3/2}$, which is curved. For Fig. 3.11 only the data up to 120 mN load of [20] were used, the higher values were omitted. It is clearly seen that the left plot (exponent 3/2) exhibits a curvature, while the right-hand plot (exponent 2) gives a straight line. The details of the nanoindentations (up to 10 mN) and the question of the cut at the F_N axis are effectively obscured at that scale. Therefore, the theoretically deduced square relation appears to hold only in microindentations. This view is further supported by the analogous treatment of the microindentation to crystalline SiO_2 (001) and sapphire [20], which give similarly good plots with the square relation for the same load range (not depicted).

The reason for the switch of the exponent from 3/2 to 2 when going from nanoindentation to microindentation must be sought in finite tip radii or flat truncated ends, which become only negligible if the indentation depth exceeds about ten times the radius or diameter. This is hardly reached in nanoindentations. If we try on the basis of Hertzian theory (3.14) must be consulted in this respect: the sphere would expect $m = 1.5$, or the flat punch with $m = 1$ may be thought averaged with the effective cone with $m = 2$ for the nanoindentation region. There is a preference of the latter possibility for pyramidal tips due to the very strong influence of the effective cone angle (Fig. 3.1). When the more likely flat truncation (usually termed radius of curvature) becomes negligible at deep microindentation the cone feature with $m = 2$ prevails. However, the exponent 3/2 may have a deeper physical meaning if Hertzian theory is not applicable to nanoindentation.

Interestingly, there are opposing claims of varying Meyer exponents ($n =$ 1.65–1.9) for microindentations with Vickers indenters [4], and there may be ambiguities with the microindents of hard metal (Vickers, up to 2.5 N) [17], which provide the $F_N - h^{3/2}$ trial plots better than the $F_N - h^2$ trial plots in particular in the high range of the load. Also microindents with Al (Berkovich, up to 120 mN) [20] and W (Berkovich, up to 120 mN) [20] provide ambiguity. Both types of plot exhibit kinks, and an experimental assessment which plot should be preferred does not appear possible. The exponent 3/2 would "see" kinks at 40 and 20 mN, respectively (low load part less steep), the exponent 2 at 35 and 30 mN, respectively (low load part steeper). But these kinks cannot indicate phase transitions. These occur already in the nanoindentation range (Sect. 3.7). Rather an intermediate exponent would give reasonable straight lines. It is certainly hard to judge if additional effects may come to the fore at the very high loads of micro- and macroindentation. We return now to nanoindentations that give more direct unambiguous information.

A crucial test for the new $F_N \sim h^{3/2}$ law in nanoindentations are highly elastic and fully elastic load/unloads on fused quartz with Berkovich and spherical indenter. The former exhibits the kink due to phase transition under pressure at about 105 nm penetration and the square plot is curved. Interestingly, the fully elastic loading and unloading curves with the conospherical indenter ($R = 1\,\mu m$, several cycles) on fused quartz are also linear in the very first part of the $F_N - h^{3/2}$ plot (Fig. 3.12). This depth region corresponds to the probing of the hydration layer (cf. Fig. 3.10b). In both cases the square trial plots fail as they are markedly curved.

The $F_N \sim h^{3/2}$ law is also valid for nanoindentations on metals. Only $F_N - h^{3/2}$ plots (but not $F_N - h$ or $F_N - h^2$ plots) give straight lines that cut close to the origin with aluminum if the Berkovich $\{k = 0.2565\ (\mu N/nm^{3/2})$, Fig. 3.13$\}$ or the conospherical indenter (Fig. 3.14) is applied. The very minor positive cuts at the F_N-axis reflect the oxide layer on the surface at ambient atmosphere (Figs. 3.13, and 3.14). The linear plots for the indentations on aluminum (Figs. 3.13, 3.14, and Fig. 1 in [7]) do not exhibit kinks up to the high tested load in the nanoregion. There are almost pure plastic responses of the soft material and no phase transitions.

The $F_N \sim h^{3/2}$ proportionality is also obtained with a conosphere indenter ($R = 1\,\mu m$) on aluminum ($H = 0.35\,GPa, E_r = 66\,GPa$) (Fig. 3.14). Nevertheless, at very low load up to $50\,\mu N$ a larger slope of the 1.5 exponential plot

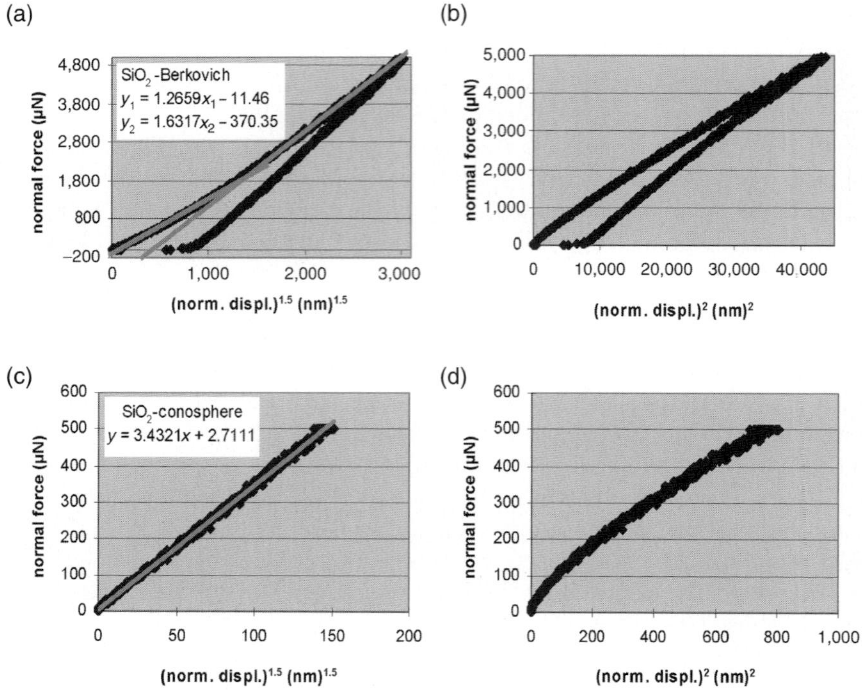

Fig. 3.12. Linear $F_N - h^{3/2}$ plots and $F_N - h^2$ trial plots of highly elastic and fully elastic loading and unloading curves on fused quartz; (**a**), (**b**) with Berkovich and (**c**), (**d**) with conosphere ($R = 1\,\mu m$) indenters indicating the validity of the exponent 3/2 and excluding the exponent 2

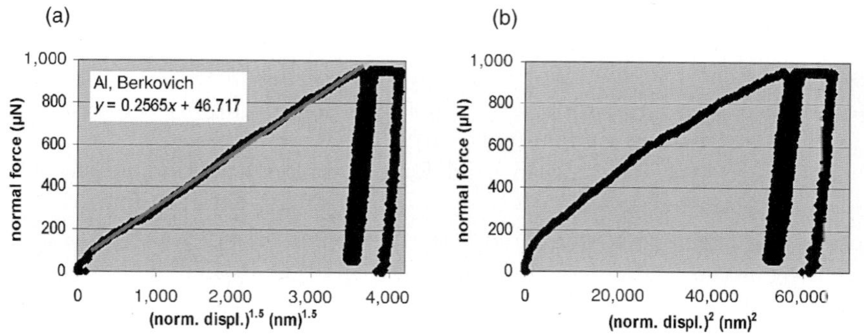

Fig. 3.13. Loading/unloading curves on aluminum with a Berkovich indenter; (**a**) $F_N - h^{3/2}$ plot with the slope that represents the nanoindentation parameter, the deviation at the beginning derives from the hydrated alumina layer that is unavoidable on aluminum in air; (**b**) $F_N - h^2$ invalid curved trial plot

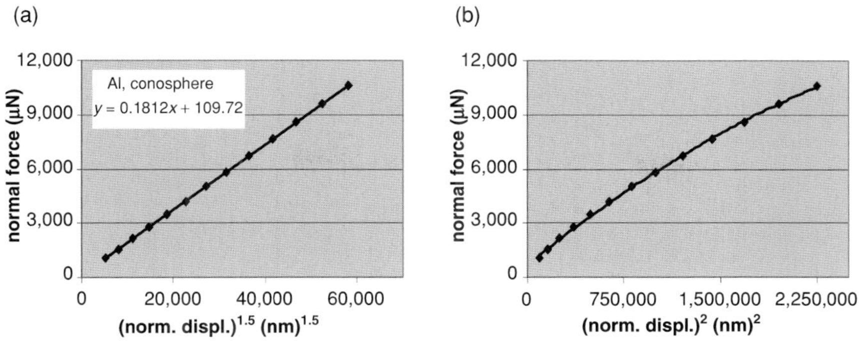

Fig. 3.14. Nanoindentation loading curves on aluminum with conosphere indenter ($R = 1\,\mu\text{m}$); (**a**) $F_N - h^{3/2}$ plot with the slope (k) that represents the nanoindentation parameter; (**b**) invalid $F_N - h^2$ trial plot with curved shape

($1.042\,\mu\text{N nm}^{-1.5}$; not depicted) was obtained with the conospherical indenter, which is followed by pop-ins after about $80\,\mu\text{N}$ load and must be attributed to the unavoidable alumina layer on the Al surface in air.

Gold (polycrystalline) ($H = 3.3$; $E_r = 105\,\text{GPa}$) is harder than aluminum and has a higher elastic modulus. The marked differences are also reflected in the different total works of the indentation (cf. Sect. 3.8) (the total work of indentation for Al up to $1,000\,\mu\text{N}$ load of the conosphere is 3.74 times larger than that for Au). Also the nanoindentation coefficients (slopes, k-values) are much higher for gold. Contrary to aluminum, gold gives a clearly kinked $F_N - h^{3/2}$ plot when the load of the conosphere ($R = 1\,\mu\text{m}$) nanoindentation is followed up to $10,000\,\mu\text{N}$ (Fig. 3.15). The low force indentation plot up to $200\,\mu\text{N}$ has a considerably higher slope (k-value) than at the higher forces. This points to a pressure-induced phase transition of gold (for example, face-centered cubic into hexagonal close-packed or body-centered cubic phase) [41] (Sect. 3.7 and 3.8). It appears extremely valuable that the simple analysis of the loading $F_N - h^{3/2}$ plots of nanoindentations allows for tracing of these events. Such behavior is completely obscured in microindentations. The $F_N - h^2$ trial plots are again curved, which excludes their validity.

The linear plot for the 3/2 power of the penetration depth was also found with silicon [7]. Clearly, the 1.5 exponential relationship (normal force) ~ (normal displacement)$^{3/2}$seems to be universal irrespective of the material. This is also shown in Sect. 3.10 and 3.11 for anisotropic organic and polymeric materials. Furthermore, it is very important to extend the study of loading curves to anisotropic crystalline oxides such as α-quartz and salt crystals such as strontium titanate. It was termed favorable that the Mohs standards "did not exhibit significant anisotropies" [42]. However, the slope of $F_N - \text{h}^{3/2}$ plot in anisotropic crystals must depend on the crystal face. This is dealt with in Sect. 3.8 after a more detailed discussion of the phase transitions in Sect. 3.7.

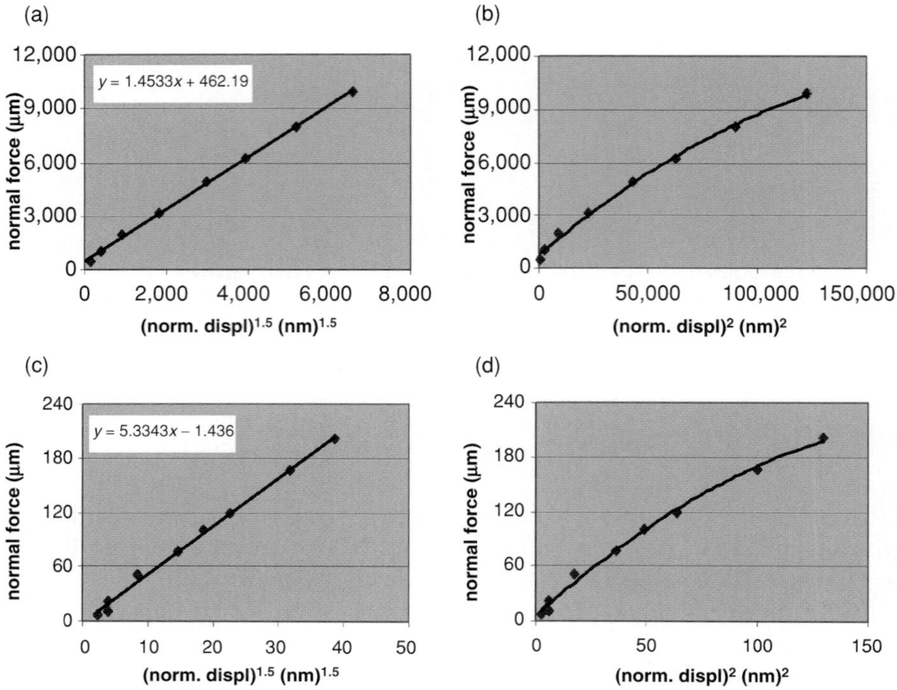

Fig. 3.15. Loading curves on polycrystalline gold of conosphere tip nanoindentations; (**a**,**c**) $F_\mathrm{N} - h^{3/2}$ plots at high and low load range with very different slopes (representing the nanoindentation parameters) due to a phase transition; (**b**,**d**) curved invalid $F_\mathrm{N} - h^2$ trial plots

3.7 Phase Transitions

It has been argued that the average pressure under a "geometrical similar" tip such as cone or pyramid does not depend on the load or penetration [43]. However, in nanoindentations the tips are not geometrically similar as they have significant radii usually in the 100 nm range, which cannot be neglected. According to elastic theory [1] the mean pressure p_av under a sphere is described by (3.19) where "a" is the radius of the projected area and R the radius of the sphere.

$$p_\mathrm{av} = (4E_\mathrm{r}/3\pi)(a/R). \tag{3.19}$$

This means that the average pressure increases initially and that enormous mean pressures are reached very soon in the early nanoindentation range. For example, they reach values of 0.368 E_r if h_c is at half of the radius R. It is only at microindentations where the mean pressure for a (pseudo) conical indenter is almost correctly described by (3.20) [43] or similar equations that take into account corrections for the $h_\mathrm{max}/h_\mathrm{c}$- ratio [11].

$$p_\mathrm{av} = E_\mathrm{r}/2\mathrm{tg}\alpha \tag{3.20}$$

It should be noted that long-range pressure relief would decrease the calculated pressure values. Such long-range pressure relief may be phase transitions along highly occupied lattice axes of crystals. Any pressure-induced phase transition is expected to exhibit in precisely analyzed load-displacement curves of nanoindents as a kink because the phase-transformed material behaves differently. Furthermore, if nanoindents actually start with a more or less spherical or truncated pyramidal (mean of flat and effective conical punch) geometry the linear $F_N - h^{3/2}$ plots (but not $F_N - h^2$; cf. Sect. 3.6) would be explained on the basis of Hertzian theory. The correct linear plot (3.18) can only trace the kink when the pressure increased sufficiently to induce the phase transition in the core zone.

The load does not appear in (3.19). Rather the modulus and the radii determine which pressures will be reached. The "effective cone angle" of the pyramidal tip plays its role if the influence of the tip radius or of the flat end diameter can be neglected at depths larger than about 10 R or 10 d. Very large average pressures occur at high values of the elastic modulus. In many cases such average pressures exceed the pressures required for transitions from low- to high-pressure crystal or glass modifications. For example, the average pressure at $h_c = \mathrm{R}/2$ calculates to 46 GPa for trigonal α-quartz on (0001) ($H = 10.3$, $E = 124$ GPa, $\nu = 0.077$) and to 27 GPa for fused quartz ($H = 9.5$, $E = 73$ GPa, $\nu = 0.17$) according to (3.19). These average pressure values are well above the phase transition pressures of α-quartz (>2.2 GPa into monoclinic coesite and >8.2 GPa into tetragonal stishovite) [44] or fused quartz (amorphous–amorphous transition at ca. 3 and 5 GPa) [40]. The average pressure of corundum, sapphire, or ruby (α-Al_2O_3 on (001), $H = 30$–35 GPa, $E = 441$ GPa, $\nu = 0.234$) under the same conditions calculates to 162 GPa according to (3.19). The phase transition (into the Rh_2O_3-structure) requires 79 GPa [45]. Interestingly, the reference indentation data of sapphire [21] exhibit a kink (at about 230 μN load and 30 nm depth) when plotted according to (3.18). Furthermore, the fully elastic indentation that has been reported for tungsten [20] ($H = 6.0$ GPa, $E = 410$ GPa, $\nu = 0.280$) provides a kink at about 120 μN load and 30 nm penetration when plotted according to (3.18), and there are the well-known interchangeable α-W and ß-W modifications that are close in energy content.

Kinks are to be expected in the nanoindentation loading curves because the different modifications are different materials with different mechanical properties. The high-pressure form governs therefore the first part of the unloading curves in common nanoindentation experiments. It is hard to predict if the high-pressure forms are rapidly reversible or not. As the theory of plastic deformation and "indentation hardening" is not well developed it is not yet possible to predict at which load the kink will appear. It is therefore very important to realize that such kinks are only expected if the correct exponent 3/2 is used to h, as this is now the reliable description of the loading curve for nanoindentations. Such kinks are obscured when using a faulty exponent (for example of 2), or by curve fitting, or in microindentations. The rather low pressure values for the phase transitions of quartz are the reason for early kinks in the $F_N - h^{3/2}$ plots. This is unfortunate, as fused quartz is the most frequently used calibration standard. Its curve fitting hampered the quantitative analytical description of the nanoindentation loading curves. Phase transitions in indentations are not unusual and can be studied further by Raman spectroscopy [46]. However, these are indirect if the permanent impressions are investigated and the high-pressure phase does not survive the unloading. The rather troublesome example of multiple silicon phases is discussed in connection with the more distinct detection of phase transitions by nanoscratching in Sect. 4.5.3 (Chap. 4). The practical use for finding pressure-induced phase transitions from the nanoindentation loading plot with most easy experimentation must be stressed. Such knowledge is widely unavailable for organic molecular crystals, in particular drugs (Sect. 3.10), and may be of high importance in organic polymer properties (Sect. 3.11).

3.8 The Work of Indentation and its Anisotropy

The integral under the loading curve gives the total work W_{Ntot} necessary for creating the impression up to the final load F_N (3.21).

$$W_{Ntot} = \int F_N \mathrm{d}h \quad (\mu\mathrm{N\ nm}) \tag{3.21}$$

The quantity W_{Ntot} can be planimetrically obtained or by use of an integrator. The unloading curve separates the elastic and plastic part of the area, which allows for separation of W_p (the permanent plastic work) and W_e (the elastic work). These quantities have found some interest for the characterization of materials, but most applications are still focused on the values of H and E_r. It has been found that W_p is linearly related with $F_N^{3/2}$ for macroscopic loads to aluminum ($1-50$ N), and the so-called "true hardness" (see Sect. 3.5) was determined from linear $W_p - F_N^{3/2}$ plots [47]. Such linear plots for W_{Ntot} and W_p were also reported for ceramics [36]. The specific work of permanent indentation was defined by dividing W_p through the volume of the indent. This gives the energy required to create a unit residual volume of indentation. However, the formal deduction used an $F_N \sim h^2$ relation [36], which is not valid for nanoindentation and the volume of the impression is not precisely obtainable from the contact area A_{hc} and h_c for a nonideal pyramid or cone. "Work-of-indentation hardness" values can be graphically determined. These values are different from Meyer hardness H but their detection does not require estimation of penetration depth, contact areas, or contact volumes, which can be inaccurate [48]. Thus hardness H can also be defined according to (3.22) from the plastic work and the residual volume.

$$H_{true} = F_{Nmax}/A_{pr} = W_p/V_p = 24.56\ F_N^3/9\ W_p^2 \quad \text{(for a Berkovich)}. \tag{3.22}$$

This technique reduces the errors of the standard procedure due to pile-up. The method breaks down at very low loads and small penetration depths.

This was attributed to deviations from the ideal tip geometry. However, these deviations may also derive from the special effects that occur under these circumstances as detectable by the kinks in the $F_N - h^{3/2}$ plots.

A plot of the "elasticity index" (H/E_r) values against $(1 - W_p/W_{Ntot})$ has been found to be approximately linear for metals and ceramics (3.23) [49]

$$H/E_r = \text{const}(W_{Ntot} - W_p)/W_{Ntot}. \tag{3.23}$$

Equation (3.23) relates plastic and elastic properties with a common dimensionless constant, which is valid for a given indenter angle. The value H/E_r can therefore be easily obtained by the work measurements. Interestingly, this empiric relation allows for a new determination of H and E_r when using F_{Nmax} and the initial slope of the unloading curve but without requirement of the projected area as expressed in (3.24) and (3.25).

$$H = \text{const}^2\ \pi\ (W_{Ntot} - W_p)^2(\mathrm{d}F_N/\mathrm{d}h)^2/W_{Ntot}^2 4\ F_{Nmax}, \tag{3.24}$$

$$E_r = \text{const}\ \pi\ (W_{Ntot} - W_p)(\mathrm{d}F_N/\mathrm{d}h)^2/W_{Ntot} 4\ F_{Nmax}. \tag{3.25}$$

However, the problems with the uncertainty of the slope determination (see Fig. 3.8; Table 3.2) remain. The indentation ductility W_p/W_{Ntot} may be linearly plotted against h_{res}/h_{max} [50]. This improves the assessment of the h_{res}/h_{max} ratio, however, it has two regions of linearity (kink at $h_{res}/h_{max} \approx 0.4$). It would therefore appear more useful to use the nanoindentation parameter k as defined in (3.18) for the characterization of materials. Corresponding formulas for the determination of H and E_r with the W_p/W_{Ntot} ratio using spherical indenters have been deduced [51, 52]. These are equally independent of sinking-in or pile-up. Clearly, what has been found and deduced for W_p [36, 47] must also be valid for W_e and W_{Ntot}, the total work of indent. It should be particularly noted that W_{Ntot} does not depend on assumptions about the contact area (see Fig. 3.6) or any iterations. Therefore, comparison of the work of indentation of different materials at the same load F_N is possible. For example, 1,000 µN conosphere ($R = 1$ µm) indents require the W_{Ntot} of 5,950 (µN µm) for Al and 1,590 (µN µm) for gold, which is a highly useful characterization of practical importance.

If different conical or pyramidal indenters are used, these can also be compared if $W_{Ntot}\mathrm{tg}\alpha$ is used as a new parameter, where α is the effective cone angle (3.26).

$$W_{Ntot}\mathrm{tg}\alpha = \text{const}\ {F_N}^{3/2}. \tag{3.26}$$

The dimension of the empirical constant for extrapolations to different loads is (µm µN$^{-1/2}$). Equation (3.26) has been applied for fused quartz with various indenter tips (Table 3.3) [9].

The performance of (3.26) is reasonably good if it is considered that no corrections for the tip radii and other deficiencies of the tips have been made and that a material has been under investigation that exhibits a pressure-induced phase transition. But there are the uncovered influences of the indenter shape

Table 3.3. Comparison of indentations on fused quartz at 700 μN load with pyramidal indenter tips

indenter	H (GPa)[a]	E_r (GPa)[a]	α	tg α	h_{max} (nm)	W_{Ntot} (μNμm)	W_{Ntot} tgα
60° pyramid	10.5	70.4	24.45°	0.4547	213	57.95	26.4
cube corner	10.2	69.9	42.28°	0.9093	143	37.79	34.4
Berkovich	9.1	69.8	70.30°	2.7928	66	17.34	48.4

[a] H and E_r values refer to 1,000 μN load at nanoindentations

on the work of nanoindentation that may be less important in microindentations. The variations in h_{max}, p_{av}, and W_{Ntot} are very large, but the parameter W_{Ntot}tgα is reasonably constant and the latter may be used for extrapolations to other tip geometries using additional empiric correction factors. The results suggest comparisons of different compounds at the same load if the same indenter tip is used. Furthermore, (3.26) allows for the calculation of W_{Ntot} at different loads.

W_{Ntot} may be experimentally subdivided into the plastic part W_p and the elastic part W_e by separate integration of the area between the loading and unloading curves as well as the area under the unloading curve, respectively, but the creep problem will decrease the reliability. This may nevertheless be useful for characterizing the anisotropy of different crystallographic faces of a crystalline material. High-grade natural low temperature- or α-quartz and strontium titanate have been indented on various faces and the anisotropies of H, E_r, k, W_p/W_e, and W_{Ntot}tgα are listed in Table 3.4.

Table 3.4. Various plastic and elastic parameters of α-quartz from 30/10/30 s cube corner nanoindentations up to 5,000 μN load on five main faces of crystalline α–quartz ($P3_22_1$) and 30/10/30 s Berkovich nanoindentations on strontium titanate ($Pm3m$) up to 3,000 μN load

compd.	face	h_{max} (nm)	H[a] (GPa)	E_r[a] (GPa)	k_1 (μN nm$^{-3/2}$)	k_2 (μN nm$^{-3/2}$)	W_p/W_e	W_{Ntot} tgα (μN μm)
SiO_2	(10-10)	201	15.3	109.0	1.956	1.590	1.07	420.8
SiO_2	(01-10)	191	16.2	119.7	2.145	1.728	1.09	378.3
SiO_2	(01-11)	179	17.4	133.6	2.730	1.844	1.26	379.1
SiO_2	(10-11)	193	16.5	105.0	2.256	1.668	1.17	404.7
SiO_2	(1-100)	193	16.6	109.4	2.303	1.656	1.05	395.6
SrTiO3	(100)	102	11.7	236	2.754	3.536	1.53	329.7
SrTiO3	(110)	103	12.0	254	2.462	3.390	2.04	331.1
SrTiO3	(111)	102	11.1	246	2.317	3.096	2.08	355.7

[a] The generally recommended 20–95% fit to the unloading curve was used

The data for crystalline SiO_2 show little variation in H and E_r except on the (01-11) face. But the total work of indentation is considerably different on the different faces. A correlation with the other quantities, including the

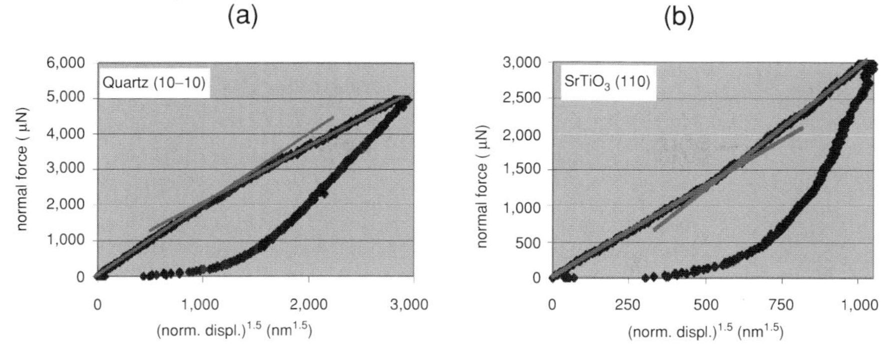

Fig. 3.16. $F_N - h^{3/2}$ plots on (100) of (**a**) crystalline quartz (cube corner) and on (**b**) (110) of crystalline strontium titanate (Berkovich)

"plasticity index" W_p/W_e, cannot be seen. The latter changes but not in line with the total work. The k-values are characteristic for the observed anisotropy, which reflects different atom densities on the different faces.

The data for crystalline $SrTiO_3$ show even less variation in the more common mechanic parameters. However, the indentation coefficients k and the W_p/W_e values vary more pronouncedly than the total work of indentation and provide the better means for the characterization of the anisotropy. Overall, the anisotropy found by indentation is relatively minor. This contrasts the appearance of the very different AFM images in Fig. 3.5. The discussion of the crystal packing that gives rise to the differences is executed in connection with the nanoscratching results, which disclose the anisotropy more distinctly (Chap. 4). The value of the k parameter from nanoindentations for the detection of pressure-induced phase transitions is again demonstrated with $F_N - h^{3/2}$ plots in Fig. 3.16, where there are kinks at low penetration. It should be noted that pop-ins [11], which can be initiated by some mechanical disturbance or noise from the environment, must be avoided. Different materials and different faces have different sensitivities toward noise. Most sensitive is the (111)-face of $SrTiO_3$ that usually requires several repeated cautious measurements for obtaining traces free of pop-ins. Only these were used in Table 3.4. Additionally observed steep slopes at very low penetration (<15 nm) of the linear plots must be attributed to surface hydrates of the oxide and of the salt. The initial slope (before the kink) may be smaller or larger than the second slope, depending on the material (Fig. 3.16).

It is well known that trigonal α-quartz experiences phase transformations at high pressure and also strontium titanate exists at low temperature in a different modification:

trigonal α-quartz → monoclinic coesite(> 2.2 GPa) → tetragonal stishovite(> 8.2 GPa)	cubic $SrTiO_3$($Pm3m$); tetragonal (I4/mcm)?

3.9 Recent Approaches to Nanoindentation at Highest Resolution

The ultimate description of nanoindents is, of course, atomistic. Actually, atomistic [53–55] and final element simulations [56] are being performed. However, there is a difficulty, which seems to be accounted for only recently. The usual load-controlled nanoindents force the tip constantly down, so that the nanoscopic events cannot be resolved and any discontinuities are smeared out in the loading curve. This is totally different if displacement-controlled indents are measured at ultralow loads. In these experiments the normal displacement is increased at constant rate and the normal force monitored. For example if the soft aluminum crystal (100) was indented by this technique with a sharp tip down to 100 nm depth 24 sudden load drops and recoveries, 12 of them down to zero force were detected (final load was 90 μN) [3]. Most importantly, at least five discernible load drops were observed in the first 10 μN of loading within a 113 nm displacement region. Thus, discontinuous yielding can result in sudden full stress relaxation even at ultralow loads. Sudden load drops have also been reported for gold single crystals under displacement control [57–59]. Other intriguing aspects are occasional load drops that start from a load lower than the starting load of the previous load drop and of load drops during displacement hold periods. These are basic data for the theoretical treatment of the indentation process. They indicate with highest resolution (also in the time domain) transient material behavior and a time lag (kinetic factor) in burst events so that the indenter loses contact to the surface and thereafter reloads (predominantly) elastically. The presence and the formation of intermediate lower load states (that are not resolved under load control) is certainly the basis for atomistic or finite element treatments of what is termed "nucleation of dislocation propagation" or of "dislocation-like discontinuities" (but there may be different mechanisms with other classes of materials, and there is also the possibility of anisotropic phase transformations; Sect.3.7) in order to deduce the empirically found power law of the loading curve in nanoindentations with the new indentation parameters irrespective of the plastic/elastic ratio or the various types of solid materials. Therefore, a general theory may be envisaged that might replace Hertzian theory.

3.10 Nanoindentations with Highly Anisotropic Organic Crystals

Only few organic crystals have yet been studied by nanoindentation [60]. They behave differently from inorganic materials because they exhibit only weak intermolecular van-der-Waals and/or eventually hydrogen bridge bonds between the isolated molecules. This feature leads to a higher risk of bursts and cracks or fissures in the loading curve of nanoindents. Furthermore, due to the anisotropic shape of most organic molecules their crystals exhibit highly pronounced anisotropies. Only very few ball-shaped organic molecules are known. However, nanoindentation can, of course, be applied to organic crystals. Anisotropic deformations upon mechanical stress by (nano) indentation of organic crystals have only lately been reported [60]. The shapes of the indents on organic crystals may be very diverse. Examples are given in Figs. 3.17 and 3.18 for the molecular crystals of the compounds **1**, **2**, and **3** on the indicated faces. **1** ($P2_1/n$) has a 3D-interlocked lattice, **2** ($P2_1/c$) exhibits double layers on (001) and **3** ($P2_1/c$) has skew cleavage planes under (110). Further highly anisotropic indenter imprints to α-cinnamic acid on (010) with vertical cleavage planes are shown in Fig. 4.43 in connection with nanoscratching results. Clearly, we have very different conditions and the indents are indeed very different.

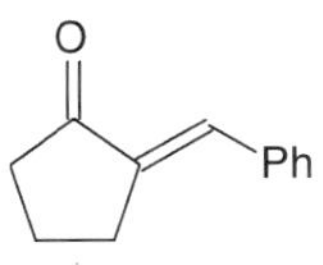

1

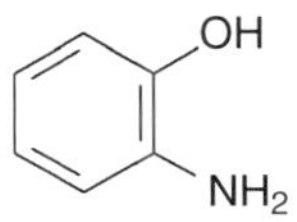

2

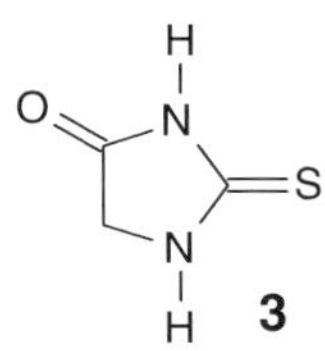

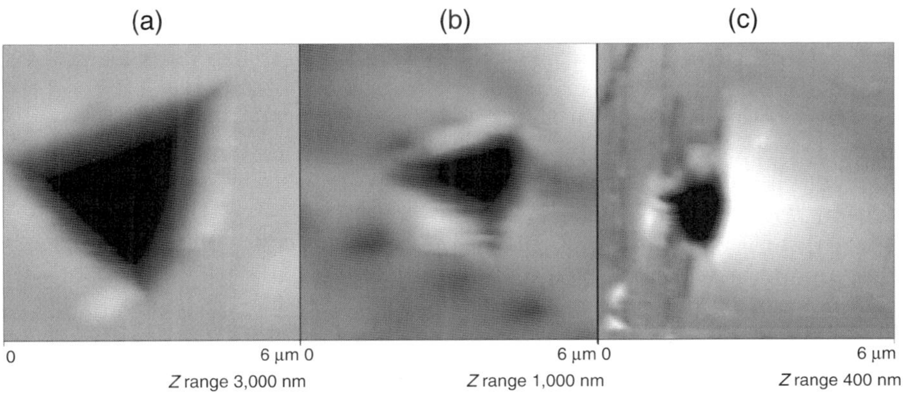

Fig. 3.17. AFM topologies of deep cube corner indents on crystals of (**a**) **1** (010), (**b**) **2** (001), and (**c**) **3** (110)

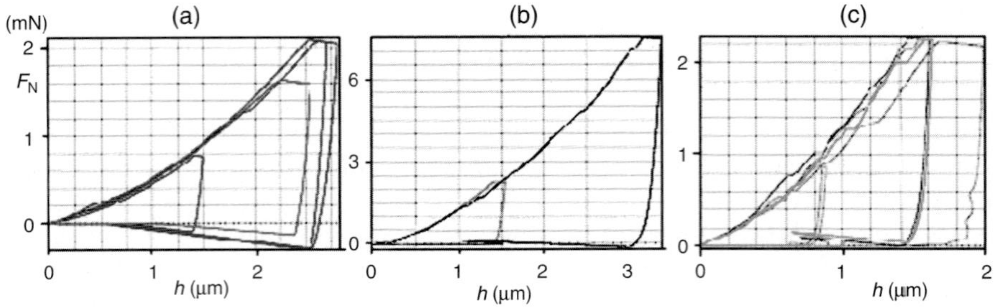

Fig. 3.18. Series of force–displacement curves for the nanoindents on (**a**) (010) of 2-benzylidene-cyclopentanone (**1**), (**b**) (001) of o-aminophenol (**2**), and (**c**) (110) of thiohydantoin (**3**)

The cube corner impression of 2-benzylidene-cyclopentanone (**1**) on the (010)-face is rather symmetric and only minor pileup can be traced by the AFM (Fig. 3.17a). However, the load displacement cycle curve (Fig. 3.18a) indicates creep and adhesion by the depression of F_{Nmax} and with the negative values in the unloading branch. Indents may not reproduce the shape of the cube corner indenter tip as in the case of 2-aminophenol (**2**) (Fig. 3.17b) even though the load displacement cycle looks quite normal, except for some minor pop-ins (Fig. 3.18b). It appears that elastic recovery is more pronounced in the vertical y-direction. A most interesting case is obtained with the cube corner impression in thiohydantoin (**3**): the shape of the indenter is hardly recognized in Fig. 3.17c. Fissures are seen on the left side but extended regular pile-up occurs to the right at the given crystal/tip orientation. The force displacement curves (Fig. 3.18c) show several steps (crack/fissure formation). The unloading branch does end above the zero line. H and E_{r} values as obtained by the standard iteration routine (Sect. 3.4) without taking care of apparent deviations from the assumptions of Fig. 3.6 with respect to the shape of the contact are therefore not precise [60]. The need for less ambiguous materials parameters is particularly important here. There are anisotropies, cracks, fissures, pileup, creep, adhesion, and unusual long-range effects.

The deep nanoindent on **3** (Fig. 3.17c) piles up the displaced material on one side of the calibrated cube corner indenter. The migrated material cannot be debris as it has the same standard hardness and elasticity as the original crystal. The piled-up volume (0.298 μm^3) is measured larger than the volume of the depression (0.170 μm^3), a fact that certainly relates to the cracking step formation (vertical to the long crystal edge) and to tip/sample convolution, which underestimates the volume of the sharp impression. But these values clearly indicate that no material is lost by abrasion. The unusual anisotropy is apparently caused by molecular migrations under the (110)-surface coming above its level, which occur only in one direction. The piled-up hill has a plane of symmetry with an ascend angle at its steep side of $<28°$ and extending over more than 3 µm with a uniform slope of 1.5–2° [60]. An angle of 45° is reached at the right-hand slope of the depression in Fig. 3.17c, while the slope at the left side (up to 29°) is smaller, because the molecules could not exit there. The crystal packing of **3**($P2_1/c$) [61, 62] on the (110)-face indicates parallel monolayer sheets that lie in the (10-2)-plane and are 66° skew under the (110)-face. Two types of hydrogen bonds form the infinite sheets. However, there are no hydrogen bonds between the sheets. Thus, the crystal may be cleaved parallel to these sheets and thiohydantoin crystals (**3**) exhibit clear-cut packing anisotropies. Hence, molecules or small hydrogen-bridged aggregates of molecules migrated upward along the skew cleavage planes. These anisotropic molecular migrations are even more distinctly revealed by nanoscratching experiments (Chap. 4, Figs. 4.30, 4.31). The detection of pressure-induced molecular migrations within crystals is of considerable practical importance for the foundation of new environmentally benign solid-state chemical syntheses, which require such molecular migrations within reacting crystals in order to release local pressure [63, 64]. All experimental evidence points to molecular (or small hydrogen-bridged molecular aggregates) migrations along the cleavage planes that are inclined to the right and thus open to the right-hand side in the given orientation of the crystal (Fig. 3.17c). The response to the pressure imposed by the indenter under increasing load is molecular migrations in this case. These molecular migrations are observed over more than 3 μm at the present crystal orientation and load only to the right side. Ditches are only at the left side. Clearly, anisotropic materials transport with potential for surface modeling can be obtained by mechanical pressure on the (110)-face of **3** [60]. These and further molecular migrations are more thoroughly treated in Chap. 4, Sect. 4.5.7, as nanoscratching provides much more detailed results.

When molecular crystals are mechanically deformed the molecules are not destroyed but only intermolecular interactions are broken and reformed (only solid organic polymers and explosives suffer from mechanical bond cleavages). The marked anisotropies, pronounced adhesion effects, creep, and the molecular migrations along cleavage planes are not bothersome for the analysis of loading curves of molecular crystals, though in the absence of severe pop-ins by crack/fissure formation. It is highly gratifying that the new power law for nanoindentations proves also valid for organic molecular crystals. Despite the weak intermolecular bonds the $F_N \sim h^{3/2}$ relation (3.18) holds again [7, 9]. This has been shown for anthracene (**4**, planar, no H-bonds) in Fig. 3.19, for tetraphenylethene (**5**, 3-dimensional, crowded, no H-bonds) in Fig. 3.20, for thiohydantoin (**3**, essentially planar with H-bonds) in Fig. 3.21, and for ninhydrin (**6**, 3-dimensional with H-bonds) in Fig. 3.22.

The slope of the linear plot for the 1.5th power of the penetration depth, i.e., the indentation parameter k with dimension (μN nm$^{-3/2}$) (3.18) represents the geometric mean of the work done by contact area and the contact pressure imposed by the indenter tip for the nanoindentation range. That range is of particular interest for submicrotreatment of surfaces and for wear applications.

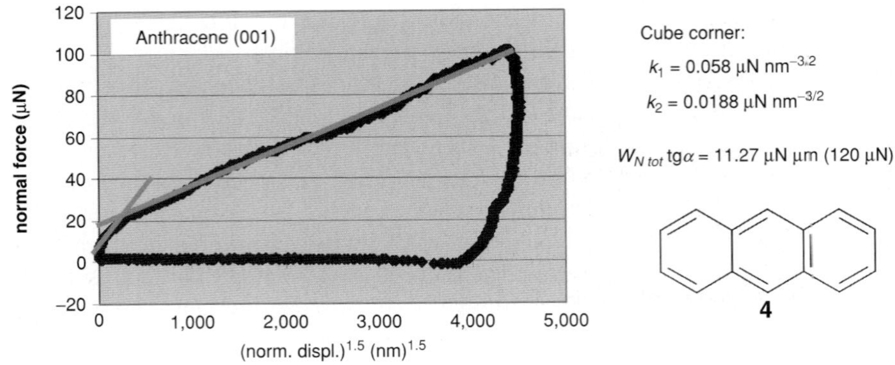

Fig. 3.19. $F_N - h^{3/2}$ plot of the loading and unloading curve of a cube corner nanoindentation on (001) of anthracene (**4**) with the slopes of the straight lines that represent the nanoindentation parameters

Anthracene (**4**) is a planar aromatic hydrocarbon with a layered crystal structure. It provides a pretty rough loading curve, which has an initial high slope followed by a lower slope in the $F_N - h^{3/2}$ plot on its (001) face (Fig. 3.19). The unloading curve starts with a negative slope (in the absence of a hold period) indicating creep. It would be difficult to obtain reliable stiffness contact area results for H and E_r determinations. Therefore, the nanoindentation parameters k_1 and k_2 or the total work of indentation W_{Ntot} (Sect. 3.8, (3.21), (3.26)) are more useful for the characterization of the mechanical properties on that face. The kink points to phase transformation, and indeed it is known that anthracene experiences pressure-induced phase transitions at 2.4 and above 3 GPa pressure [65].

The hydrocarbon tetraphenylethene (**5**) is a bulky unsaturated molecule. The phenyl groups are out of a common molecular plane due to internal steric hindrance. A rough loading curve ensues on its natural (10-1)-face, which has a kink in the $F_N - h^{3/2}$ plot, indicating pressure-induced phase transition (Fig. 3.20). The slopes k_1 and k_2 and the total work of the indentation W_{Ntot} are found as indicated. Again the unloading curve starts with a negative slope (in the absence of a hold period) indicating creep.

The almost planar heterocycle thiohydantoin (**3**) with its skewed layered structure on the natural (110) face, which experiences anisotropic molecular migration (Fig. 3.17), appears to be another candidate for a system with pressure-induced phase transition. The slopes k_1 and k_2 and the W_{Ntot} value are given in Fig. 3.21. The unloading curve is more regular, but the roughness of the loading curve does not indicate a well-defined contact depth for the determination of S, H, and E_r. Nevertheless, iterated values from the 20% to 95% range of the unloading curve were of some value for the comparison of essentially equal hardness and elasticity of pristine material and long-range "pile-up" by far reaching molecular migration (Fig. 3.17) [60].

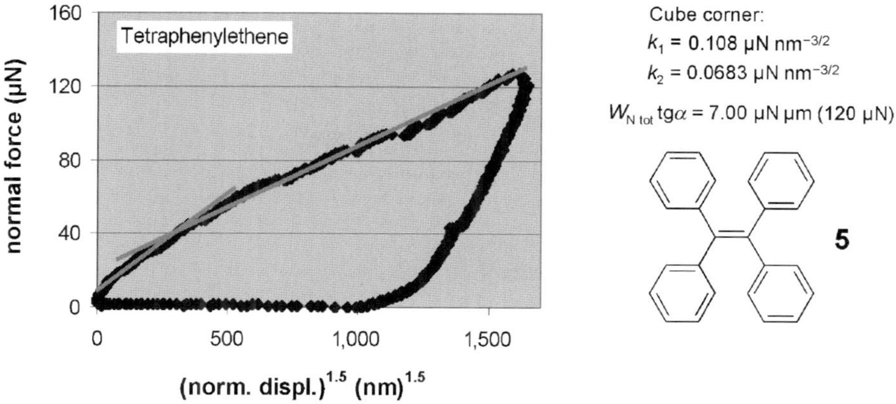

Fig. 3.20. $F_{\mathrm{N}} - h^{3/2}$ plot of the loading and unloading curve of a cube corner nanoindentation on (10-1) of a tetraphenylethene crystal (**5**) with the slopes of the straight lines that represent the nanoindentation parameters

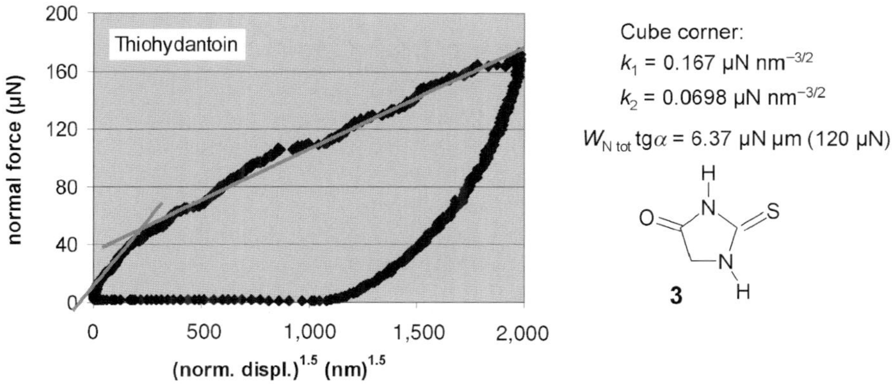

Fig. 3.21. $F_{\mathrm{N}} - h^{3/2}$ plot of the loading and unloading curve of a cube corner nanoindentation on (110) of thiohydantoin (**3**) with the slopes of the straight lines that represent the nanoindentation parameters

The well-known reagent ninhydrin (**6**) is a 3D molecule exhibiting multiple hydrogen bonds. The indentation on its (110) face tends to produce steps. There are two different linear regions in the $F_{\mathrm{N}} - h^{3/2}$ plot (Fig. 3.22), which again suggests a pressure-induced phase transition. The unloading curve has a negative slope at the beginning, which indicates creep, and the parameters k_1 and k_2 as well as the W_{Ntot} value are most appropriate for the mechanical characterization.

If the k parameters (μN nm$^{-3/2}$) (which represent the geometric mean of work per contact area and contact pressure) and the W_{Ntot} values (μN μm)

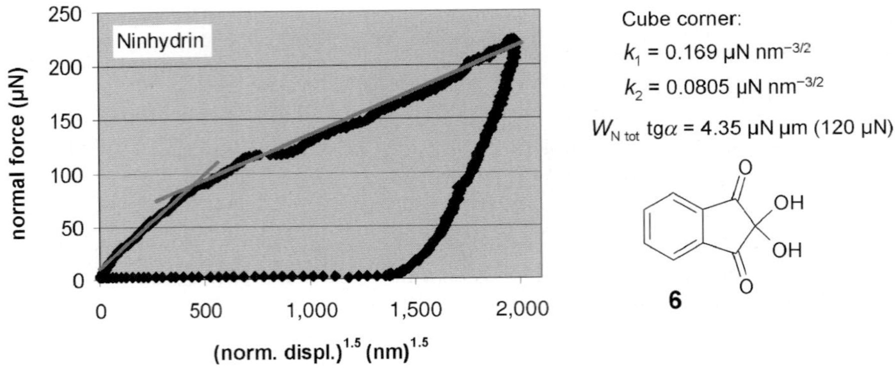

Fig. 3.22. $F_{\mathrm{N}} - h^{3/2}$ plot of the loading and unloading curve of a cube corner nanoindentation on ninhydrin (**6**) on (110) with the slopes of the straight lines that represent the nanoindentation parameters

Table 3.5. Comparison of cube corner indentation on various organic molecular crystals **1**–**6** at the chosen final load

Compd. (face)	F_{N} (µN)	H^{a} (GPa)	$E_{\mathrm{r}}^{\mathrm{a}}$ (GPa)	h (nm)	W_{Ntot} (µN µm)	W_{Ntot} tgα (µN µm)	k_1 (µN nm$^{-3/2}$)	k_2 (µN nm$^{-3/2}$)
4 (001)	120	0.21	10.9	236	12.40	11.27	0.058	0.0188
4 (110)	120	0.22	9.2	247	14.10	12.82	0.061	0.0224
5 (10-1)	120	0.48	11.9	132	7.70	7.00	0.108	0.0683
1 (010)	120	n. m.	n. m.	311	20.6	18.7	0.028	
1 (010)	2000	0.11	4.9	2440	1931	1756	0.0141	0.0171
3 (110)	120	0.57	12.5	112	7.00	6.37	0.167	0.0698
3 (110)	2000	0.55	13.7	1360	1116	1014	n. m.	n. m.
2 (001)	2000	0.31	12.4	1225	1102	1002	0.0380	0.0460
6 (110)	120	0.73	17.9	145	4.78	4.35	0.169	0.0805

[a]at 200 µN load; obtained from the ISO 14577 standard data treatment

(which relate the total work of indentation to the load) of the different organic crystals or their faces are compared, mechanical orders are obtained that should be more reliable than the unloading H and E_{r} values obtainable from the standard data treatment (assuming comparable behavior with respect to the standard materials). Table 3.5 compares the organic compounds **1**–**6** at the chosen normal loads F_{N}.

The data of Table 3.5 reveal that little can be derived from the small standard H (molecular crystals, intermolecular forces) and standard E_{r} values (mostly plastic behavior). However, there is clear-cut correlation of the k-values and the total work of indentation with the bonding properties of the molecular crystals. Furthermore, the two linear regions in the $F_{\mathrm{N}} - h^{3/2}$ plots point to pressure-induced phase transformations that may be of interest for further high-pressure studies. Such easy detection of pressure-induced phase transformations may be of particular use for pharmaceutical industries, which will need to know about all existing polymorphs of their drugs (even more precise detection in nanoscratching: Chap. 4).

Anthracene (**4**) has the molecules stand in layers on their short axis in two different directions on the (001) face. Conversely, on (110) of **4** the molecules stand and lie on their long axis in rows separated by skew cleavage planes [63, 64]. These packing differences influence the mechanical properties. Thus, the k-values are smaller on (001), the penetration is smaller than on (110), even though the H and E_r values cannot predict that sequence. The total work of indentation is smaller on (001) because the flat layers can be easily shifted laterally while the tip penetrates.

Tetraphenylethene (**5**) is a bulky molecule with considerably interlocking layers on the vertical (010) cleavage plane under (10-1) [66]. Therefore, the k-values are larger than in **1** (lower penetration) because the shifting of molecules upon indentation is more difficult. W_{Ntot} is much smaller due to lower penetration.

Benzylidene-cyclopentanone (**1**) is a considerably bent 3D molecule with the possibility for partial internal rotations. The molecules are packed in an interlocking lattice that exhibits channels along [100] under (40-1). As there are only van der Waals' interactions in the molecular crystal **1** it is rather soft and yielding (low packing density; $d = 1.185\,\mathrm{g\,cm^{-3}}$) [64]. Therefore, the k-values are low and the total work is relatively high.

Thiohydantoin (**3**) has an almost planar ring but the methylene-H deviates from the plane. Furthermore, its crystal exhibits hydrogen bonds within molecular monolayers [64] that have to be broken and rearranged if molecules are to be displaced. This increases the k-values on the natural (110) face with steep layers under it. The penetration is relatively low and so is the total work.

2-Aminophenol (**2**) is essentially planar and the crystal exhibits hydrogen bonds solely in zigzag double layers that are flat on the natural (001) face. Due to the easy shifting aside of flat layers the k-values are small, and the total work is lower than in the case of **1** despite the hydrogen bonds in **2**.

Ninhydrin (**6**) is a 3D molecule and it exhibits hydrogen bonds, which connect the molecular layers on (100) to double layers [64] that may thus be shifted but less easily because of higher thickness and as there is some interlocking between these double layers. It gives therefore k-values that are similar to those of **3**, but the total work is less due to the double layers as compared to monolayers in **3**.

The new parameters k ($\mu\mathrm{N\ nm^{-3/2}}$) (3.18) and $W_{\mathrm{Ntot}}\mathrm{tg}\alpha$ ($\mu\mathrm{N}\ \mu\mathrm{m}$) (3.26) at a specified load with reference to the indenter type (here cube corner) are well suited for the mechanical characterization of organic crystals. They appear more important as there are severe doubts concerning the precision of the iterated values for H and E_r in these molecular crystals. Various types of organic materials have been compared using the same indenter, and the anisotropies on different faces of the crystals **4** have been exemplified. However, directional anisotropies on particular crystal faces (that could be less distinctly revealed using Knoop indenter) require the 2D facilities of the nanoscratching (Chap. 4).

3.11 Nanoindentations with Organic Polymers

Organic polymers are soft and experience much creep, sagging, and even plastic flow. This is evident from Fig. 3.3 that shows a sequence of unloading/reloading curves. Clearly, the hysteresis loops are an indication of the viscosity properties and rapid phase transformations due to mechanical breaking of very long main chains to give macroradicals that disproportionate (to give end alkane and end alkene chains) until these are too short for further mechanical chain breaking. H and E_r determinations are therefore very difficult to assess with the standard ISO 14577 technique [20]. However, the generally applicable $F_N - h^{3/2}$ plots of the loading curves in nanoindentations are also successful in polymers, giving straight lines in smooth loading curves of nanoindentations. This is shown and compared to invalid $F_N - h^2$ trial plots in Figs. 3.23–3.25 for injection-molded polycarbonate (PC; polycarbonate Macrolon® for optical lenses), injection molded polymethylmethacrylate (PMMA), and molded cross-linked CR39 [from diethyleneglycol bis(allylcarbonate)], respectively.

PC is a polymer with benzene rings, isopropyl- and carbonic ester groups. A well-developed kink indicating the plastic–viscoelastic transition under pressure is obtained at rather high (about 550 µN) load in its $F_N - h^{3/2}$ plot (Fig. 3.23). Clearly, the square plot is again invalid in the nanoindentation and it would totally obscure the plastic–viscoelastic transition feature. There is creep and a straight line for the unloading as well. The k-values (slopes of the loading plot) are 0.441 and 0.295 µN nm$^{-3/2}$. High values of k indicate a low depth of penetration and a low total work of the indentation (Table 3.6). The rather high load required for the phase transition of PC due to breaking of linear chains is remarkable.

Table 3.6. Comparison of cube corner mechanical parameters of nanoindentations on glassy polymers

Polymer	F_N (µN)	H^a (GPa)	E_r^a (GPa)	h (nm)	W_{Ntot} (µN µm)	W_{Ntot} tgα	k_1 (µN nm$^{-3/2}$)	k_2 (µN nm$^{-3/2}$)
PC	900	3.11	17.9	178	71.02	64.57	0.441	0.295
PC	400	3.16	21.9	93	15.28	13.89		
PMMA	400	0.34	5.25	395	71.69	65.19	0.0714	0.0536
CR39	400	0.285	2.3	500	87.71	79.75	0.0489	0.0339

[a] Obtained by the ISO 14577 standard data treatment

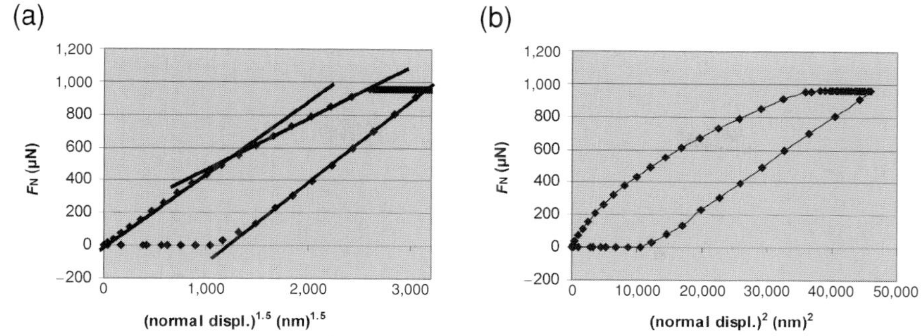

Fig. 3.23. Loading and unloading plots of a cube corner nanoindentation on PC; (**a**) $F_N - h^{3/2}$ plot showing two straight lines and a kink in the loading curve that is not seen in the invalid $F_N - h^2$ trial plot (**b**)

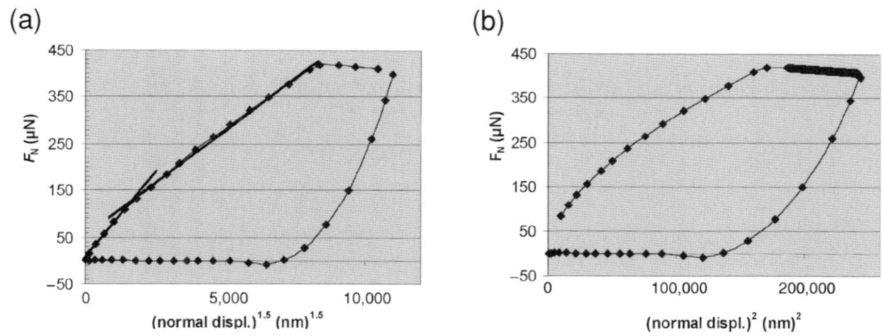

Fig. 3.24. Loading and unloading plots of a cube corner nanoindentation on PMMA; (**a**) $F_N - h^{3/2}$ plot showing two straight lines and a kink in the loading curve that is not seen in the invalid $F_N - h^2$ trial plot (**b**)

PMMA is an aliphatic chain polymer with methoxycarbonyl side groups. Its nanoindentation (Fig. 3.24) gives the pronounced kink for the plastic viscoelastic transition in the $F_N - h^{3/2}$ plot at about 130 μN load, which would be totally obscured in the $F_N - h^2$ trial plot. There is creep and sagging. The slopes ($k_1 = 0.0714$; $k_2 = 0.0536$) are about 6 times lower than in the case of injection-molded PC, and the total work is much higher. As discussed for Fig. 3.3, the reason for the phase transition derives from mechanical breaking of very long main chains of the polymer. Evidence for radical production under various kinds of mechanical destruction of PMMA is available from spectroscopic [67] and chemical analyses [68].

The cross-linked CR39 is a purely aliphatic polymer with a high percentage of carbonic ester and ether groups. The kink again indicating the plastic-viscoelastic transition under mechanical stress is clearly detected at about 70 μN load already (Fig. 3.25). Again the $F_N - h^{3/2}$ proportionality is nicely

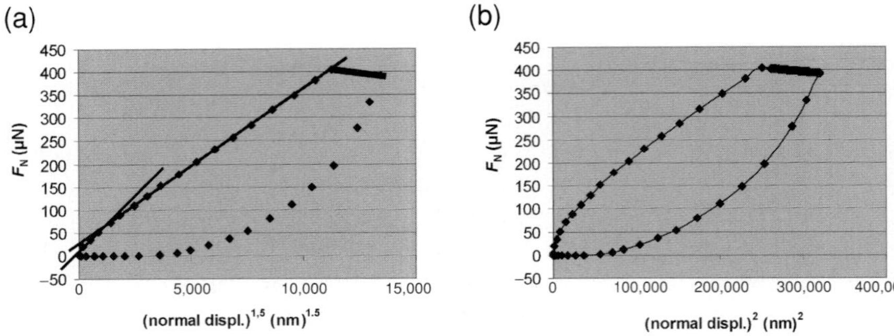

Fig. 3.25. Loading and unloading plots of a cube corner nanoindentation on CR39; (**a**) $F_N - h^{3/2}$ plot showing two straight lines and a kink in the loading curve that is not seen in the invalid $F_N - h^2$ trial plot (**b**)

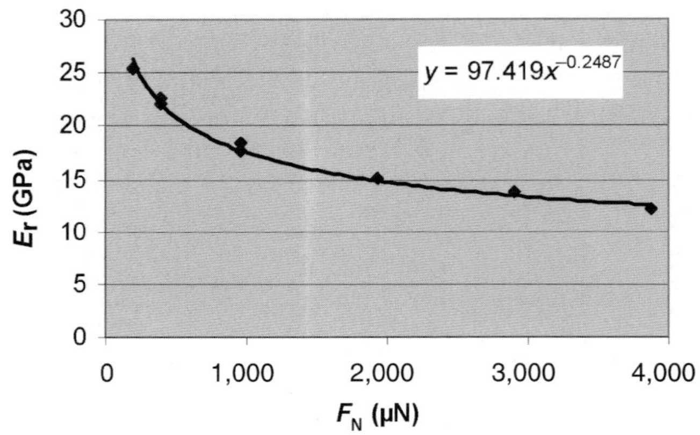

Fig. 3.26. Dependence of the elasticity modulus E_r of injection molded PC (Macrolon®) from the load F_N in cube corner indentations

shown by the straight lines but not in the invalid $F_N - h^2$ trial plot. The k-values ($k_1 = 0.0489$; $k_2 = 0.0339$) are smaller and the total work is higher than in PC and PMMA. There is creep and sagging and a strongly curved unloading curve. Importantly, polymer bonds are broken and after disproportionation of thus formed macroradicals the cross-linking decreases. The polymer data are compared in Table 3.6.

The variations in Table 3.6 reflect the mechanical properties of the polymers in the injection molded (PC, PMMA) or molded (CR39) state. It is also seen that the standard H and E_r values for PC (Macrolon®) are considerably higher than the ones of PMMA and CR39. However, H and E_r as obtained by the standard procedure may be very dependent on the load in glassy polymers. Interestingly, the E_r-values of PC (as determined by the iterative standard technique) decrease according to a power function when increasing the load (Fig. 3.26) whereas the H-values decrease linearly from 3.18

(200 μN) to 2.88 GPa (4000 μN). Such relations may be of theoretical interest and can serve to eliminate single deviating measurements from consideration. An exponential behavior was not found with PMMA and CR39 where both standard H and E_r decrease linearly from 200 to 1250 μN load: $H = 0.346$–0.328, E_r $5.28 - 5.24$ for PMMA and $H = 0.286 - 0.284$, $E_r = 2.5 - 2.1$ GPa for CR39. The assumption of independence of E_r from the load for the detection of S, H, E_r (3.9, standard procedure) is far from being obeyed, particularly with injection-molded PC.

Major differences of the polymers may be the brittleness of the cross-linked example and the type of curling of the chains in PC and PMMA. In all cases the shifting of material requires breaking of C–C or C–O or C–N bonds if applicable and this must be the major reason for the plastic–viscoelastic transitions.

The nanoindentation analysis using the generally applicable $F_N \sim h^{3/2}$ proportionality opens up new possibilities of polymer characterization by tracing of pressure-induced plastic–viscoelastic transitions just by using the correct relation (3.18). This was clearly not possible with the previous assumption of a square relationship that is definitely not applicable for nanoindentations. Various sources of errors in the standard determination of H and E_r (and their dependence on the depth) for viscoelastic polymers have been exhaustively described by Briscoe et al. [69].The difficulties with the measurement of H and E_r make the k-parameter and the total work of the indentation more valuable even though they rely on the rate of indentation with the viscoelastic materials and the correct shape of the indenter. The viscoelasticity of these polymers is also indicated in the hysteresis loops of the unloading/reloading curves as depicted in Fig. 3.3.

It should be noted that H and E_r determinations according to the standard technique can nevertheless be of practical importance for the local distinction of different materials in polymer blends or in organic materials if the absolute values are of less importance and cannot be assessed, as these are presently in widespread use. A striking medicinal application was reported with submicroscopic deposits in used blood bags that were identified with H and E_r and then located with SNOM [70] (Chap. 2, Sect. 2.9.3).

3.12 Concluding Remarks

The power laws of the loading curves differentiate nanoindentation and microindentation. The new $F_N \sim h^{2/3}$ relation for nanoindentations (as opposed to the theoretically supported $F_N \sim h^2$ relation for microindentations) is valid independent of the type of the chemical bonds that must be broken upon plastic deformation. These are covalent, ionic, metallic, intermolecular (organic molecular crystals), and mixed (for example polymers). It is highly rewarding that the $F_N - h^{3/2}$ plots of the loading curves are linear despite the many odd effects that do not agree with the assumptions made in the depth-sensing theory for isotropic materials. The latter make the determination of hardness H and elastic modulus E_r bothersome. The major reason for the difference of nanoindentation and microindentation is the final end-radius R of all diamond indenter tips in the 100 nm ranges, which may be also a flat truncation of comparable diameter. Both cannot be neglected up to penetration depths of 10 R or 10 d. In view of the strong influence of the effective cone angle (Fig. 3.1) the flat truncation is preferred over the sphere proposal. The standard procedure (ISO 14577) tries to handle all difficulties by three-parameter iteration of the stiffness S and by eight-parameter iteration of the contact area A_{hc}, the physical meaning of which does not sound very convincing. It does not use the benefits of the correct loading curve analyses in nanoindentations.

It appears very important that the same $F_N \sim h^{2/3}$ relationship is obtained irrespective of the chosen indenters and irrespective of elastic, or elastic–plastic, or plastic response. The proportionality factor (slope) of the linear plot of loading curves of nanoindentations is to be termed as *indentation parameter* and has the dimension μN nm$^{-3/2}$. It depends on the indenter tip, but it is a characteristic materials property for specified indenter tips that is most easily and assumptionless obtained. This indentation parameter should be tabulated with respect to the experimental conditions.

It is unlikely that the exponent 3/2 is related to the variable Meyer parameter (exponent n) that ranged between 1.65 and 1.9 with the Vickers indenter on glass and could only be determined at very high forces (100–1,000 mN) from doubly logarithmic normal force versus displacement curves that were obtained for various loading regimes [4]. In microindentations (load: 10-2,000 mN) the fine details of nanoindentation are not recognizable. Neither are the possible phase transitions recognizable from the microindentation loading curves and unrelated phenomena may occur. It should be particularly stressed that nanoindentation loading curves follow the $F_N - h^{3/2}$ relation both in the purely elastic and in the almost purely plastic case and in between. Therefore attempts to subdivide the loading curve in a fully elastic initial part up to a critical load F_{Nc} and thereafter a purely plastic part or to theoretically treat loading curves as purely plastic does not seem to make much sense in attempts for simulations (e.g., [1]). The coincidence with experimental results is deficient for both loading and unloading curves in nanoindentations, and if a simulation gives a reasonable loading curve it may be that the actual kink due to phase transition has been overlooked.

The now-proposed nanoindentation parameters k (μN nm$^{-3/2}$) and W_{Ntot} (μN μm) are obtained without assumptions as to the contact geometry (Fig. 3.6) and without the multiparameter iterations that are typical for the determination of H and E_r. They are also related to the indenter shape, but there are means to characterize the tips. Only nanoindentation but not microindentation is able to trace pressure-induced phase transitions by kinks in the linear plots of the loading curve.

The organic crystals and polymers deserve special concern due to unusual effects that are not taken care of in the data evaluation for inorganic standard materials. Highly anisotropic organic crystals of various types give cracks, fissures, rough loading curves, creep, adhesion, considerable pile-up, and molecular migrations. This detracts from validity of the assumptions made for the contact geometry for H and E_r determinations. Therefore such values cannot be easily assessed, and their quantitative significance may be doubtful. Furthermore, organic polymers are often very soft and experience much creep, sagging, and even plastic flow. It is therefore particularly important to use the now-proposed parameters k (μN nm$^{-3/2}$) and W_{Ntot} (μN μm) for the mechanical characterization. A consistent treatment of organic solids is possible on that basis.

Nevertheless, too many questions are open with the indentation technique and a better test procedure is required as both the elastic and plastic deformations are in principle anisotropic and as more precise information about pressure-induced molecular migrations and phase transitions are desirable. Nanoscratching is more precise, more rewarding, and more straightforward not only in this respect, but also for the question: where does the plastically displaced material go?

Another technique for obtaining hardness and modulus values is ultrasonic atomic force microscopy [71–73]. It is not described and compared here, as the theory of the dynamic AFM processes (including piezo-mode AFM) is far from being fully understood and there is still no broad application, but this may change in the future.

References

1. A.C. Fischer-Cripps: Nanoindentation, Mechanical Engineering Series. Springer, New York (2002)
2. O.L. Warren, J.F. Graham, P.R. Norton: Tapping mode imaging with an interfacial force microscope. Rev. Sci. Instrum. **68**, 4124–4131 (1997)
3. O.L. Warren, S. A. Downs, T.J. Wyrobek: Challenges and interesting observations associated with feed-back-controlled nanoindentation. Z. Metallkd. **95**, 287–296 (2004)
4. H. Meinhard, P. Grau, G. Berg, S. Mosch: Hardness and flow behavior of glass in the nanometer range. An interpretation of the load dependence of the hardness. Glass Sci. Technol. **70**, 333–339 (1997)
5. H. Hertz: Über die Berührung fester elastischer Körper. J. Reine Angew. Math. **92**, 156–171 (1882)
6. I.N. Sneddon: The relation between load and penetration in the axisymmetric Boussinesq problem for a punch of arbitrary profile. Int. J. Eng. Sci. **3**, 47–57 (1965)
7. G. Kaupp, M.R. Naimi-Jamal: Nanoscratching on surfaces: the relationships between lateral force, normal force and normal displacement. Z. Metal. **95**, 297–305 (2004)
8. T. Chudoba, F. Richter: Investigation of creep behaviour under load during indentation experiments and its influence on hardness and modulus results. Surf. and Coat. Technol. **148**, 191–198 (2001)

9. M.R. Naimi-Jamal, G. Kaupp: Quantitative evaluation of nanoindents: Do we need more reliable mechanical parameters for the characterization of materials? Z. Metallkd. **96**, 1226–1236 (2005)
10. S.E. Bresler, S.N. Zhurkov, E.N. Kazbekov, E.M. Saminskii, E.E. Tomashevskii; Investigation of macroradicals arising during mechanical destruction of polymers. Zhurnal Sakharnoi Promyshlennosti **29**, 358–364 (1959).; Chem. Abstr. 108586 (1959)
11. B. Wolf, A. Richter, M. Günther: Approaches of quantifying the entire load-depth curve in terms of hardness. Z. Metal. **94**, 807–812 (2003)
12. G. Y. Chin, M.L. Green, L.G. Van Uitert, W.A. Hargreaves: Hardness anisotropy of strontium fluoride, barium fluoride, sodium chloride, and silver chloride crystals. J. Mater. Sci **8**, 1421–1425 (1973)
13. H. Li, R.C. Bradt: The microhardness indentation size–load effect (ISE) in hard ceramic materials. J. Hard Metals **3**, 403–419 (1992)
14. T.Y. Tsui, W.C. Oliver, G.M. Pharr: Indenter geometry effects on the measurement of mechanical properties by nanoindentation with sharp indenters. Mater. Res. Soc. Symp. Proc. **436**, 147–152 (1997)
15. M.I. Petrescu. A possible interpretation of the exponent in the Meyer Law for microhardness in a group of sulfidic minerals. U. P. B. Sci. Bull. Series **B 63**, 51–56 (2001)
16. K. Herrmann, N.M. Jennett, S. Kuypers, I. McEntegaart, C. Ingelbrecht, U. Hangen, T. Chudoba, F. Pohlenz, F. Menelao: Investigation of the properties of candidate reference materials suited for the calibration of nanoindentation instruments. Z. Metal. **94**, 802–806 (2003)
17. T. Chudoba: Depth sensing hardness measurements. A new model for the interpretation of results. Materialprüfung **39**, 357–361 (1997)
18. K.M. McElhaney, J.J. Vlassak, W.D. Nix: Determination of indenter tip geometry and indentation contact area for depth-sensing indentation experiments. J. Mater. Res. **13**, 1300–1306 (1998)
19. A Bolshakov, G. M. Pharr: Influences of pileup on the measurement of mechanical properties by load and depth sensing indentation techniques. J. Mater. Res. **13**, 1049–1058 (1998)
20. W.C. Oliver, G.M. Pharr: An improved technique for determining hardness and elastic modulus using load and displacement sensing indentation experiments. J. Mater. Res. **7**, 1564–1583 (1992)
21. TriboScope® Users Manual of Hysitron Inc. (1999)
22. G.M. Pharr, W.C. Oliver, F.R. Brotzen: On the generality of the relationship among contact stiffness, contact area, and elastic modulus during indentation. J. Mater. Res. **7**, 613–617 (1992)
23. B. Wolf, C. Deus, P. Paufler: Comparative morphology study of indentations in quasicrystalline AlPdMn and single crystalline NiAl investigated by atomic force microscopy. Surf. Interf. Anal. 25, 561–568 (1997)
24. M. Kempf, M. Goeken, H. Vehoff: Nanohardness measurements for studying local mechanical properties of metals. Appl. Phys. **A**: **A66**, 843–846 (1998)
25. B. Bhushan: Nanotribology and its applications to magnetic storage devices and MEMS. NATO ASI Ser., Ser. **E286**, 367–395 (1995)
26. J. Thurn, R.F. Cook: Simplified area function for sharp indenter tips in depth-sensing indentation. J. Mater. Res. **17**, 1143–1146 (2002)

27. K. Herrmann, N.M. Jennett, W. Wegener, J. Meneve, K. Hasche, R. Seeman: Progress in determination of indenters used for nanoindentation. Thin Solid Films **377–378**, 394–400 (2000)
28. N.M. Jennett, A.S. Maxwell, K. Lawrence, L.N. McCartney, R. Hunt, J. Koskinen, J. Meneve, W. Wegener, T. Muukkonen, F. Rossi, N. Gibson, Z. Xu, A.J. Bushby, S. Brookes, A. Cavaleiro, K. Herrmann, B. Bellaton, R. Consiglio, F. Augereau, O. Kolosov, D. Schneider, T. Chudoba: Indicoat Final Report: Determination of Hardness and Modulus of thin Films and Coatings by Nanoindentation. NPL Report MATC (A) 24 (2001)
29. T.Y. Tsui, W.C. Oliver, G.M. Pharr: Nanoindentation of soft films on hard substrates: The importance of pile-up. Mate. Res. Soc. Symp. Proc. Vol **436**, 207–212 (1997)
30. B. Bhushan (Ed.), Springer Handbook of Nanotechnology, Chapter 22.4, pp. 707–711, Springer, Berlin (2004)
31. S.V. Hainsworth, H.W. Chandler, T.F. Page: Analysis of nanoindentation load–displacement loading curves. J. Mater. Res. **11**, 1987–1995 (1996)
32. S.V. Hainsworth, T.F. Page: Mechanical property data for coated systems – the prospects for measuring "coating only" properties using nanoindentation. Mater. Res. Symp. Proc., Maters. Res. Soc. **436**, 171–176 (1997)
33. T.F. Page, G.M. Pharr, J.C. Hay, W.C. Oliver, B.N. Lucas, E. Herbert, L. Riester: Nanoindentation characterization of coated systems: P: S2 – a new approach using the continuous stiffness technique. Mater. Res. Soc. Symp. Proc. **522**, 53–64 (1998)
34. J.L. Loubet, J.M. Georges, J. Meille: In *Microindentation Techniques in Materials Science & Engineering*, (Eds. P.J. Blau, B.R. Lawn) ASTM, Philadelphia, pp. 72–89 (1986)
35. K. Zeng, C.H. Chiu: An analysis of load–penetration curves from instrumented indentation. Acta Mater. **49**, 3539–3551 (2001)
36. M.T. Attaf: A unified aspect of power-law correlations for Berkovich hardness testing of ceramics. Mater. Lett. **57**, 4627–4638 (2003)
37. M. Sakai: The Meyer hardness: a measure for plasticity? J. Mater. Res. **14**, 3630–3639 (1999)
38. J.L. Hay, W.C. Oliver, A. Bolshakov, G.M. Pharr: Using the ratio of loading slope and elastic stiffness to predict pile-up and constraint factor during indentation. Mater. Res. Soc. Symp. Proc.Vol. **522**, 101–106 (1998)
39. T. Chudoba: Interpretation of depth sensing hardness measurements. Forschungszentrum Rossendorf e. V., Bericht **FRZ-129**, 49–53 (1996)
40. K. Trachenko, M. Dove: Intermediate state in pressurized silica glass: Reversibility window analogue. Phys. Rev. B **67** 212203/1–212203/3 (2003)
41. R. Ahuja, S. Rekhi, B. Johansson: Theoretical prediction of a phase transition in gold. Phys. Rev. **B63** 212101/1–212101/3 (2001)
42. U.D. Hangen: A comparison of nano-hardness and scratch resistance on Mohs minerals. *Z. Metallkd.* **92**, 1074–1077 (2001)
43. N.A. Stilwell, D. Tabor: Elastic recovery of conical indentations. Proc. Phys. Soc. London **78**, 169–179 (1961)
44. Matthes: *Mineralogie (3)*. Springer, Berlin, p. 55 (1990)
45. T. Mashimo, K. Tsumoto, K. Nakamura, Y. Noguchi, K. Fukuoka, Y. Syono: High-pressure phase transformation of corundum (α-Al_2O_3) observed under shock compression. Geophysical Research Letters **27**, 2021–2024 (2000)

46. A. Kailer, K.G. Nickel, Y.G. Gogotsi: Raman microspectroscopy of nanocrystalline and amorphous phases in hardness indentations. J. Raman Spectr. **30**, 939–946 (1999)
47. M. Sakai: Energy principle of the indentation-induced inelastic surface deformation and hardness of brittle materials., Acta Metall. Mater. **41**, 1751–1758 (1993)
48. J.R. Tuck, A. M. Korsunsky, S. J. Bull, R. I. Davidson: On the application of the work-of-indentation approach to depth-sensing indentation experiments in coated systems. Surf. and Coati. Technol., **137**, 217–224 (2001)
49. Y.T. Cheng, C.M. Cheng: Relationships between hardness, elastic modulus, and the work of indentation. Appl. Phys. Lett. **73**, 614–616 (1998)
50. Y.T. Cheng, Z. Li, C.M. Cheng: Scaling relationships for indentation measurements. Philos. Mag. **A 82**, 1821–1829 (2002)
51. J. Malzbender: Energy dissipated during spherical indentation. J. Mater. Res. **19**, 1605–1607 (2004)
52. W. Ni, Y.T. Cheng, C.M. Cheng, D.S. Grummon: An energy-based method for analyzing instrumented spherical indentation experiments. J. Mater. Res. **19**, 149–157 (2004)
53. C.L. Kelchner, S.J. Plimpton, J.C. Hamilton: Dislocation nucleation and defect structure during surface indentation. Phys. Rev. B **58**, 11085–11088 (1998)
54. J.A. Zimmerman, C.L. Kelchner, P.A. Klein, J.C. Hamilton, S.M. Foiles: Surface step effects on nanoindentation. Phys. Rev. Lett. **87**, 165507/1–165507/4 (2001)
55. K.J. Van Vliet, J. Li, T. Zhu, S. Yip, S. Suresh: Quantifying the early stages of plasticity through nanoscale experiments and simulations. Phys. Rev. **B 67**, 104105/1–104105/15 (2003)
56. For example A. Muliana, R. Steward, R.M. Haj-Ali, A. Safena: Artificial neural network and finite element modeling of nanoindentation tests. Metal. Mater. Trans. **A**: **33**, 1939–1947 (2002)
57. J.D. Kiely, R.Q. Hwang, J.E. Houston: Effect of surface steps on the plastic threshold in nanoindentation. Phys. Rev. Lett. **81**, 4424–4427 (1998)
58. J.D. Kiely, J.E. Houston: Nanomechanical properties of Au (111), (001), and (110) surfaces. Phys. Rev. B **57**, 12588–12594 (1998)
59. J.D. Kiely, K.F. Jarausch, J.E. Houston, P.E. Russell: Initial stages of yield in nanoindentation. J. Mater. Res. **14**, 2219–2227 (1999)
60. G. Kaupp, J. Schmeyers, U.D. Hangen: Anisotropic molecular movements in organic crystals by mechanical stress. J. Phys. Org. Chem. **15**, 307–313 (2002)
61. L.A.Walker, K. Folting, L.L. Merritt: Crystal structure of 2-thiohydantoin. Acta Crystallogr. Sect. B **25**, 88–93 (1969)
62. G. Kaupp, J. Schmeyers: Gas/Solid Reactions of Aliphatic Amines with Thiohydantoins: Atomic force microscopy and new mechanism. Angew. Chem. Int. Ed. Engl. **32**, 1587–1589 (1993)
63. G. Kaupp: Solid-state reactions, dynamics in molecular crystals. Curr. Opin. in Solid State and Mater. Sci. **6**, 131–138 (2002)
64. G. Kaupp: Solid-state molecular syntheses: Complete reactions without auxiliaries based on the new solid-state mechanism. CrystEngComm. **5**, 117–133 (2003)
65. J.M. Leger, H. Aloualiti: X-ray study of anthracene under high pressure. Solid State Commun. **79**, 901–904 (1991)
66. G. Kaupp, J. Schmeyers: Gas/solid reactions with nitrogen dioxide. J. Org. Chem. **60**, 5494–5503 (1995)

67. R.A. Graham, P.M. Richards, R.D. Shrouf: Direct evidence for formation of radicals in a shock-loaded polymer. J. Chem. Phys. **72**, 3421–3422 (1980)
68. N.K.Baramboim, I.I. Nikitina: Mechanical degradation of unsaturated three-dimensional polyesters. Izv. Akad. Nauk Kazak. SSR, Ser. Khim. **23**, 77–79 (1973); Chem. Abstr. 479583 (1973)
69. B.J. Briscoe, L. Flori, E. Pelillo: Nano-indentation of polymeric surfaces. J. Phys. D. **31**, 2395–2405 (1998)
70. G. Kaupp, A. Herrmann, M.R. Naimi-Jamal, S. Stepanenko, V. Stepanenko: Near-field microscopy for organic and biological surfaces. Research report of the federal ministry for education and research (BMBF), FKZ 13N7519, 30.6.2003, 1–33 (2003); available upon request at TIB-UB Hannover, Germany: www.tib.uni-hannover.de
71. L. Jia, M. Mandanas, C. Miyasaka, B.R. Tittmann, G.L. Messing. Analysis of organic binder distribution in spray dried granules by ultrasonic–atomic force microscopy, Proc. SPIE **3585**, 270–281 (1999)
72. U. Rabe, M. Kopycinska, S. Hirsekorn, W. Arnold. Evaluation of the contact resonance frequencies in atomic force microscopy as a method of surface characterization, Ultrasonics **40**, 49–54 (2002)
73. W. Arnold, S. Amelio, S. Hirsekorn, U. Rabe. Quantitative contact spectroscopy and imaging by atomic–force acoustic microscopy, Mater. Res. Soc. Symp. Proc. **591**, 183–194 (2000)

4

Nanoscratching

4.1 Introduction

Scratch tests are of enormous practical importance. They used to be applied on a macroscopic, microscopic, and nanoscopic scale, the latter with the atomic force microscope (AFM). The most recently emerging development is nanoscratching, using nanoindenters equipped with a two-dimensional (2D) or three-dimensional (3D) transducer. Scratching of surfaces is closely related to the field of tribology, but greasing is to be avoided for original materials characterization. Scratching experiments lead to abrasive wear in the majority of cases.

Applications were primarily in macroscratching with foundations in microscratching. The industrial importance of macro- and microscratching has been widespread and various different approaches exist for the definition of scratch hardness in the various fields. Thus, the Mohs hardness is a nonlinear scratch hardness still in extended use (DIN EN 101 and DIN EN ISO 1518) and sclerometers provide relative macroscopic scratch hardness values (DIN EN 438). Architects use grated scratch devices for probing the quality of cement flooring. The macroscopic applications compare either the depth of a scratch at a predefined load on varied scratching tools or the limiting load that is needed for a scratch to emerge. But there are also sound emission analyses (upon failure after linear increase of the normal load or during plowing with constant load) in scratch tests. More recently two different definitions of scratch hardness H_s have been put forward: $H_s = F_L/A_N$ [1] and $H_s = F_N/A_L$ [2] where A_N and A_L are the projected normal and lateral areas, respectively. The scratch resistance Erichsen, which is used to judge long-term stability, is prescribed in DIN 53799/10. Instrumentation for the study of microscratching wear of metallic materials at high temperatures (room temperature to $1,000°$ C) has also been developed [3]. Several mechanisms have been distinguished primarily on the basis of microscratching. Thus, grooving wear is subdivided into various factors that depend both on the materials properties and on the indenter tip geometry (attack angle). The term *plowing* has been coined for largely plastic formation of a permanent ditch with abrasion, pileup to both sides and eventually slip lines away from the scratch direction. The volume of pileup does normally not correspond with the ditch volume as microchipping or brittle removal of dust loses material and these processes may strongly prevail. Additionally, adhesional loss of material (removal of scales from the surface) may be discussed as well as microfatigue upon repeated plowing and cross plowing. The mechanism may change at phase boundaries. In composites with different hardness of the components the softer partner might be abraded while the harder partner may crack out and produce fissures due to pressure enhancement that builds up around it. Also fissuring or microcracking with detachment of larger particles can be very important. These distinctions are well developed and of high importance in metalworking with work hardening where also the influence of excessive speed in turning and drilling may be prevailing. For example, it is well known that higher speed favors chipping over plowing by increasing the brittleness. A review on microscratching has been published [4]. Abrasion and wear were the major issues.

Macro- or microscratching only rarely revealed that anisotropic materials give anisotropic scratch responses. It was recognized that microscratching on KCl, copper, and austenic single crystals in different directions gave different cross sections (Q) of the permanent ditches and different F_L/Q values under constant normal load (F_N) [3]. However, these results did not find widespread use and concern. There exist less than 50 reports on microscratching of inorganic crystals since 1967. They covered the salts NaCl, LiF, CaF_2; the oxides MgO, SiO_2, Al_2O_3, ZrO_2, Al_2SiO_5, $LiNbO_3$, $Li_2B_4O_7$, Bi-germanate; the chalcogenides CdS, InS, PbS, Sb_2S_3, $TbSbSe_2$, CdTe, ZnCdTe; nitrides/phosphides/arsenides TiN, InP, GaAs; carbides SiC, WC; the elements graphite, diamond, Si; the metals Cu, Al, Pb, W, Cr, Mg; some alloys such as steel, CuAl, MnZn, NiAl, AlCuCoSi; ceramics, ferrites; and the Mohs hardness standards.

Industrial applications of scratch tests on coatings largely prefer microscratching devices (tip radii of 20 or 200 µm) and evaluate diverse properties such as mechanical performance, adhesion, delamination, abrasion, mar resistance, critical load, etc. Of course, special problems are encountered when scratching coatings and thin films on supports. The major problems here are the separation of the film, the support properties from a composite indentation response, and the requirement of ultrasmall-depth indents well below film thickness. The basics are still being explored [5, 6]. Adhesion strength measurements with coatings are performed with linearly increasing normal load while the indenter moves over the surface (or the surface under the indenter). Acoustic detection is practical here as cracking or fissuring at the critical load emits acoustic signals. The critical load for detachment is characteristic for the adhesive strength and such techniques are therefore in wide industrial use.

These data are basic for the judgment of abrasive wear properties. But it will be necessary to use nanoscratching for the evaluation of submicrotools or for better characterizing long-term stability of coated surfaces that are in long-term sliding use as the applied loads are much smaller than in microscratching. However, the basic features of nanoscratching are just being explored. It is to be expected that its application will be helpful and also provide a deeper insight in the macroscopic industrial applications. The increased precision allows for quantitative treatments and the evaluation of basic laws.

While the applied forces in nanoscratching are much lower than in microscratching the various mechanisms may be similar but not identical. Sharper indentation tips with larger attack angles can be used and more detailed information is obtained. Obvious practical applications of nanoscratching are to be expected in the field of minerals, construction (pipelines, floor coverings, ceramics, etc.), coatings (heavy duty tools, varnishes, lacquers, etc.), nanoparticle composites (scratch resistant finish for cars and flooring, UV protection in sunscreens, UV stabilization in plastics and clothes, self-cleaning facades, etc.), thin films (scratch-resistant eyeglasses, polymers, microtools, self-assembled monolayers, etc.), and ultra-thin films of nanoparticles. Ultrathin coatings with a wealth of industrial applications (e.g., nanoparticles, lubricants, antistiction layers, adhesion layers, molecular glues, coupling agents, etc.) require submicron or nanotechniques for their analyses. Self-assembled monolayers may be superlubricants: they decrease wear by withstanding higher normal load than the plain support during sliding of an AFM tip (nominal "curvature" radius 100 nm, cantilever stiffness 30 N m^{-1}, load range 20–60 μN) (pp. 853–856 in [5]).

There are some further reports of nanoscratching with cantilevered AFM tips at increased load. However, regular AFM tips at 10^{-8} to 10^{-6} N force are not able to achieve controlled scratching of surfaces or to provoke long-range molecular migrations with formation of surface features. Efficient abrasion of material at about 10 nN force may only be obtained with off-axis cantilever "scratchy" tips (Chap. 1, Fig. 1.14) on organic crystals and only if these exhibit 3D interlocking, which immobilizes the molecules [7], [8]. Standardization of such nanoscratching would be virtually impossible with micromachined cantilever tips. In the AFM scratching experiments the applied force is not normal to the surface. But reliable mechanical nanotesting requires normal force and thus the setup of a nanoindenter with 2D transducer providing linearly variable forces. Very little is known about 3D nanoscratching, which would also detect the forces in the y-direction at right angle to the lateral scratch direction. The new type of nanoscratching has a big advantage over nanoindentations (Chap. 3), because mechanical properties and anisotropies are more easily and distinctly studied. For example, the indentation hardness (H) depends on the depth by the so-called "indentation size effect" [9] and on numerous inconsistencies in its determination (Chap. 3), which makes it difficult to assess them. Furthermore, asymmetries in the remaining tip impression are hard to interpret. Also the fate of the displaced material has been of little concern in "plastic" indentations. Hence establishing a materials balance was not usually tried, but many studies in metalworking deal with the ratio of chipping and plowing under various conditions. Anisotropy was seen [10] by using the two axes of the Knoop indenter with long and short diagonal length ratios of $a : b = 7 : 1$. These can be applied at orthogonal orientations and the lengths of the remaining diagonals of the impression are measured in order to calculate and compare Knoop hardness and Young's modulus. However, the amount of information is limited. Nanoscratching is able to determine anisotropies much more distinctly. Its increased sensitivity is also helpful for much more detailed studies of face anisotropies than previously and angle anisotropies can be tackled for the first time.

Surprisingly, it was not particularly acknowledged in the previous studies that bonds must be broken in all infinite covalent crystals and materials if they are indented or scratched. With cross-linked organic polymers there is also bond breaking (Sect. 3.11) and the electron spin resonance spectra of the free radicals formed as the result of chain disruption found much interest (Sect. 4.6). The new nanoscratching adds abrasive phenomena to plasticity or elasticity, and it adds molecular migrations within molecular crystals as pressure is at work in front and to the sides of the translating tip [11]. Also phase transformations under pressure are easily and distinctly detected in nanoscratching (Sect. 4.5.3).

Clearly, nanoscratching using a nanoindenter with a 2D transducer is a versatile technique for advanced mechanic testing of solid surfaces in addition to nanoindentation (Chap. 3). The basics are still not exhaustively worked out, but residual scratch resistance measurements [12] and very pronounced anisotropies of the nanoscratches in nonisotropic materials [11, 13] are highly promising.

4.2 Foundations of the Nanoscratching Technique

Few theoretical analyses of nanoscratching are known. A rather detailed report [14] described the "friction coefficient" (ratio of lateral force over normal force) in microscratching "as the resistance to abrasion over the whole period of the scratching process" and derived equations for the highly complicated changes of that ratio in different load regimes for ramp experiments with Si, Al, float glass, and fused silica. Unfortunately the "friction coefficient" was not constant and double logarithmic plots were required for obtaining linear plots for such "friction coefficient" versus normal force in microscratching up to 300 mN force but only in some of the investigated cases [14]. A Vickers microscratching study on various metal alloys and hard phases used the same definition for the "scratch coefficient" at constant load for various temperatures as basis for "grooving wear rates" [1]. Conversely, the simple 1.5 exponential relationship of the nanoindentation process at low-to-moderate forces for nanoindents with all types of materials (Sect. 3.6) and the now-available nanoscratching experiments reveal useful quantitative relations [13] (Sect. 4.5.1). The lateral force (F_L) is defined here as the force necessary for pushing the indented tip horizontally forward at a constant rate while the normal force (F_N) is held at a constant value. Some authors use the term tangential force (F_T) for that motion and define "lateral force" orthogonal to it, but not consistently [15,16]. Lateral force (F_L) in constant-load–constant-rate nanoscratching for very diverse solid materials does not linearly relate with the normal force (F_N). The general quantitative relation is described by $F_L \sim F_N^{3/2}$ [13]. The materials studied include isotropic glasses, semiconductors, crystalline oxides, salts, and various types of molecular crystals. The quantitative relation allows obtainment of characteristic materials constants describing mechanic properties, but the so-called "friction coefficient" cannot be constant in nanoscratching and remains a doubtful construct. In ramp scratching (F_N is constantly increased while the indenter moves horizontally at constant rate) there occurs sliding followed by plowing. We have tried to define friction coefficients for both parts and sum them up to a total friction coefficient in ramp experiments [16], but the plowing part does not conform to the 1.5 exponential relation both in edge- or face-in-front constant force scratching [13].

Previous interest in nanoscratch experiments derived from the study of residual scratch resistance (scratch work over excavated volume) that was tried to relate with hardness. However, it is very hard to obtain reliable measurements of the volume for various technical reasons (Sect. 4.5.6). Thus, better and unambiguous parameters are specific scratch work, full scratch resistance, phase transitions, facial and angular anisotropy, and anisotropic molecular migrations in molecular crystals [13]. These practical features have not been addressed at all in very detailed theoretical considerations [18,19].

The constant-load scratching may be performed with a pyramidal edge in front or with a pyramidal plane in front. But horizontal ditches are not always formed (Fig. 4.1a). The reasons for the building of upward (Fig. 4.1b) or downward ditches (Fig. 4.1c) have not been exhaustively elucidated. They may be sought in different types of abrasion mechanism as sketched in Fig. 4.1. The actual behavior depends on the brittleness or adhesion of the materials scratched. All depth values are averaged in the central 67% range, as there are clear distortions at the beginning and at the end of the scratch. This helps also in eliminating roughness of the depth line due to fissures and cracks that should be kept to a minimum by proper choice of the experimental conditions in any particular case.

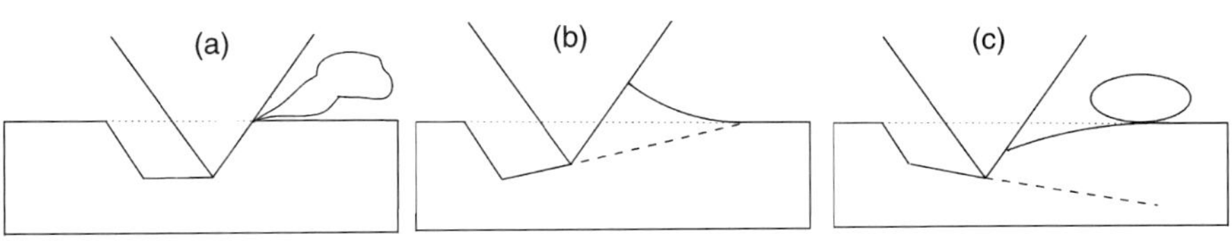

Fig. 4.1. Sketch of various types of abrasive nanoscratch appearance: (**a**) horizontal at smooth abrasion with a cloud of fine dust, (**b**) upward in the adhesive case, (**c**) downward if blocks of material are ejected

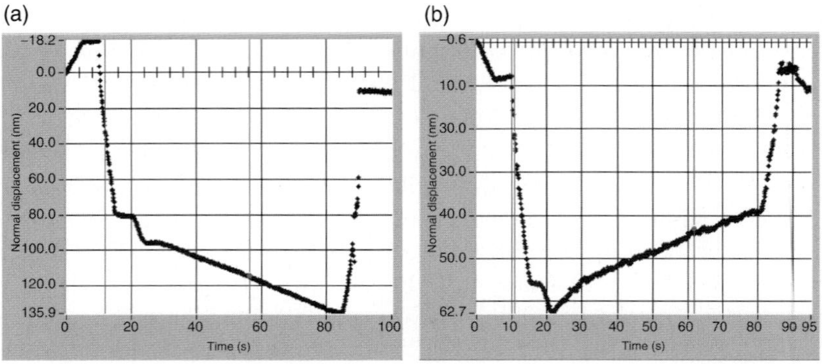

Fig. 4.2. Normal displacement (h) curves upon constant load cube corner scratching; (**a**) on fused quartz (side in front, 3 µm, 1,484 µN normal load) and (**b**) on strontium titanate (111) (edge in front, 3 µm, 993 µN normal load); the scratch time is 60 s

The occurrence of downward and upward scratching is instrumentally monitored and an average value for the normal displacement for the range of interest has to be taken if it is not neutral. Figure 4.2 shows examples for two of the possibilities in Fig. 4.1. The ranges of analysis can be chosen freely but should cover the inner 67% if possible.

In ideal scratch plowing, e.g., on polymers with Berkovich indenter at the beginning of the microscratching range, the material is pushed upward to both sides and there may be higher elastic recovery than in indentations (Chap. 3). A rapid elastic recovery at the rear of the indenter during scratching is assumed. Abrasion is enhanced at cross double scratching under these conditions and very flat scratch profiles have been reported over as much as 300 µm length in a polymer at about 1.5 µm depth [20].

4.3 Equipment

Nanoscratching on solid surfaces is a rather new technique in the mechanical testing. The capacitive 2D-transducer is commercially available for the Triboscope® or Triboindenter® of Hysitron, Inc. since 1998. Different techniques use Nano Test® from Micro Materials, Nanoindenter XP® of MTS Systems Corporation, Nanoscratch Tester® of CSEM Instruments, or UMIS® of CSIRO and others. Acoustic detectors may be added. Most commercial instruments work at normal force control, which is suitable for uniform materials. However, the capability for depth control is important for composite materials such as hard metals and the like in order to achieve a constant cross section of the ditch and monitor the site-specific variations of the normal force F_N [21]. All suitable nanoindenters (Chap. 3) require a double-axis transducer with two additional force–displacement sensors in order to monitor and control the position in the x-direction. The Triboscope® system controls and measures the displacements capacitively. The lateral force transducer performs both indentation (load control) and scratch testing (displacement control). The operator specifies length and duration of the scratch. The transducer is firmly connected to any SPM stage by its mounting in place of the AFM head in order to enable SPM measurements before (surface roughness) and directly after the scratch with the same indenter tip under 3D-piezo control. Calibration and analysis software is usually provided by the supplier or may be adapted from commercial software packets such as ELASTICA of ASMEC. The AFM part may be the NanoScope III or one of many other brands. The necessary equipment must not be very complicated (Fig. 4.3).

In the ramp experiment, the indenter scratches at constant rate for a predetermined distance while the normal load is linearly increased [14]. Conversely, the constant load experiment runs at constant rate for a predetermined distance without changing the normal force. The lateral force that ensues is recorded by the additional sensors of the 2D transducer and provides also

Fig. 4.3. A basic nanoscratching workplace by combination of a Triboscope® with a Nanoscope III® showing suspension mount on rubber bands, the transducer on top of the AFM piezo, leveling device and turn ring (detailed in inset), a light source, a telescope, a transparent cover for noise protection, with three screens (of two computers), and control box (under the right screen); the AFM measurements are controlled and visualized with the left and central screens

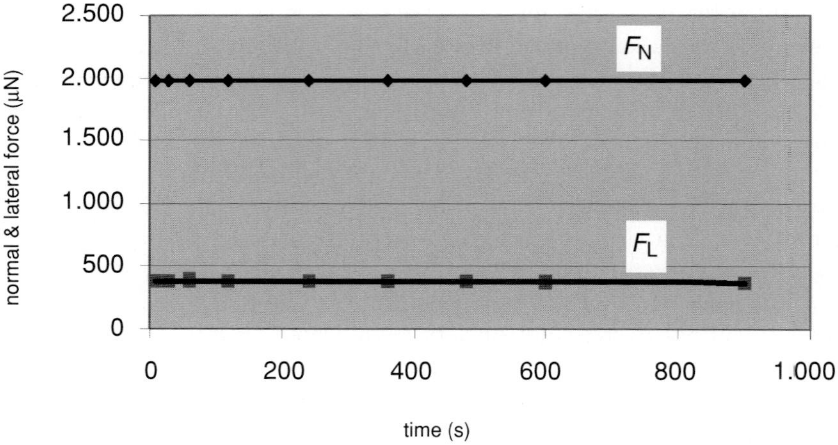

Fig. 4.4. Scratching on fused quartz with a Berkovich indenter at 2 mN load and 3 µm length at various scratch durations

the scratch work. The scratching process is normally not very sensible to the scratching rate. Figure 4.4 gives an example for 3 µm Berkovich scratching on fused quartz. It can be seen that F_L does not change if F_N is kept constant when the scratch time is varied from 10 to 900 s. Usually 30 or 60 s are chosen for 3–10 µ m scratches [22].

The volume of the excavated ditch can be measured by AFM and the scratch resistance calculated from these values [12]. There is, however, some tip–sample convolution if elastic recovery is low and successive filling of the ditch with abraded dust upon repeated AFM traces. Such dust may also occur as loose debris piled up in part at the rims of the scratch. This drawback is not removed by replacing the AFM tip with a sharp one and sometimes tedious search of the impression, but there are also instances where no dust debris occur, for example in the case of molecular migrations at the complete expense of abrasion (Sect. 4.5.7). For precise measurements of angular dependences of scratches a goniometer device is required that allows for adjusting sample tilt and adjustment of the rotation axis of a sliding bearing for transducer rotation. Such an addition has been constructed and is available for the Triboscope®. After instrument calibration the sensors of the 2D transducer record the values of normal force and lateral force. In all constant load experiments the lateral force is averaged over the central regions of the scratches (about 67%) in order to eliminate distortions at the start of the scratch and at the unloading end. Steps, peaks, or depressions that may have arisen due to single fissures or cracks (that can be minimized by slower scratching if necessary) are neglected. Typical scratch rates are 30 or 60 s for a 3 µm scratch length. But usually there is no rate dependence on the F_L and $F_N^{3/2}$ ratio from

3 to >900 s per 3 or 5 μm scratch length except for viscoelastic samples such as organic polymers. Pyramidal indenters may be oriented using the shape of their triangular indents. The nanoscratches can be performed edge-in-front or side-in-front and this has to be stated as well as the type of the tip. The quantitative data depend on geometry.

4.4 The Appearance of Nanoscratches

The appearance of symmetric nanoscratches with different materials may be very different. This points to different mechanisms of the abrasion as sketched in Fig. 4.1. Some examples are shown in Fig. 4.5 for ramp and constant load nanoscratches. Abrasion of material may be complete, or there may be pileup in front or at the sides. As pressure is not only exerted laterally in scratch direction but also laterally sidewise there may be elastic recovery and plastic deformations at the sidewalls and it is usually not possible to assess the amount of material abraded. Furthermore, it is not surprising that nanoscratches tend to exhibit rough borders due to side fissures and cracks, which have to be minimized by proper choice of conditions. Depending on the crystal structure it is even possible to produce asymmetric scratches with anisotropic

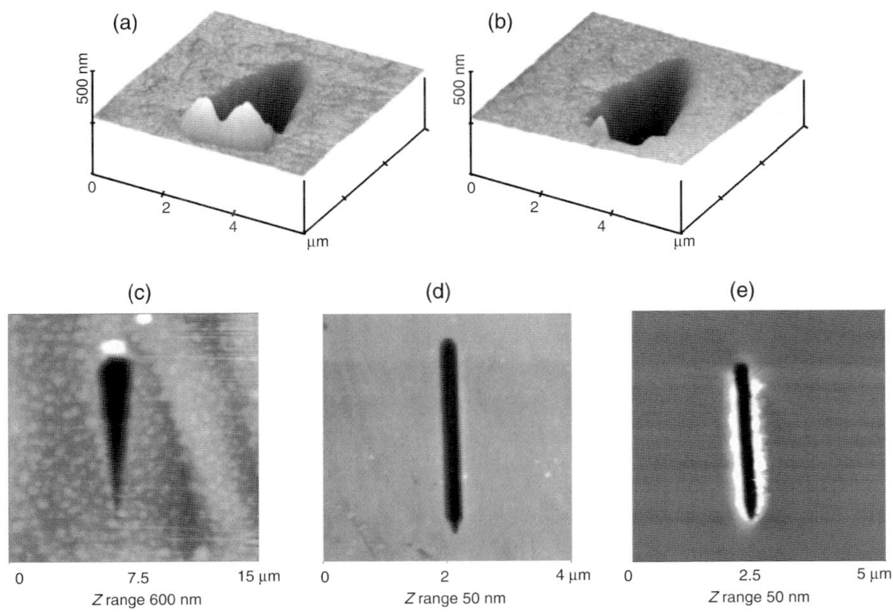

Fig. 4.5. Various appearance of abrasive ramp **(a, b, c)** and constant load nanoscratches **(d, e)** with and without pileup; the scratching was on single crystals of thiohydantoin (10–2) (a, b), 2-benzylidene-cyclopentanone (**c**), fused quartz (**d**), and strontium titanate (100) (**e**)

molecular migrations particularly with organic molecular crystals (Sect. 4.5.7). Importantly, AFM measurements of the steep ditches with the same indenter tip are charged with tip–sample convolution errors that may be quite high. However, looking for the tiny nanoscratches with regularly sharp AFM tips is very time consuming or impossible and bears the risk of partly filling the ditch with abraded dust. Generally, the first ditch volume measurement provides the largest reading but this decreases in consecutive AFM scans when also some of the pileup material has shifted. Nevertheless the imaging of the nanoscratches gives important information, even if it appears advisable to develop scratch resistance parameters that do not require the uncertain volume measurement. Microscopic or electron microscopic techniques are not able to improve the assessment with respect to the ditch volume. Importantly, abrasive nanoscratches are less sensitive to greasing impurities that may be present undetectably on the surface than in sliding friction measurements. Therefore natural surfaces with moderate roughness can be safely studied but the tip should be normal to the surface in narrow limits (better than 1° or less) in order to avoid severe shear force effects.

Most of the lateral force is in the scratch direction, which is measured by the 2D transducer, but there is also normal lateral force (to the sidewalls of the ditches), which can be measured by emerging 3D transducers. While Fig. 4.5 does not allow to directly "see" the effects of the sidewise lateral force this will be evident in anisotropic scratching of molecular crystals with molecular migrations in Sect. 4.5.7.

4.5 Quantitative Treatment of Nanoscratching

4.5.1 The Relation of Lateral and Normal Force

There is a widespread use of the so-called "friction coefficient" ($\mu = F_L/F_N$) in sliding and abrasive scratching [16]. However, such "friction coefficient" would indicate a direct proportionality of lateral force and normal force. Such direct proportionality has never been demonstrated experimentally and has apparently never been deduced theoretically. Rather "size effects" have been assumed. However, knowledge of the dependence of the lateral force F_L (in direction of the scratching) from the normal force F_N is most important for all quantitative treatment of nanoscratching. It determines the depth of the ditch under load for the material under investigation. It is therefore highly gratifying that an apparently general exponential relation between F_L and $F_N^{3/2}$ was experimentally found and secured for virtually all types of solid materials in constant load nanoscratching at constant rate (4.1) [13, 17].

$$F_L = K F_N^{3/2}. \tag{4.1}$$

The constant K of dimension $[1/\sqrt{\ }\text{force}]$ has to be termed as "scratching coefficient", which is characteristic for the material on particular faces in particular directions and depends also on the indenter geometry. This experimental result may be derived on the basis of the standard formulas of the indentation process (Chap. 3), but the uncertainties of these (assumptions and iterative relation to a standard with totally different behavior) do not lead to useful formulas for the extraction of hardness from K (4.9) at the present state of theory. However, the values of K ($\mu N^{-1/2}$) are useful on an experimental basis for the quantitative description of nanoscratching. Their constancies (unlike the so-called "friction coefficients") should also be helpful for advanced descriptions of ramp scratching.

The experimental evidence for (4.1) has been demonstrated for oxides, elements, metals, salts, organics [13,17], and polymers (Sect. 4.6). For example, fused quartz is an isotropic glass with SiO_4 tetrahedrons randomly connected by strong O–Si–O bonds. Many of these covalent Si–O bonds have to be broken when excavating ditches. Scratches at different constant normal force values are plotted versus different powers of the lateral force required (Fig. 4.6). It is found that only a plot of lateral force versus (normal force)$^{3/2}$ is both linear and cuts at the origin as is necessary (Fig. 4.6b). Conversely, the trial plots for direct proportionality (exponent 1) and square relationship (exponent 2) fail as they miss the origin (Figs. 4.6a,c). Only the 1.5 exponential relation is valid. The slope of the line in Fig. 4.6b (K) is to be termed as scratch parameter for the used tip geometry. The very good linearity is found up to 1,500 μN normal load in the 1.5 exponential plot.

The failure of a direct proportionality (Fig. 4.6a) is remarkable in view of the common approach for the description of the tribological behavior by a "friction coefficient," which is defined by the ratio of lateral force over normal force [13]. Therefore, the generality of the new result has to be demonstrated under various conditions and with different materials. In fact, the 3/2 exponential relation (4.1) has been secured for virtually all different kinds of materials on various faces and directions as well as with various indenters [13, 17]. Strontium titanate ($Pm3m$) is a typical example for salts

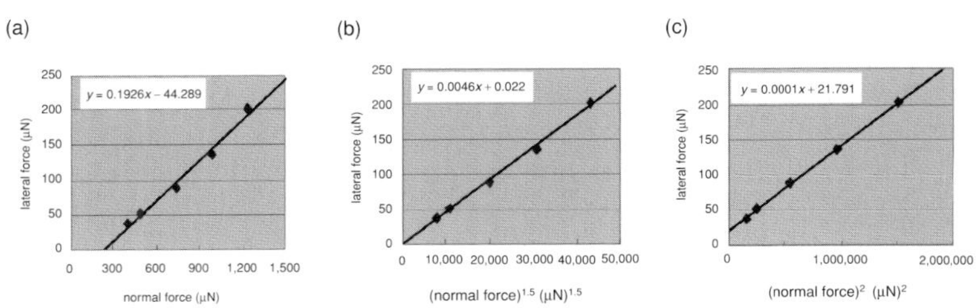

Fig. 4.6. *Fused quartz*: plots of the lateral force vs. normal force upon scratching with a cube corner indenter (edge-in-front) showing a linear plot through the origin only at the 1.5 exponent (**b**) but not in the trial plots with the exponents 1 (**a**) or 2 (**c**)

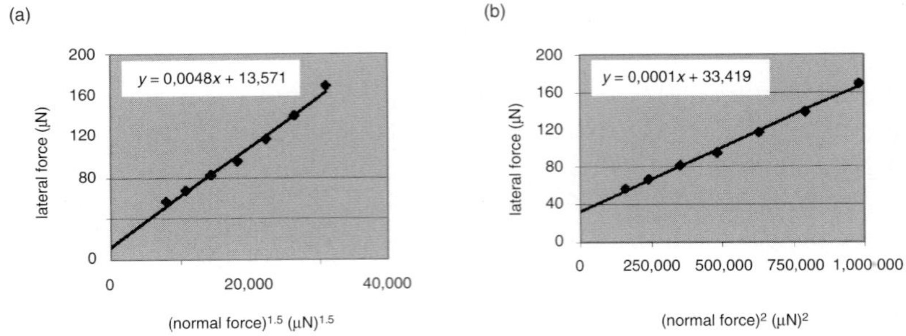

Fig. 4.7. $SrTiO_3$ on (100): plots of the lateral force vs. normal force upon edge-in-front cube corner scratching parallel to the crystal edge: (**a**) with exponent 1.5; (**b**) trial plot with exponent 2 clearly missing the origin

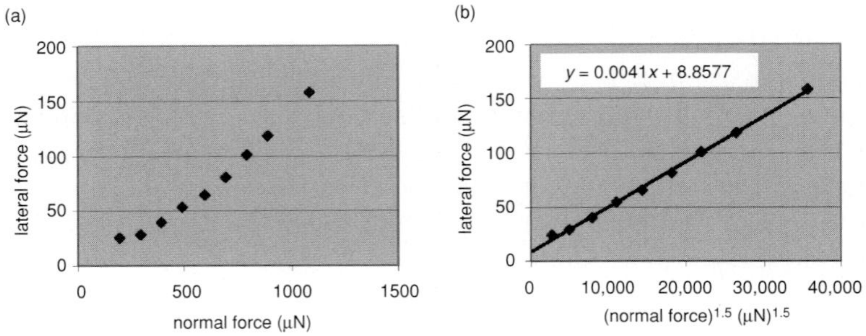

Fig. 4.8. Silicon on (100) (SiO_2 layer 10 nm): plots of the lateral force vs. normal force upon cube corner edge-in-front scratching parallel to the cleavage direction; (**a**) curved trial plot with exponent 1; (**b**) linear plot with exponent 1.5 (it does not perfectly cut at zero due to the hydrated SiO_2 layer)

(Fig. 4.7). The cut of the straight line in Fig. 4.7a misses the origin but slightly, probably because of the water layer on its surface. However, the trial square plot of Fig. 4.7b is far off and thus invalid.

A similar problem with surface layer is encountered with silicon, which has a hydrated layer of silica on it in ambient air. But the experimental evidence in terms of (4.1) (Fig. 4.8b) is again convincing. The straight line in Fig. 4.8b cuts reasonably close to the origin, whereas the trial plot in Fig. 4.8a with exponent 1 is strongly curved. A further feature with the F_L–F_N plot of silicon is a kink that occurs at higher normal force. It is discussed in Sect. 4.5.3 (Fig. 4.18).

Importantly, even organic molecular crystals of various kinds and polymers (Sect. 4.6) obey (4.1). For example, thiohydantoin (**1**) and ninhydrin (**2**) have different arrays of hydrogen bonds, whereas tetraphenylethene molecules (**3**) (Fig. 4.9) experience only van der Waals' forces in their crystals. All of these

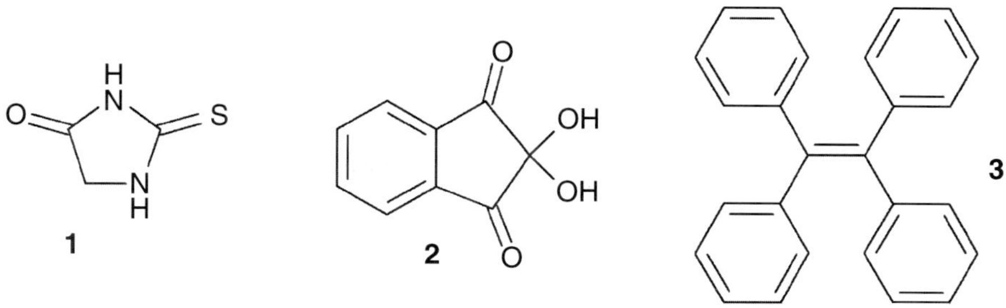

Fig. 4.9. Molecular formulas of thiohydantoin (**1**), ninhydrin (**2**), and tetraphenylethene (**3**)

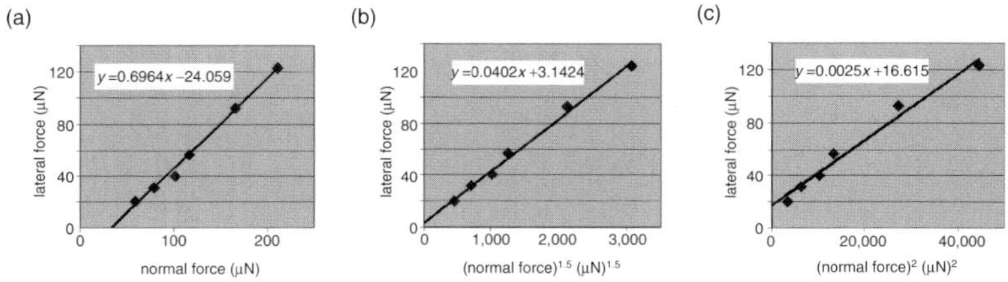

Fig. 4.10. Thiohydantoin (**1**) on (110) at the 180° orientation: plots of the lateral force vs. normal force upon edge-in-front scratching with the cube corner; the trial plots (**a**) with exponent 1 and (**c**) with exponents 2 do not cut at the origin; only plot (**b**) with exponent 1.5 is valid

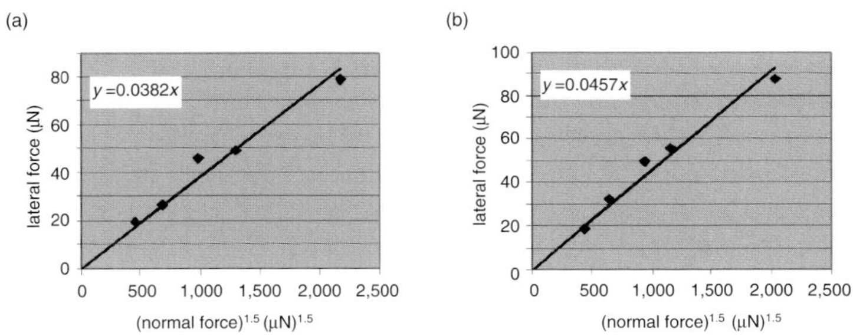

Fig. 4.11. Thiohydantoin (**1**) on (110) at the 270° orientation (migration to the left): comparison of (**a**) edge-in-front and (**b**) side-in-front scratching with the cube corner securing the 1.5 exponential relationship in both cases (**a, b**) but with slightly different slopes

obey (4.1) upon scratching on different faces in different directions and both in pyramidal edge- or face-in-front orientation, as evidenced in Figs. 4.10–4.12.

Thiohydantoin (**1**) is a flat molecule with a layered crystal structure that has all hydrogen bonds within the double layers. It was scratched edge-in-front on a (110) surface at the 180° orientation giving abrasion (0° with respect to

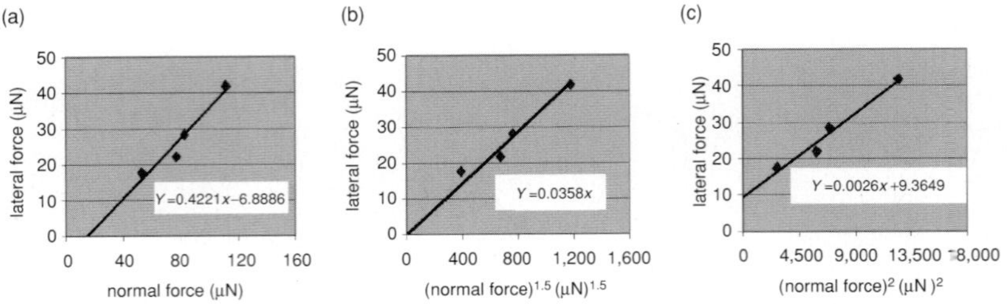

Fig. 4.12. Thiohydantoin (**1**) on (10–2) parallel to the long crystal edge upon cube corner side-in-front scratching: plots of the lateral force vs. normal force with the exponents 1 (**a**), 1.5 (**b**) and 2 (**c**)

the long crystal axes gives molecular migrations to both sides; Sect. 4.5.7). All three plots of Fig. 4.10 with the exponents 1, 1.5, and 2 seem to be close to linear; however, only the 1.5 exponential relationship has its intercept at the origin and is thus valid (Fig. 4.10b). The straight lines in the trial plots (Figs. 4.10a,c) are to be qualified as tangents. The slope of 0.0402 $\mu N^{-0.5}$ (K-value) is characteristic to this material at this face and scratch direction, representing the response to the mechanical abrasive plowing with the cube corner indenter.

As the anisotropy is very pronounced with thiohydantoin (**1**) it is of interest to explore the validity of the 1.5 exponential relationship also for the molecular migration case (Sect. 4.5.7). Figure 4.11 shows the results at the 270° orientation (when molecules migrate only to the left side). Again the 1.5 exponential plots (a,b) hit at the origin. It does not matter for the linearity if the edge (a) or the side of the cube corner (b) is in front; however the scratch coefficient (slope) is considerably different. The constant K of 0.0382 $\mu N^{-0.5}$ from the edge-in-front scratching (a) (Fig. 4.11a) is smaller (by 5%) than in the abrasion case of Fig. 4.10b. This is significant and may indicate differences in the scratch mechanism at the different angles.

The results with the very different indenter orientations indicate again that the indenter geometry is not the reason for the exponential law of (4.1). Clearly, the 1.5 exponential relationships are also retained if the side of the indenter is in front. This finding is of highest importance, as it demonstrates only a low sensitivity of the scratch coefficient K with respect to the tip shape. The difference of the K-values in Fig. 4.11 is 16%.

The validity of the 1.5 exponential relationships at side-in-front scratching was also secured with thiohydantoin (**1**) on its (10–2) cleavage plane where the molecules are arranged flatly in infinite hydrogen bridged layers (the nanoscratching on that face results in exclusive abrasion; Sect. 4.5.7, [11, 13]). Only the 1.5 exponential plot (b) in Fig. 4.12 represents the valid relationship, because it hits the origin unlike the trial plots (a) or (c). The slope of Fig. 4.12b has to be compared with the side-in-front value of Fig. 4.11b, of course. The value of 0.0358 on a (10–2) surface is considerably (22%) smaller than the value of 0.0457 on a (110) surface at the 270° scratch direction. This indicates a lower resistance for the abrasion on the (10–2) surface of thiohydantoin.

It is important to secure the 1.5 exponential relationship of lateral force and normal force with further compounds. Ninhydrin (**2**) is a dipolar bicyclic compound with a three-dimensional structure and forming infinitely hydrogen-bridged double layers [23]. Figure 4.13 shows, that upon scratching only the 1.5 exponential plots (b) and (e) hit the origin, unlike the (a) or (d) and (c) or (f) plots that are based on the exponents 1 and 2, respectively. Again it does not matter if edge-in-front or side-in-front scratching is performed with the cube corner. The only variation is in the scratch coefficient. The difference in slope is again comparably small (17%). However, scratch coefficients are recommended for edge-in-front scratches, as these should be less sensitive for tip changes, if a 60° pyramid, a Berkovich, a Vickers, or a conosphere indenter should be compared with the cube corner.

Tetraphenylethene (**3**) is a bulky unpolar hydrocarbon that does not exhibit hydrogen bridges. It is of interest to note if the 1.5 exponential relationship still applies in the case of exclusive van der Waals' crystal forces. The results in Fig. 4.14 show that this is indeed the case. Only the 1.5 exponential plot (b) intercepts at the origin unlike the trial plots with the exponents 1 (a) and 2 (c). The slope reaching the value of $0.0598\,\mu N^{-0.5}$ is larger in this case when compared to the hydrogen-bridged crystals of **1** and **2**. The higher energetic effort derives from the interlocking of the bulky molecules (Fig. 4.44).

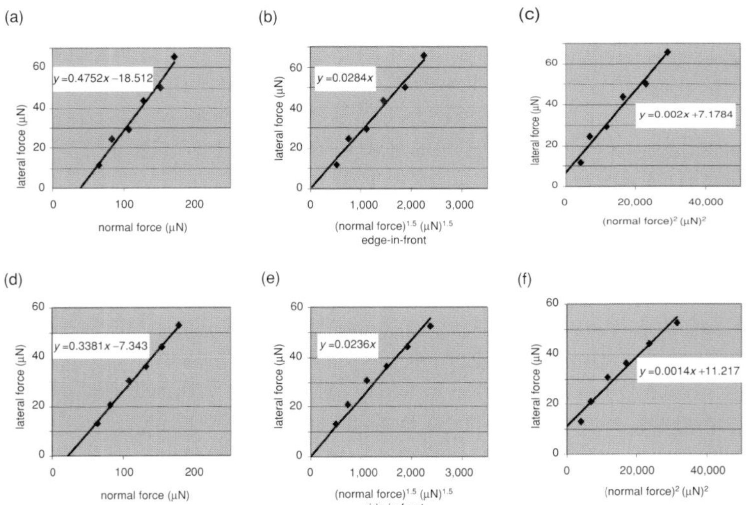

Fig. 4.13. Edge-in-front (**a**–**c**) or side-in-front (**d**–**f**) cube corner scratching of ninhydrin (**2**) on an (1–10) surface along the c-axis: plot of the lateral force vs. (normal force)^n with the exponents 1 (**a,d**), 1.5 (**b,e**) and 2 (**c,f**), showing the 1.5 exponential power law obeyed in both scratch modes

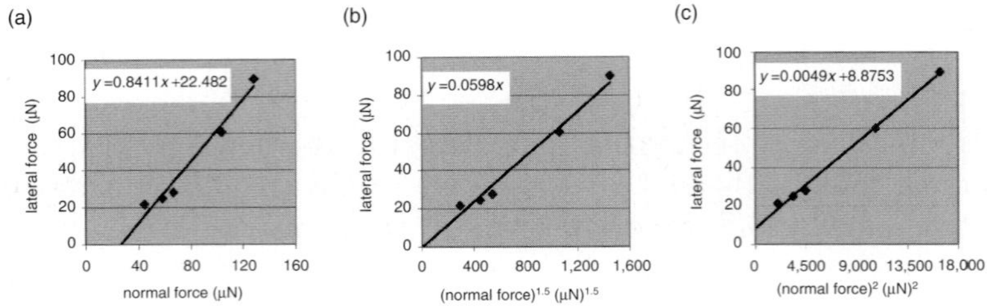

Fig. 4.14. Tetraphenylethene (**3**) on (10–1): plots of the lateral force vs. normal force in the side-in-front cube corner scratching along the longest of the crystal edges; (**b**) with the exponent 1.5; (**a**) with 1, and (**c**) with 2; the trial plots (**a**) or (**c**) with the exponents 1 or 2, respectively, do not hit the origin

Nanoscratching on diverse materials (oxides, salts, covalent semimetals, molecular crystals, polymers) consistently reveals proportionality of (lateral force) $\sim$ (normal force)$^{3/2}$. This relationship is valid for amorphous (fused quartz) and crystalline materials such as silicon, silica, strontium titanate, and the polar molecular crystals of thiohydantoin and ninhydrin (**2**) with hydrogen bonds, or pure van der Waals' crystals of tetraphenylethene (**3**). It is of highest interest that the 3/2 exponential relationship does not change with the bonding situation.

4.5.2 Scratch Work Considerations and Mathematical Justification of the $F_L \sim F_N^{3/2}$ Relation

Per definition, the scratch work is given by the lateral force (F_L) and scratch length (l)

$$W_{sc} = F_L \cdot l. \tag{4.2}$$

By substitution in (4.1) it follows the obvious proportionality of scratch work and (normal force)$^{3/2}$ (4.3):

$$W_{sc} = K \cdot l \cdot F_N^{3/2}. \tag{4.3}$$

The slopes of nanoscratch work (lateral force · scratch length) vs. (normal force)$^{1.5}$ are linear and cut at the origin, as expected. This may be exemplified with the plots in Fig. 4.15b. The slope value ($K \cdot l$) has the dimension (μm · μN$^{-0.5}$). Clearly, the trial plot (a) with the exponent 1 fails. It appears that (4.3) is practical for extrapolation purposes.

Furthermore, for the comparison of different materials and for the qualification of face and angle anisotropies of a material (Sect. 4.5.6/4.5.7) it is necessary to define the specific nanoscratching work (spec W_{sc}) under specified load. For practical reasons this should be defined for 1 μm nanoscratching length. Therefore, the specific nanoscratch work is the lateral force (at uniform rate) times 1 μm as in (4.4), which contains the factor 1 μm for correct dimension.

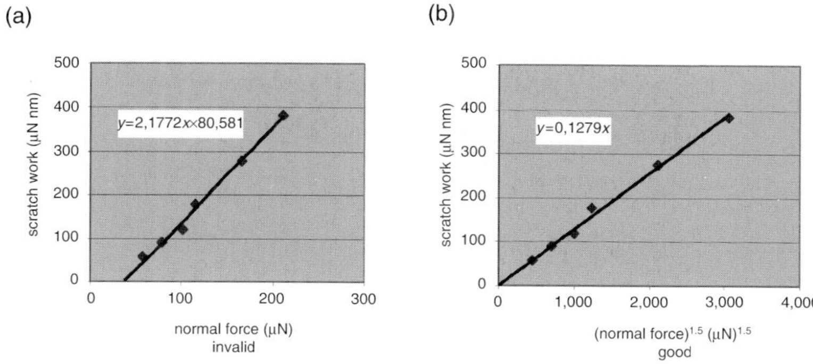

Fig. 4.15. Plots of the scratch work W_{sc} vs. force upon edge-in-front scratching on a (110) surface of thiohydantoin (**1**) at the 180° orientation (abrasion, Sect. 4.5.7); (**a**) invalid vs. normal force and (**b**) vs. (normal force)$^{1.5}$ (with cut at the origin); the scratch length was 3 μm in all experiments

$$\text{spec } W_{\text{sc}} = F_{\text{L}} \cdot 1 = K \cdot F_{\text{N}}^{3/2} \cdot 1\,(\mu\text{N }\mu\text{m}). \tag{4.4}$$

Such specific nanoscratch work or the nanoscratch coefficient $K(\mu\text{N}^{-0.5})$ (also 4.1) should be used instead of the dimensionless "friction coefficient," which was previously defined as lateral force over normal force, a ratio that cannot be constant in view of (4.1) and (4.4).

The comparison of work of nanoindentation (Sect. 3.8) and the nanoscratch work may serve as a theoretical rationalization of the $F_{\text{L}} \sim F_{\text{N}}^{3/2}$ relation on the basis of previous theories for micromechanics if these are assumed to be applicable to the nanotechniques. The indentation work W_{N} and the scratch work W_{sc} at the same penetration depth h are expressed and compared. W_{N} can be written as the product of mean pressure p_{m} and volume $V = A\ h/3$ (pyramids and cones), where A is the projected area. W_{sc} as the product of lateral force (F_{L}) and scratch length (l) is taken at the same constant normal penetration depth (h). For a conical or pyramidal indenter with effective half-cone angle α (Berkovich: 70.2996°; cube corner: 42.28°; 60° pyramid: 24.45°), the average pressure may be taken from the formula of Stilwell and Tabor as $p_{\text{m}} = E_{\text{r}} \cot \alpha/2$, where E_{r} is the reduced elasticity modulus [24]. This and the definition of hardness $H = F_{\text{N}}/A$ leads to (4.5) for the work of indentation

$$W_{\text{N}} = p_{\text{m}} V = (E_{\text{r}} \cot \alpha/2) \cdot (Ah/3) = (E_{\text{r}} \cot \alpha/6) \cdot (F_{\text{N}} h/H). \tag{4.5}$$

The standard formula for E_r (Sect. 3.4) is given in (4.6), where $\text{d}F_{\text{N}}/\text{d}h$ is usually taken as the unloading stiffness of the indent:

$$E_r = (\sqrt{\pi}\sqrt{H}/2\sqrt{F_{\text{N}}}) \cdot \text{d}F_{\text{N}}/\text{d}h. \tag{4.6}$$

Integration for F_{N} from 0 to F_{N} leads to an expression for h as in (4.7)

$$h = (\sqrt{\pi}\sqrt{H}/2E_{\text{r}}) \int F_{\text{N}}^{-1/2} \text{d}F_{\text{N}} = (\sqrt{\pi}\sqrt{H}/E_{\text{r}}) \cdot \sqrt{F_{\text{N}}}. \tag{4.7}$$

If h is substituted in (4.5) one obtains (4.8)

$$W_{\mathrm{N}} = (\cot \alpha \sqrt{\pi}/6\sqrt{H}) \cdot F_{\mathrm{N}}^{3/2}. \tag{4.8}$$

Division of W_{N} (4.8) through W_{sc} (4.2) (both at the penetration depth h) gives the proportionality of F_{L} and $F_{\mathrm{N}}^{3/2}$ in the form of:

$$(W_{\mathrm{N}} l) \cdot \mathbf{F}_{\mathrm{L}} = (W_{\mathrm{sc}} \cot \alpha \sqrt{\pi}/6\sqrt{H}) \cdot \mathbf{F}_{\mathrm{N}}^{3/2}. \tag{4.9}$$

Thus, the $F_{\mathrm{L}} \sim F_{\mathrm{N}}^{3/2}$ relation is inherent in the standard formula. The analytical precision of (4.8) and (4.9) is certainly poor in view of the many assumptions in the basic expressions used and the unfortunate iterations of terms with respect to the standard fused quartz (Sect. 3.4). Furthermore, the present deduction includes the assumption that no phase transformation occurs because this would produce different responses in indentation and scratching. As pressure-induced phase transitions seem to be more frequent than previously expected (Sect. 4.5.3), (4.8) and (4.9) are not generally usable for the absolute determination of the nanoindentation hardness (H) from nanoscratching experiments. However, their mathematical deduction shows that the empirical $F_{\mathrm{L}} \sim F_{\mathrm{N}}^{3/2}$ relation is not contradictory to common theories and that more advanced approaches to the mean pressure and the contact area would give similar equations with more reliable parameters that would allow for absolute determinations of the nanoindentation hardness H and elasticity modulus E_{r} (without dependence on multiparameter iterations) from measurements of nanoscratching forces and works. Until then, only the indentation parameter k (μN/nm$^{3/2}$) (3.18) and the total work of the indentation $W_{\mathrm{N\ tot}}$(μN μm) (3.26) provide a quantitative description of the nanoindentation experiment (Sect. 3).

4.5.3 Phase Transitions

The medium scratching pressure that builds up (p_{sc}) is given by the lateral force (F_{L}) divided by the cross-section (Q)

$$p_{\mathrm{sc}} = F_{\mathrm{L}}/Q. \tag{4.10}$$

Such pressure spreads in forward directions depending on the crystal structure and anisotropy. Slide or shear bands, or work hardening or phase transition ensue under the influence of very high pressure and are followed by failure. Usually, the pressures reach values in the multi GPa range and are more guided in the straight direction at face-in-front rather than edge-in-front scratching. Also the attack angle influences the actual pressure as the abrasion is more impeded if the tip face is more inclined. Therefore, phase transformations may easily occur, and wearing of the diamond tip by hard materials may also occur. Notwithstanding, the $F_{\mathrm{L}} \sim F_{\mathrm{N}}^{3/2}$ relation proved valid for all kinds of materials in nanoscratching irrespective of the very different events and mechanisms (Sect. 4.5.1). But if a phase transition occurs under load (see also Chap. 3.7) there must be a response in the value of the scratch coefficient K because the phase-transformed material is different from the original. For example fused quartz undergoes an amorphous-to-amorphous phase transition [25]. It is not seen in Fig. 4.6b at relatively low F_{N} values, but a kink occurs above that loads and a second linear region with different slope starts (also for the cube corner, not shown) due to phase transition. This effect can only be seen in nanoscratching but not in microscratching. Such kinks are depicted for nanoscratches with the 60° pyramid in Fig. 4.16b and with the Berkovich tip in Fig. 4.16c. A simple 1 by 1 proportionality is again excluded, as the plot in Fig. 4.16a (60° tip, similarly for the Berkovich) is curved and does not hit at the origin. The evident dependencies of the various K-values (slopes) on the phase and on the indenter geometry are clearly seen.

Cubic strontium titanate ($Pm3m$) on polished (100) is a good representative for a crystalline salt, which undergoes a pressure-induced phase transition (perhaps into the tetragonal low-temperature modification). This is most easily seen in Fig. 4.17b with a very pronounced kink and very different characteristic scratch coefficients (slopes) K. The intercept of the first part in Fig. 4.17b is close to zero unlike the curves with exponent 1 (Fig. 4.17a) or 2 (not shown). The slight deviation from zero may derive from surface hydration of the salt. This deviation was found larger (but still small) in independent experiments that were undertaken with the same sample about one year later and may reflect the influence of ambient humidity. As there was no visible cracking up to 1,500 μN load, the kink secures the phase transition, which changes the response against the nanoplowing by virtue of a different slope. Actually, further crystal modifications are known for strontium titanate [26,27]. The measured slope Δ (lateral force) / Δ {(normal force)$^{3/2}$}(μN$^{-0.5}$) is the constant of the material phase for the cube corner indenter with the edge-in-front, when the scratch is parallel to the edge for the crystal packing on the (100) surface.

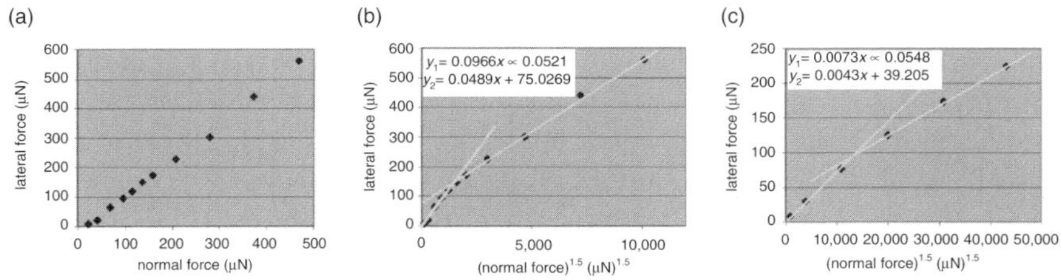

Fig. 4.16. Nanoscratches on fused quartz: (**a**) invalid trial $F_{\mathrm{L}} - F_{\mathrm{N}}$ plot (60° tip; similarly with Berkovich tip) as it does not hit at the origin; (**b**, **c**) $F_{\mathrm{L}} - F_{\mathrm{N}}^{1.5}$ plots exhibiting kinks by intersection of two linear ranges (different slope values K), which are due to the amorphous-to-amorphous phase transformation with 60° tip (**b**), and Berkovich tip (**c**), although the early kink in (b) might point to a different process

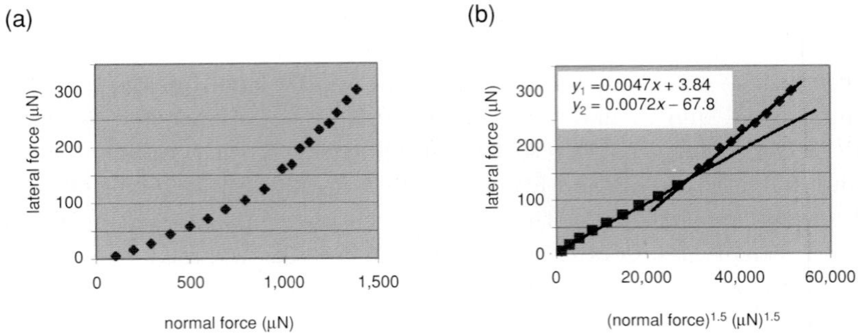

Fig. 4.17. Edge-in-front cube corner scratching parallel to the crystal edge of $SrTiO_3$ (100); (**a**) curved trial $F_L - F_N$ plot, (**b**) linear $F_L - F_N^{1.5}$ plot; the kink in (**b**) is the result of the phase transition under pressure, the scratch coefficients (slopes) are indicated

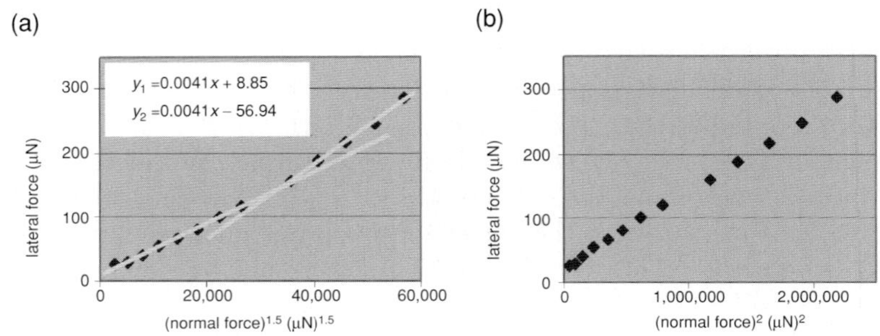

Fig. 4.18. Silicon wafer (100) (with 10 nm silica cover) upon scratching with a cube corner indenter (edge-in-front) parallel to the cleavage direction: (**a**) plot of $F_L - F_N^{1.5}$ indicating the phase transition by the kink, while the positive cut at the y-axis is due to the silica cover; (**b**) square invalid trial plot that would not indicate the phase transition

Other directions and other faces of the same material or other indenters give different values.

Silicon forms an infinite covalent lattice. It assumes a silica layer in ambient atmosphere that will cause slight deviations at the beginning of the correct nanoscratching plot. Again, linear or square (Fig. 4.18b) relationships between lateral and normal force do not hold. Figure 4.18a indicates two straight lines intersecting at a kink in the correct $F_L - F_N^{1.5}$ plot. Once more a marked phase transformation is disclosed by nanoscratching (probably diamond cubic Si into metallic superconducting tetragonal β-tin Si or to the even denser packed phases of Si).

Pressure-induced phase transitions of silicon found much interest and are long known. At 12 GPa hydrostatic pressure it forms the tetragonal β-tin polymorph, at 13–16 GPa a simple hexagonal phase, at 36 GPa a hexagonal closest packed phase, at 78 GPa a face centered cubic face that does not change up to 248 GPa; after pressure relief the body-centered cubic metastable phase is obtained [28]. Several intermediate phases have also been described [29]. Recently, the transition to the β-tin Si has been indirectly claimed by TEM and micro Raman spectroscopy from nanoindentation impressions after unloading. Instead of the β-tin Si the occurrence of metastable phases (rhombohedral Si, body-centered cubic Si, and amorphous Si) were deduced from Raman lines and it was concluded that these formed by transformations of intermediate β-tin Si (Chap. 22.3.5 in [5]), [30]. The difficulties in the construction of a microindentation device for in situ Raman microspectroscopy both for transparent and non-transparent materials are well known [31]. Clearly, the only possibility of focusing light to the nontransparent squeezed out silicon could not detect the high-pressure β-tin Si (or more dense metallic Si) phase, because the pressure is already released there even if it is ductile.

In situ measurements would be important. Clearly, nanoscratching alone does only detect the occurrence of phase transformations and it locates the phase-transformed material in front of the tip for possible in situ or (easier) ex situ investigation with a well-oriented pressure gradient. The most promising way to explore the actual phases at different distances from the tip will be scanning diffractometry with microfocus synchrotron radiation [32, 33]. Beam sizes have now reached the submicron scale reaching routinely the 100-nm width. Such measurements are being pursued and they promise to identify different phases along the pressure gradient with high resolution [34].

Ductile metals can be nanoscratched without chipping. The best and reproducible results are obtained if a blunt cube corner ("curvature" radius: 200 nm) is used, as the effective cone angle is most favorable and the penetration depth is suitable in the nanoscratching range. This is clearly demonstrated in Fig. 4.19 by smooth scratches (3 μm in 60 s) of gold, silver, and copper at 250 μN normal load [22]. The appearance of all three results is very similar. There are no irregularities by cracks or ruptures and the volumes of the pileups are found larger than the ditch in all three cases. This seems to exclude loss of material by chipping. Surface roughness is not critical.

Gold, silver, and copper crystallize in cubic face-centered lattices ($Fm3m$), and phase transformations are frequent for such packing. In fact the quantitative treatment gives similar results for the three ductile metals. In none of the three cases are the straight lines in the F_L - F_N plots valid, because they do not cut at the origin (Figs. 4.20a, 4.21a, 4.22a). Neither are square relations valid between F_L and F_N^2, as there is strong curvature in Figs. 4.20c, 4.21c, 4.22c. Again the 3/2 exponential relations are valid and the corresponding plots disclose the phase transitions by very pronounced kinks.

Yellow gold of high purity has the sharp kink at 740 μN normal load in the 1.5 exponential plot, when nanoscratched with the blunt cube corner (Fig. 4.20b) [22]. This points to the known face-centered cubic into hexagonal close-packed or body-centered cubic phase transformation [35] much more distinctly than by nanoindentation (cf. Sect. 3.7).

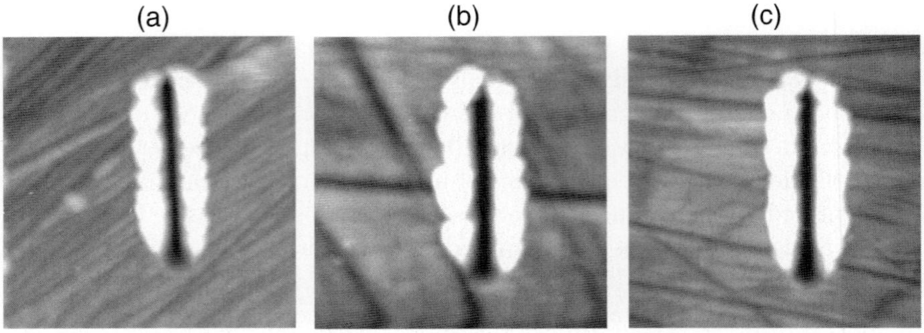

Fig. 4.19. AFM topologies (5 μm, z-scale: 400 nm) of nanoscratches with a blunt cube corner at 250 μN normal load on (**a**) gold with 60 nm depth, (**b**) silver with 80 nm depth, and (**c**) copper with 70 nm depth

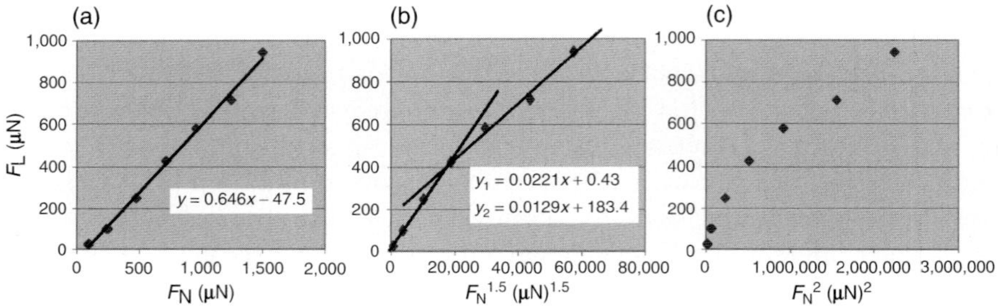

Fig. 4.20. *Gold*: cube corner edge-in-front nanoscratching; (**a**) trial F_L - F_N plot that is invalid as it does not cut at the origin; (**b**) valid F_L - $F_\mathrm{N}^{1.5}$ plot, the kink is the result of the phase transition under pressure, the scratch coefficients (slopes) are indicated; (**c**) F_L - F_N^2 trial plot that is strongly curved

Silver exhibits the kink of an analogous phase transition with the blunt cube corner in the 1.5 exponential plot at 620 μN normal load (Fig. 4.21b) [22], but the scratch coefficients are very similar to the values of gold. A hexagonal densely packed silver polymorph has been described in a mineral from northeastern USSR [36], and this is certainly a good candidate for the high-pressure form. Furthermore, a phase transition of cubic silver at 19 GPa has been described [37].

Copper has its kink of an analogous phase transition upon nanoscratching with the blunt cube corner indenter at 740 μN normal load (Fig. 4.22b) [22]. The scratch coefficients are slightly larger than with gold and silver. The pressure influence to copper in the range of 0–30 GPa has been described in [38]. Furthermore, phase transformation of copper and silver through shearing is known from [39].

The occurrence of phase transitions is most easily detected by quantitative nanoscratching, but only when using the experimentally secured $F_\mathrm{L} \sim F_\mathrm{N}^{1.5}$ re-

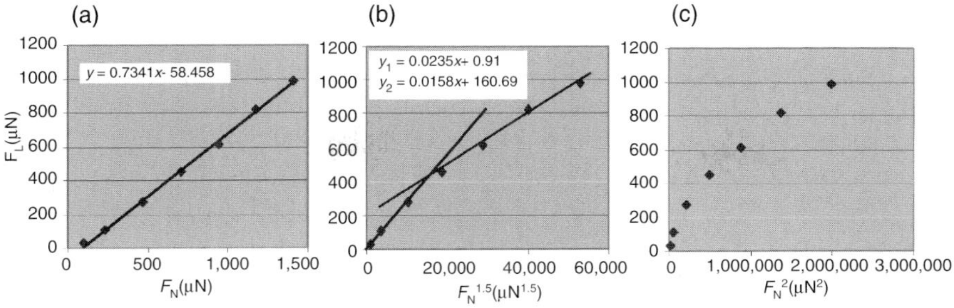

Fig. 4.21. Silver: cube corner edge-in-front nanoscratching; (**a**) trial $F_L - F_N$ plot that is invalid as it does not cut at the origin; (**b**) valid F_L - $F_N^{1.5}$ plot, the kink is the result of the phase transition under pressure, the scratch coefficients (slopes) are indicated; (**c**) F_L - F_N^2 trial plot that is strongly curved

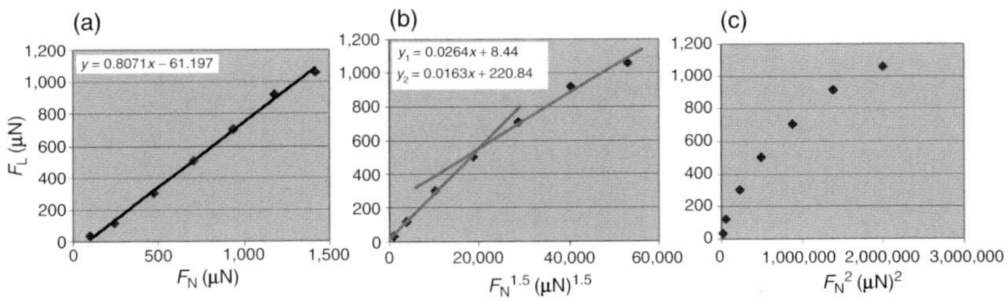

Fig. 4.22. *Copper*: cube corner edge-in-front nanoscratching; (**a**) trial $F_L - F_N$ plot that is invalid as it does not cut at the origin; (**b**) valid $F_L - F_N^{1.5}$ plot, the kink is the result of the phase transition under pressure, the scratch coefficients (slopes) are indicated; (**c**) F_L - F_N^2 trial plot that is strongly curved

lation and the corresponding plots. The new quantitative law (4.1) is therefore also valid for metals.

A practical application of the results concerns nanoscratching with the blunt cube corner on a One Eurocent coin that was previously finely polished. It was interesting to see if the coin has better mechanical properties than copper. The results are given in Fig. 4.23 [22]. Again, only the 1.5 exponential plot is valid. The kink in Fig. 4.23b occurs at 710 μN normal load. It is clearly seen that the scratch coefficients correspond closely with pure copper, so there is no improvement with respect to bulk copper. The penetration depth of the scratch was 60 nm at 250 μN and 200 nm at 1.5 mN normal load. The deeper lying alloy does not show up in the results, probably because it is less ductile than the copper coating.

An interesting field of research will stay the investigation of phase transitions of alloys and many other solid materials with nanoscratching. For example, a trigonal silver–zinc alloy can be transformed into the cubic CsCl structure at approximately 3 GPa with a finite volume discontinuity [40]. This

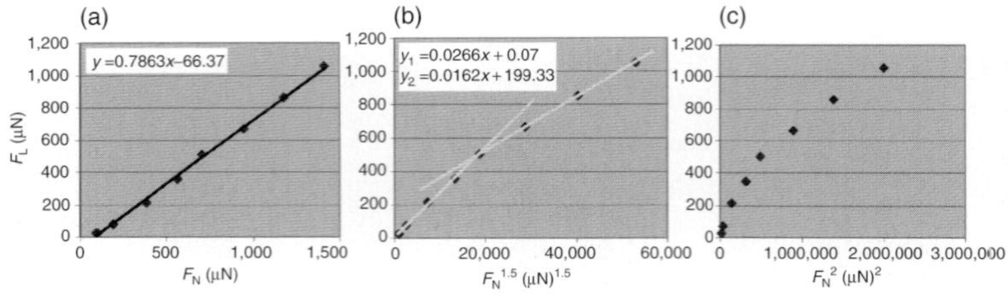

Fig. 4.23. *Polished One Eurocent coin*: cube corner edge-in-front nanoscratching; (**a**) trial F_L - F_N plot that is invalid as it does not cut at the origin; (**b**) valid F_L - $F_N^{1.5}$ plot, the kink is the result of the phase transition under pressure, the scratch coefficients (slopes) are indicated; (**c**) F_L - F_N^2 trial plot that is strongly curved

phase transition should easily show up by nanoscratching as well. These applications appear very promising and are most easily performed.

4.5.4 The Relation of Lateral Force and Normal Displacement: Consistency of the New 3/2 Exponential Relationships in Nanoscratching and Nanoindentation

The quantitative relationship of (normal force) $\sim$ (normal displacement)$^{3/2}$ in nanoindentations has been treated in Chap. 3, Sect. 3.6. This new law can be combined with the quantitative relationship of (lateral force) $\sim$ (normal force)$^{3/2}$ in nanoscratching. Hence the relation (lateral force) $\sim$ (normal displacement)$^{9/4}$ is obtained by substitution. Therefore, a simple test can secure both laws and the quality of the experimental data. Figure 4.24 shows indeed that the exponent 2.25 at normal displacement leads to the expected straight lines, which intercept at the origin within experimental accuracy. All the different nanoscratched materials of this Chap. 4 withstand the test. Figure 4.24 shows it for fused quartz, strontium titanate, silicon, and the organic crystals **1**–**3**. The test is also positive for the side-in-front scratches. This indicates again that the new empirical laws are valid for all materials and it stresses the reliability of the data. Furthermore, it becomes apparent from these linear plots if a suspicious single result in the series requires a control measurement (for example the fourth point in Fig. 4.24d).

4.5.5 Scratch Resistance

Scratch resistance was initially defined with respect to a critical load in ramp experiments (continuous increase of the load during constant rate lateral translation of the indenter) above which a permanent surface change such as plowing abrasion of material or detachment of coatings according to various failure mechanisms occurs. However, scratch resistance should also be defined for the abrasive process itself. A first approach in that directions aimed to

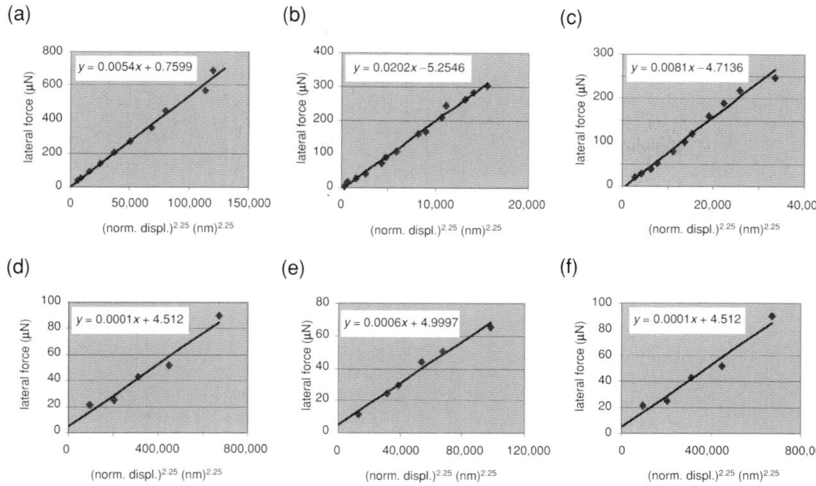

Fig. 4.24. Lateral force vs. (normal displacement)$^{9/4}$ relationship in the edge-in-front scratches of (**a**) fused quartz, (**b**) $SrTiO_3$, (**c**) silicon, (**d**) thiohydantoin (**1**), (**e**) ninhydrin (**2**) and (**f**) tetraphenylethene (**3**), providing proof for consistency of the found relationships of nanoindentation and nanoscratching and reliability of the experimental data

improve the understanding of frictional abrasive wear because Mohs hardness and other types of hardness are not fully adequate for judging macroscopic wear. Actually, no universal relations of hardness with the observed wear and its anisotropy exist. That the Mohs standards did "not exhibit significant anisotropies" was termed favorable and it was therefore tried to apply the new technique of nanoscratching to them though without taking care of anisotropies [12]. This approach defined residual scratch resistance ($R_{\mathrm{sc\ res}}$) as the scratch work (lateral work times scratch length) over the residual excavated volume (V_{res}) by the nanoscratching (4.11).

$$R_{\mathrm{sc\ res}} = F_{\mathrm{L}} l / V_{\mathrm{res}}. \tag{4.11}$$

A much better relation to the wear was demonstrated than with the Mohs scale [12]. The Mohs-scale materials revealed that the $R_{\mathrm{Sc\ res}}$ values differed significantly even at comparable nanoindentation hardness H. Unfortunately, the values tabulated for the scratch resistance depended on the load. The drop of $R_{\mathrm{sc\ res}}$ with increasing load was interpreted as a "size effect." Furthermore, a decrease of "friction forces" as the scratches become deeper was invoked as a special effect [12]. However, such "size effects" are the result of a nonlinearity between F_N and F_{L} as described in Sect. 4.5.1. A further disadvantage of $R_{\mathrm{sc\ res}}$ is the necessity to assess the precise volume excavated, which is difficult to obtain. AFM measurements with the scratching tip are highly useful but they cannot trace all of the excavated volume due to tip–sample convolution.

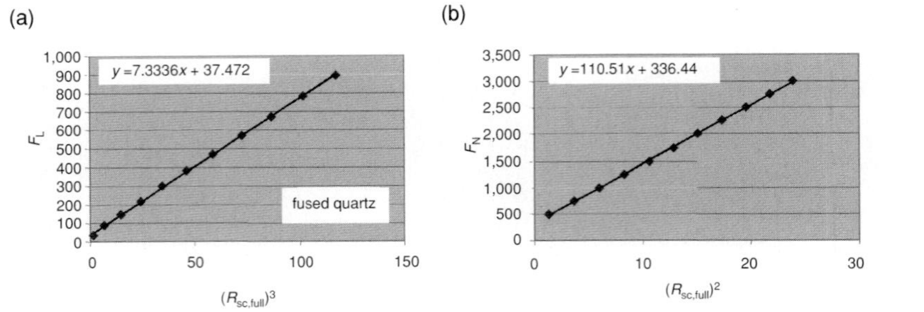

Fig. 4.25. Linear plot of (**a**) lateral force vs. $R^3_{\text{sc full}}$ and (**b**) normal force vs. $R^2_{\text{sc full}}$ from cube corner nanoscratching on fused quartz

Also the tedious search for the ditch immediately after the plowing with a sharper AFM tip does not help as dust particles are produced by the abrasive scratching that will be filled in by the scanning.

The measurement of the volume is avoided if the full scratch resistance under load ($R_{\text{sc full}}$) is defined as a new parameter. It is highly rewarding that such endeavor leads to further useful relations, including the extrapolation to different loads what may give further insights to the effects responsible for wear as well. The abrasive part $R_{\text{sc res}}$ (residual volume) is larger than the full scratch resistance $R_{\text{sc full}}$. The latter is defined by

$$R_{\text{sc full}} = F_{\text{L}}\text{l}/V_{\text{full}} = F_{\text{L}}/Q \tag{4.12}$$

The full scratch volume is equal to the indenter tip cross section (Q) below the surface times the scratch length l. In the case of a cube corner the cross section calculates equal to the basic plane (A) over $\sqrt{3}$ and (4.13) derives by substitution if the definition of the indentation hardness (H; (3.1)) is also introduced:

$$R_{\text{sc full}} = \sqrt{3}F_{\text{L}}/A = \sqrt{3}HF_{\text{L}}/F_{\text{N}}. \tag{4.13}$$

A correction considering the tip end can be neglected. By substitution with the well-founded (4.1) one obtains (4.14) and (4.15) when all the well-known constant factors are combined into one constant each:

$$F_{\text{L}} = \text{const } R^3_{\text{sc full}} \tag{4.14}$$

$$F_{\text{N}} = \text{const' } R^2_{\text{sc full}} \tag{4.15}$$

That means, the proposed plots of $F_{\text{L}} - R^3_{\text{sc full}}$ and/or $F_{\text{N}} - R^2_{\text{sc full}}$ are linear. These relations are experimentally verified in Figs. 4.25 and 4.26 with very different materials such as fused quartz and the organic material ninhydrin (**2**) by cube corner nanoscratching.

The points on the best lines in the Figs. 4.25 and 4.26 are iterated, because of the high power. They indicate the number of measurements. Nevertheless, the extrapolative power of (4.14) and (4.15) is clearly evidenced.

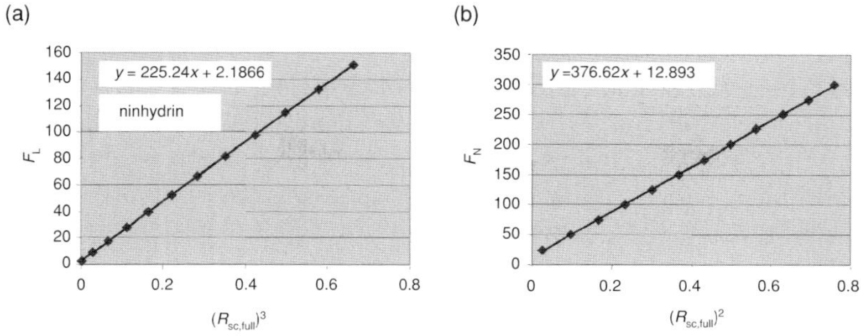

Fig. 4.26. Linear plot of (**a**) lateral force vs. $R^3_{\text{sc full}}$ and (**b**) normal force vs. $R^2_{\text{sc full}}$ from cube corner nanoscratching on ninhydrin (**2**)

Similar square and cubic dependencies of F_N and F_L, respectively, with the full scratch resistance are obtained with other materials and further indenters. The cross sections of indenters are easily obtained and the full scratch resistance ($R_{\text{sc full}}$) is therefore highly reliable. It can be used for the qualification of face and angle anisotropies in addition to the specific nanoscratch work (spec W_{sc}) according to (4.4).

4.5.6 Anisotropy in Nanoscratches

The high sensitivity of nanoscratching enables to detect and quantify anisotropies of scratch resistance ($R_{\text{sc full}}$) (4.12) and scratch work (spec W_{sc}) (4.4) on materials. The detection of anisotropic scratching with the new parameters is very important for the evaluation of the reasons for frictionally wear. It is much more precise to use $R_{\text{sc full}}$ and spec W_{sc} rather than any trials with previous residual scratch resistance ($R_{\text{s res}}$) (4.11). The latter depends too strongly on the precision of the volume measurement after the scratching, which cannot be assessed (Sect. 4.5.5). Therefore the safe parameters that avoid volume measurements should be largely preferred.

Strontium titanate ($SrTiO_3$) single crystals ($Pm3m$) have an infinite three-dimensional ionic lattice in which the titanium atoms are surrounded by six oxygen atoms in the form of octahedrons. The (100), (110), and (111) faces of $SrTiO_3$ have been studied for the demonstration of face and angle anisotropy. The projections of the crystal packing on these faces are quite different and so is the density of the ions Sr^{2+} and TiO_3^{2-} in various directions. It turns out that the measured specific scratch work (spec W_{sc}), and therefore also the proportional full scratch resistance ($R_{\text{sc full}}$), is very sensitively different on the different faces and in different scratch directions. Figure 4.27 depicts projections of the crystal packing and the measured values of the specific scratch work at three different angles on these three faces.

Under the slightly rotated (100) projection the cube structure with the strontium ion in the center can be imagined (between the titanium atoms is an oxygen atom, each). The nanoscratch work and thus also the full nanoscratch resistance values along the cube edges (at 0° and 90°) are different from the nanoscratching across the cubes at the 45° direction.

The projection on the (110) face exhibits remarkable packing differences in the 0°, 90°, and 45° directions of the nanoscratching. Therefore, nanoscratch work and thus also the full nanoscratch resistance values are found different in all chosen directions. At 0° and 90° most of the Ti–O bonds are in the scratch direction and vertical unlike the 45° direction and the lattice separation is smaller at 0° than at 90°.

In the projection on the (111) face the strontium ions are hidden behind the titanium. The nanoscratch work and thus also the full scratch resistance is considerably higher when scratching along the parallel –Ti–O–Ti–O– lines that are interconnected by the double number of oxygen bridges (0°, also 60° and 120°). The 45° and 90° scratches are off-axes.

Similar to $SrTiO_3$ there are changes of scratch resistance in crystalline α-quartz (SiO_2) as exemplified for its (1-100) face at 0°, 45°, and 90°. The crystal packing of this infinitely covalent bond crystal (*P3(1)21*) is more complicated but clearly anisotropic (Fig. 4.28). It consists of a 3D network of Si-centered tetrahedrons. All of these are linked together via their corners.

The variations in spec W_{sc} (μN μm) are not very pronounced. It is, however, seen that the *c*-direction (90°) requires the least nanoscratch work. The

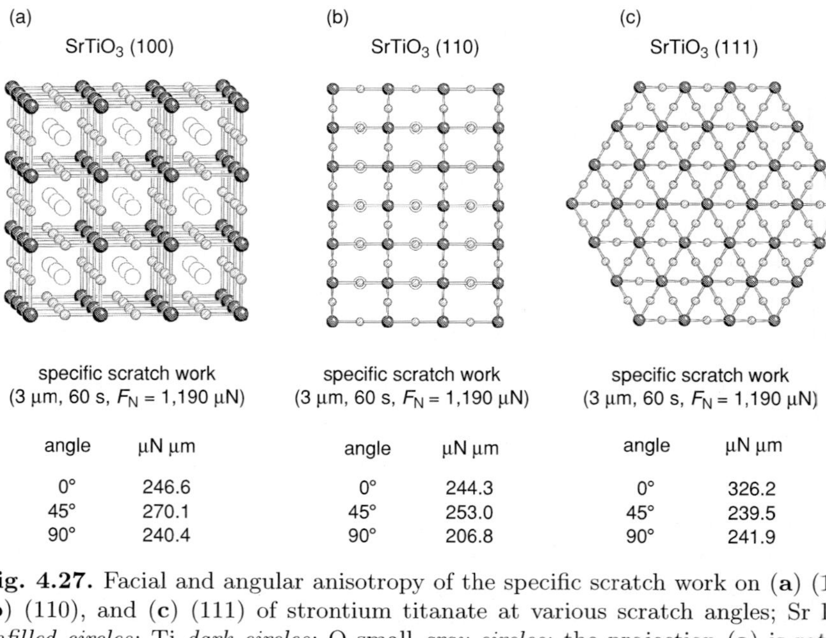

Fig. 4.27. Facial and angular anisotropy of the specific scratch work on (**a**) (100), (**b**) (110), and (**c**) (111) of strontium titanate at various scratch angles; Sr large *unfilled circles*; Ti *dark circles*; O small *gray circles*; the projection (**a**) is rotated by 5° around x and y; scratch parameters are given

spec $W_{sc} = F_L.1$ [μN μm]

Angle	μN μm (F_N = 1482 N)
90°	206
45°	223
0°	225

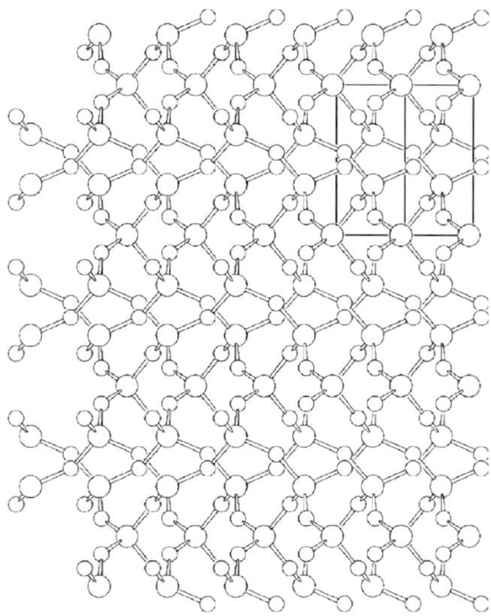

Fig. 4.28. Angular dependence of specific scratch work ($F_N = 1482\,\mu N$) on (1–100) of α-quartz upon cube corner edge-in-front scratching and the crystal packing

crystal packing in Fig. 4.28 exhibits alternating rows with very large Si–Si distances of 0.5405 nm in that direction and also the skew (10-11) cleavage plane is cutting in c-direction, which decreases the required work. On the other hand, the further directions have the atoms separated less distant on that projected face.

These fine angular anisotropies also on the four other natural main faces of α-quartz (cf. Table 3.4 in Chap. 3) [22] are not obtainable with macroscratching (Vickers, 10–100 g load) even if the critical load for microfractures was orientation dependent and marked anisotropies in the shape of the chips was observed and interpreted in terms of microfractures propagating preferentially along slip planes [41].

4.5.7 Molecular Migrations in Anisotropic Molecular Crystals

The mechanic properties of molecular crystals are just being studied [11, 42]. They behave differently from infinite covalent, ionic, or metallic crystals, as their covalent bonds cannot be broken. Only weak intermolecular bonds, including hydrogen bonds, are mechanically breakable for energetic reasons. Furthermore, organic molecules are usually far from ball shaped and crystallize in highly anisotropic crystal lattices. The possibility of long-range molecular migrations should be envisaged within molecular crystals, and these are required and occur also in their chemical reactions [43, 44]. Such molecular migrations within crystals have been unambiguously detected by atomic force and near-field optical microscopies (AFM and SNOM) [43, 44] (Chaps. 1 and

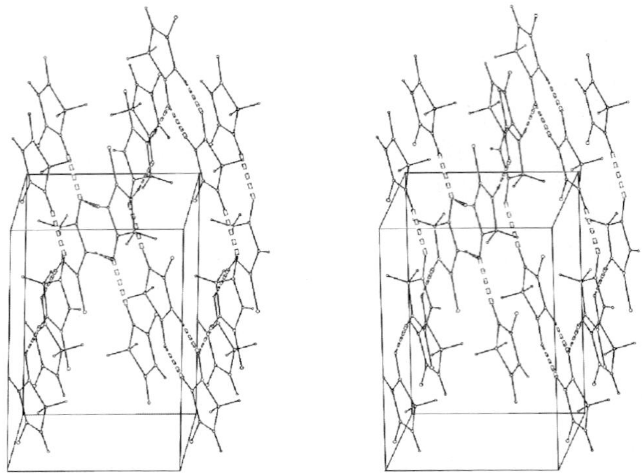

Fig. 4.29. Stereoscopic representation of the crystal packing of thiohydantoin (**1**) ($P2_1/c$) under the (110)-face in front, showing skew monolayers with cleavage planes between them; all hydrogen bonds are within the layers as shown by interrupted lines

2) and by grazing incidence X-ray diffraction (GID) [45]. These experimental results had to be additionally secured due to the generally accepted textbook denial of molecular movements within crystals for more than three decades following Schmidt's topochemistry hypothesis (strangely claiming "minimal atomic and molecular movements" in solid-state chemical reactions).

There is already clear evidence for mechanically induced molecular migrations within crystals from indentation experiments (Sect. 3.10). Furthermore, by using modern nanoscratching equipment [42] much more detailed further information is obtained. Most organic molecular crystals are highly anisotropic. Their crystal energies are small, as there are only intermolecular forces between the distinctly shaped molecules. Thiohydantoin (**1**) is an almost planar heterocyclic compound. It contains many heteroatoms and is thus dipolar and it forms hydrogen bonds. Furthermore, it crystallizes in monolayer sheets that lie in the (10-2) plane (Fig. 4.29). The monolayers are 66° skew under the natural (110) face with cleavage planes between them; all hydrogen bonds are within the layers but not connecting them [11, 42]. Such structural anisotropy must cause anisotropic nanoscratching.

The remarkable anisotropy upon nanoscratching on a (110) surface of thiohydantoin (**1**) is shown in Figs. 4.30 (ramp nanoscratching) and 4.31 (constant load nanoscratching). It is important to show that both modes lead to similar results. Every angle has its own characteristics. At 0° (along the long crystal edge) there is molecular migration to both sides and in front; the horizontal fissures indicate the direction of the cleavage planes. The constant force nanoscratching (edge in front) provides deeper scratches in the direction of the cleavage planes, as it is started after preindentation with 250 μN normal

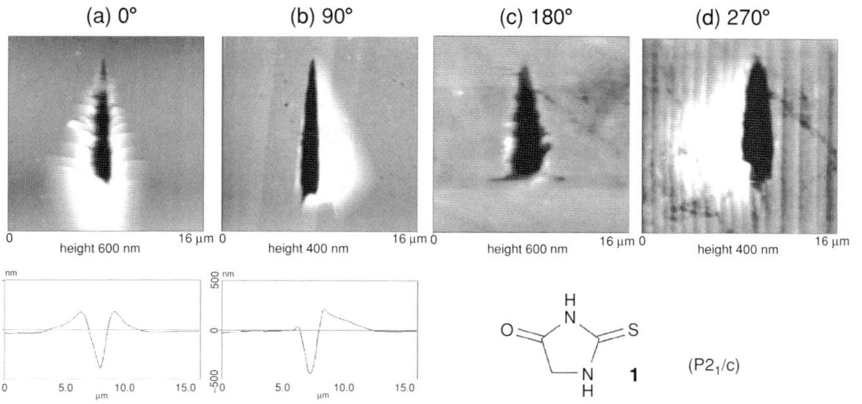

Fig. 4.30. Ramp nanoscratching (F_N up to 1 mN) of thiohydantoin on a (110) surface at 0°, 90°, 180°, 270° (by turning the crystal), showing the anisotropy of molecular migrations and exclusive abrasion in the 180° scratch [11]

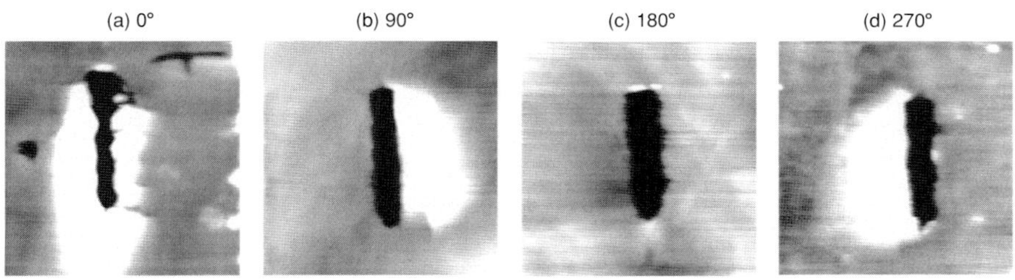

Fig. 4.31. AFM topologies (10 µm, Z-range: 200 nm) of 5 µm constant load nanoscratches in 60 s on the (110) face of thiohydantoin with a cube corner at 250 µN (for **a**,**b**, and **d**), or 200 µN (for **c**) at various directions (by turning the crystal) [42]

force. At 90° molecules migrate only to the right-hand side. Only abrasion occurs at 180°. At 270° molecules migrate only to the left-hand side. The migrated materials have about the hardness and elasticity of the initial crystal and are thus clearly distinguished by common loose pileup.

A detailed explanation of the anisotropic behavior in Figs. 4.30 and 4.31 is given in Fig. 4.32 [11, 13, 42]. The skew cleavage plane must cause the unusual piling-up of material to both sides or only at one side of the scratch. The geometric arrangement in such experiments is sketched in Fig. 4.32. Clearly, both sides and the front are open if the tip moves at 0° and it is easily comprehended that it leads to migration to both sides and in front. Conversely, the 180° direction against the skew monolayers must give abrasion. Furthermore, the left-side or right-side gaps between the sheets are blocked, as the indenter tip moves in the 90° and 270° directions, respectively [11].

As there are multiple hydrogen bonds between the thiohydantoin molecules (**1**) within the layers it cannot be claimed that only single molecules migrate.

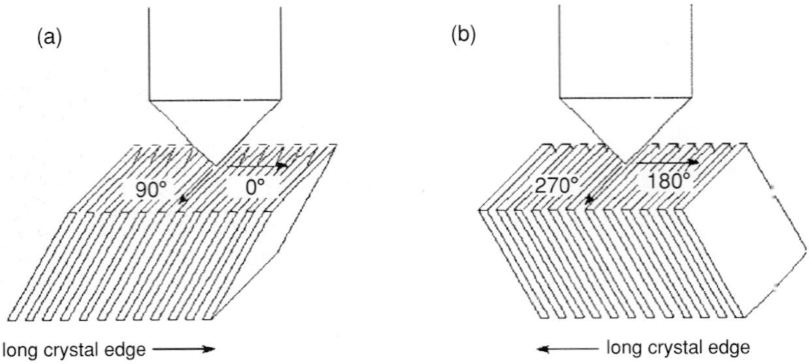

Fig. 4.32. Geometric model for the understanding of the marked anisotropies upon scratching over skew cleavage planes in four orthogonal directions of the crystal in the orientations (**a**) and (**b**)

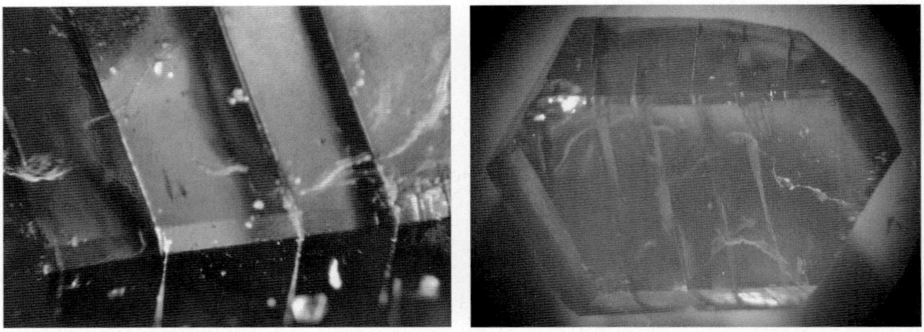

Fig. 4.33. Microscopic image of cleavages of thiohydantoin along (10–2) that are skewed under the (110) face (66°) and cut the long edge at an angle of 106° at high (microscope 100 ×) and lower magnification (32 ×) of an unusually large crystal (long edge 4.5 mm); the right-hand image has the (110) face in front [42]

Rather it appears likely that small molecular aggregates, such as almost planar hexamers that are present in the crystal structure as cyclic entities (Fig. 4.29), may be small enough for the migration. It is important to note that the different scratching mechanisms at the different angles did all obey the general (lateral force) ∼ (normal force)$^{3/2}$ relation and so does thiohydantoin (**1**) on its (10-2) cleavage plane (Sect. 4.5.1). The presence of cleavage planes, their orientation, and the ease of cleavage can also be shown with an unusually large crystal of **1**, which exhibits some cleavages (Fig. 4.33).

The crystal packing on (10–2) of thiohydantoin (**1**) is totally different from Fig. 4.29 and it is depicted in Fig. 4.34. There is no possibility for molecular migrations upon scratching in any direction on that surface but only abrasion and some shifting of fragments of the anisotropic monolayers in front of the tip that may depend on the direction.

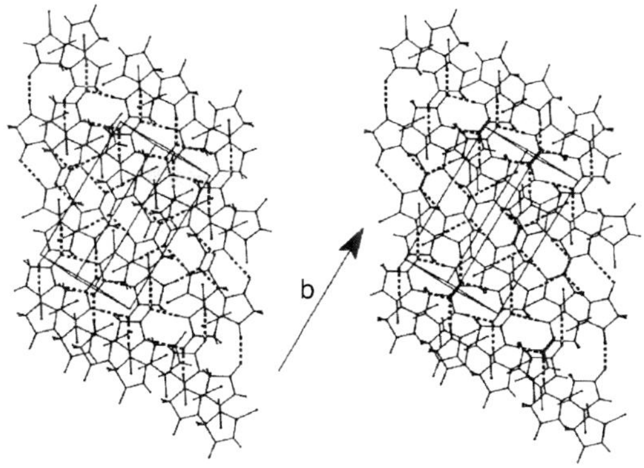

Fig. 4.34. Stereoscopic representation of the crystal packing of thiohydantoin (**1**) ($P2_1/c$) under the (10–2) face in front, showing the infinitely hydrogen-bonded flat monolayers with the flat cleavage planes between them

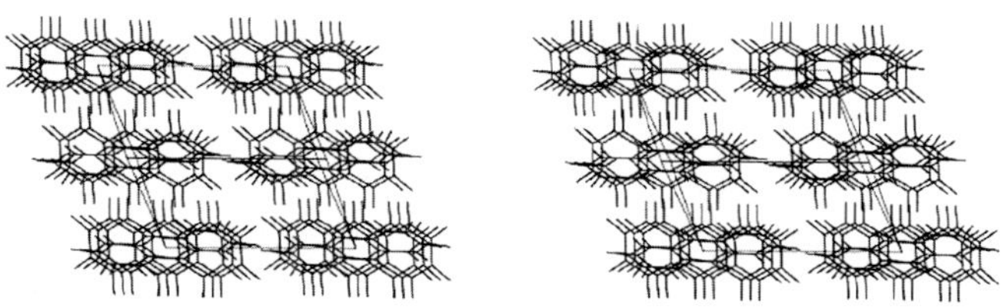

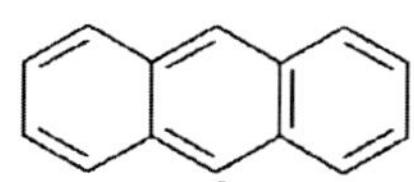

4

Fig. 4.35. Stereoscopic view of the crystal packing of anthracene (**4**) along [−110] with (110) on top, showing the skew cleavage plane direction under the surface

The expectations from Fig. 4.34 have been experimentally verified. This is shown in Figs. 4.5a,b above. The ramp scratches (0–1,000 μN) were at right angles (a) or along the long crystal edge (b) on the (10–2) cleavage face of thiohydantoin (**1**). The slight frontal pileup in Fig. 4.5a is loose debris as detected by indentation, which does not provide significant hardness values [11].

The importance of these unprecedented molecular migrations within the crystals just by application of mechanical stress requires verification also with further examples. Various principally different cases can be differentiated. Anisotropic molecular migrations are most frequently encountered with organic molecular crystals if these contain cleavage planes. Anthracene (**4**) is a planar hydrocarbon. It exhibits a skew cleavage plane direction under its (110) face, which is prominent in its prism-type crystals (Fig. 4.35). It is therefore expected that it behaves in a similar way as thiohydantoin (**1**). This is indeed the case now in the absence of hydrogen bonds, as shown in Fig. 4.36 at the depicted crystal orientations.

The volumes of migrated molecules pileup (Fig. 4.36a, c, d) and ditch, as measured with AFM, correspond closely, as do hardness and elasticity values of migrated material with that of the original (110) surface (H 0.30 GPa, E_r 11.3 GPa at 250 μN; standard technique). Conversely, the debris in front

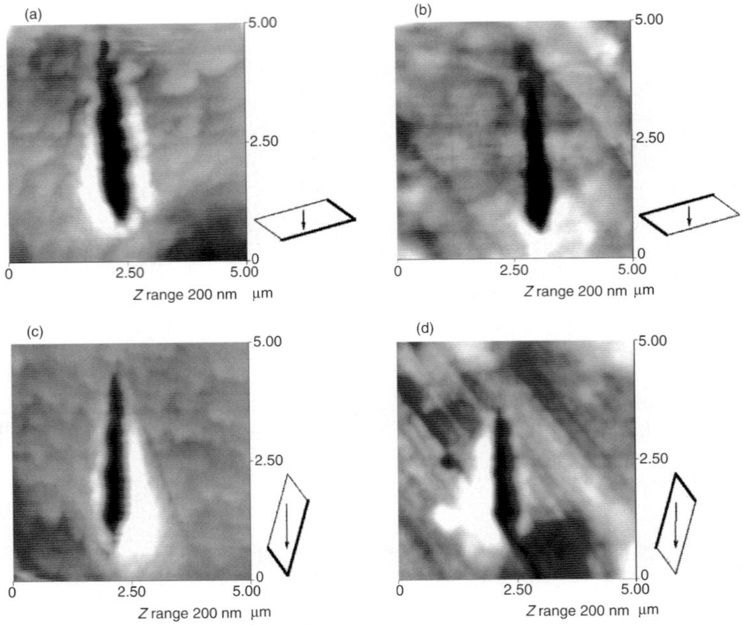

Fig. 4.36. Nanoscratching on (110) of anthracene (**4**) showing anisotropic molecular migrations (**a**,**c**,**d**) and abrasion with some debris in front (**b**), depending on the direction due to skew cleavage planes [42]

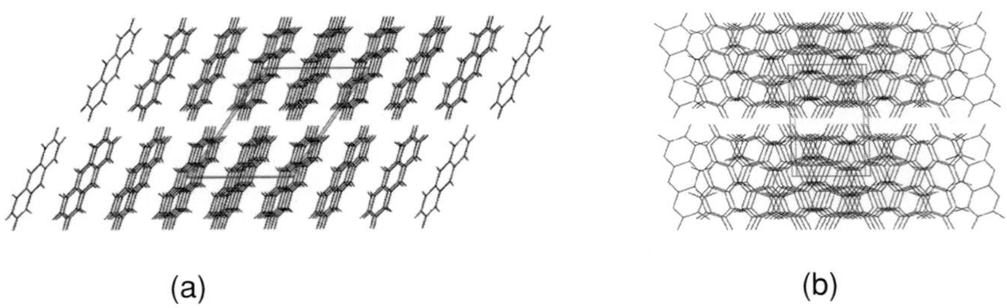

Fig. 4.37. Crystal packing of anthracene (**4**) with (001) on top: (a) along [010], (b) along [100]; both images are turned around y by 1° for better showing the inclined monolayers that are parallel under the surface

of the nanoscratch in (b) are loosely piled up. Thus, a similar analysis as in Fig. 4.32 holds true. The single organic molecules **4** experience only van der Waals interactions and are therefore not efficiently aggregated while they migrate.

In the more common scales of anthracene crystals (**4**) the (001) face is very prominent. The molecules are inclined under that face but form monolayers that are separated by the parallel cleavage planes. Figure. 4.37 depicts views from two different directions. The skew arrangement of the single molecules

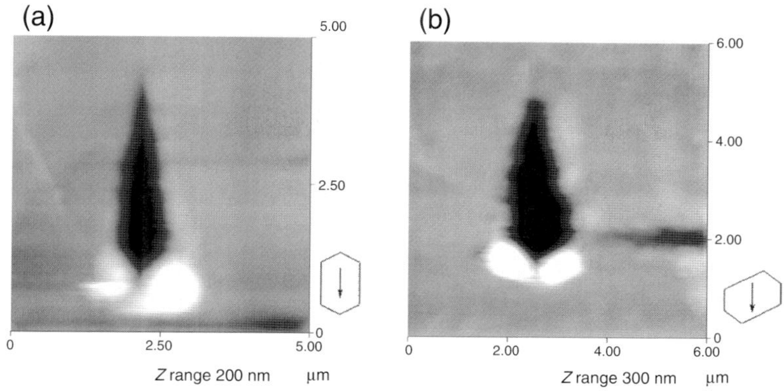

Fig. 4.38. Ramp nanoscratching (0–150 µN) on the cleavage plane of anthracene (**4**) (001) that separates monolayers, showing complete abrasion and loose pileup at the end of the scratch both in the scratch directions (**a**) and (**b**)

within the layer array does not alter the nanoscratching characteristics on these layers. Molecular migrations are impossible under that face. Therefore only abrasion and some apparent shift of layers in front of the indenter occur in any direction as shown for two in Fig. 4.38. The looseness of the debris in front was secured by indentation, which did not give reasonable hardness or elasticity values, while these values on the (001) face are measured as $H = 0.35\,\mathrm{GPa}$ and $E_\mathrm{r} = 14.8\,\mathrm{GPa}(150\,\mu\mathrm{N})$ according to the standard technique.

Thiourea crystals (**5**) on (100) offer the possibility to study planar hydrogen-bonded molecules with vertical cleavage plane arrangement. The crystal packing in Fig. 4.39 shows slightly zigzagged monolayer sheets. Most interesting are scratches along or across the direction of the sheets.

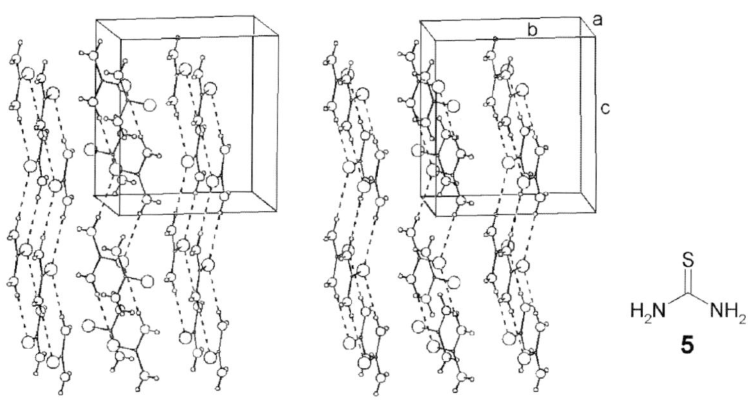

Fig. 4.39. Crystal packing of thiourea (**5**) on (100) (image rotated by 10° around x and y for a better view) showing vertical monolayer sheets separated by cleavage planes under (100) and (001) [42]

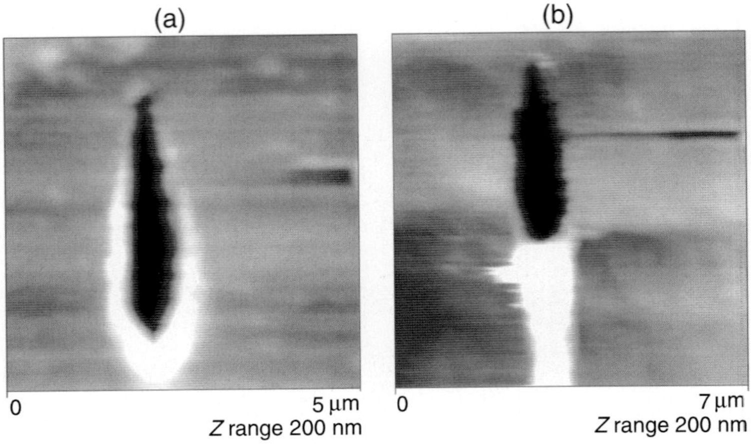

Fig. 4.40. AFM topologies on (100) of thiourea (**5**) after 4 µm cube corner scratching; (**a**) along [010], (**b**) along [001] at 0–150 µN showing in (**a**) molecular migrations to both sides and in front, and in (**b**) materials shift in front of the indenter [42]

Figure 4.40a shows molecular migrations to both sides and in front if the scratching is performed in the *b*-direction, that is crossing the layers similar to the 0° situation in Fig. 4.32. Conversely, scratching along the sheets in *c*-direction in Fig. 4.40b shifts the vertical layers in front of the tip. This is an excellent example for a slip mechanism in nanoscratching.

The unpolar cleavage plane (10–1) of α-cinnamic acid (**6**) between monolayers of hydrogen bound molecules is vertical under the most prominent (010) face along the long crystal edges (Fig. 4.41). The (010) face is another cleavage plane under (001), but the latter face is not well developed in the crystals and was therefore not investigated.

The scratching on (010) of **6** parallel to the long crystal edge (along the vertical cleavage plane direction) with the broad side of the indenter tip in front induces molecular migration in front as predicted (Fig. 4.42). Only slight effects are seen at the sides. Additionally, lots of material appear to be abraded into dust [22].

The nanoscratching information helps the interpretation of anisotropic indentation with triangular cube corner tip into α-cinnamic acid (**6**) [22] that is directly compared here. Figure 4.43 shows remarkable anisotropies evidently due to molecular migrations along the cleavage planes. These are crossed at ±60° in (a) by the two side faces of the tip while its third face is parallel orientated. Therefore, migration is to both sides. The crystal orientation angles in (b, c, d) are chosen in such a way that one of the pyramidal faces cuts the cleavage planes at right angle. Therefore, the upper side experiences more migration than the lower side in (b). Similarly, the skew migrations in

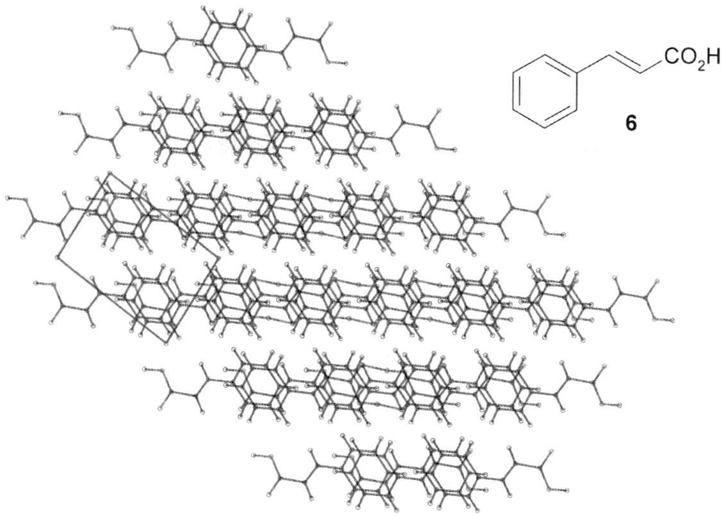

6

Fig. 4.41. Crystal packing of α-cinnamic acid (**6**) on (010) (image rotated by 10° around x and y for a better view) showing vertical monolayer sheets separated by cleavage planes under (100) and (001)

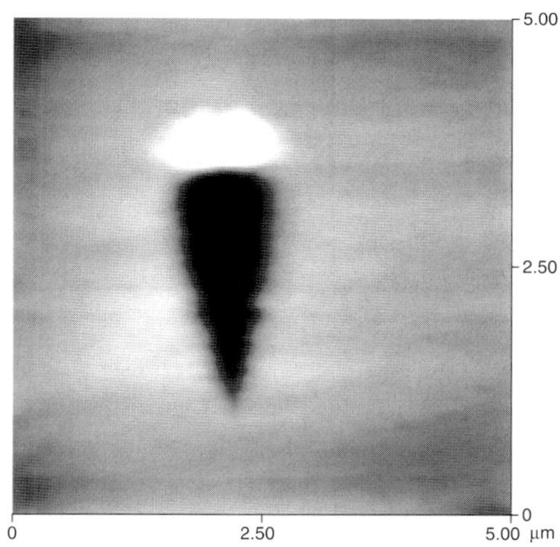

Fig. 4.42. AFM topology on (010) of α-cinnamic acid (**6**) after 4 µm cube corner ramp nanoscratching with the broad side in front; along [101] at 0–100 µN load showing molecular migrations in front of the indenter and little change to the sides [22]

(c) at 240° and (d) at 330° are more prominent to the down right and to the down left sides, respectively. The tip shape is hardly recognized and the deformations of the impression are the result of different elasticity along and across the cleavage planes (cf. Fig. 3.17).

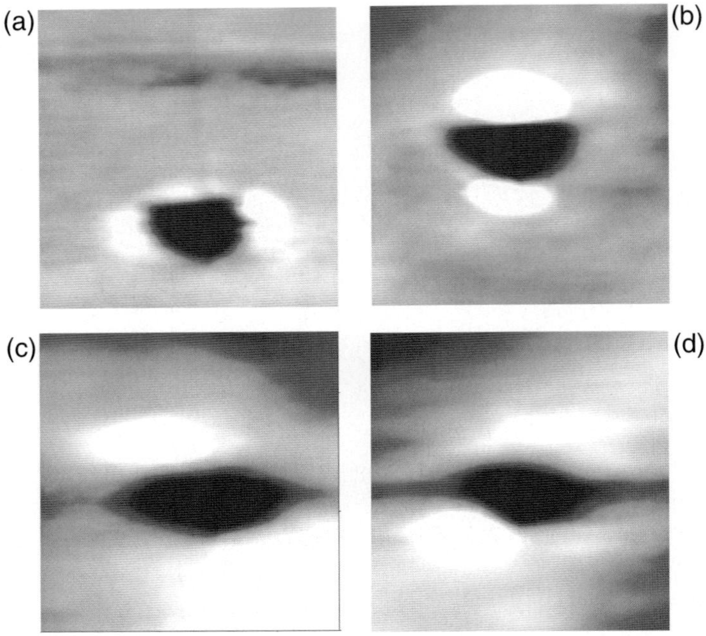

Fig. 4.43. Anisotropic molecular migrations on (010) of α-cinnamic acid (**6**) upon 200 μN load cube corner nanoindentation at different angles with respect to the triangular pyramid of the cube corner: (**a**) 0°, (**b**) 90°, (**c**) 240°, (**d**) 330° all with respect to the normal of the long crystal axis, showing the migrated material always in direction of the cleavage planes; it also shows more migration at those sides of the tip that are normal to the cleavage plane direction [22]

A very different situation is encountered if a crystal exhibits vertical monolayers of bulky 3D molecules without hydrogen bonds such as in the case of tetraphenylethene (**3**) ($P2_1$). The most prominent face is (10-1) and the plane of the layers is (010) extended along the large crystal axes. The vertical cleavage planes separating the monolayers are rather poor for migrations of molecules as there is apparent interlocking (Fig. 4.44).

Due to the considerable interlocking it is not possible to shift material in front of the tip when scratching along the layer direction (0°, Fig. 4.45a). Also the 90° scratching at right angles to planes direction does not push the molecules in one of the cleavage plane directions for upward migration and only abrasion is observed (Fig. 4.45b). However, if the crystal is turned under the scratch direction as in Figs. 4.45c, d molecular migrations only to the right or only to the left side are observed with considerable amounts in addition to some abrasion. Clearly, the molecules must get both a push from their sides to leave the layer and then also a push in the direction of the open cleavage plane for the anisotropic upward migration as observed. The efficiency of these pushes in competition to the abrasion is only partly successful and the anisotropic molecular migrations occur only in the ranges of 30°–80° (right),

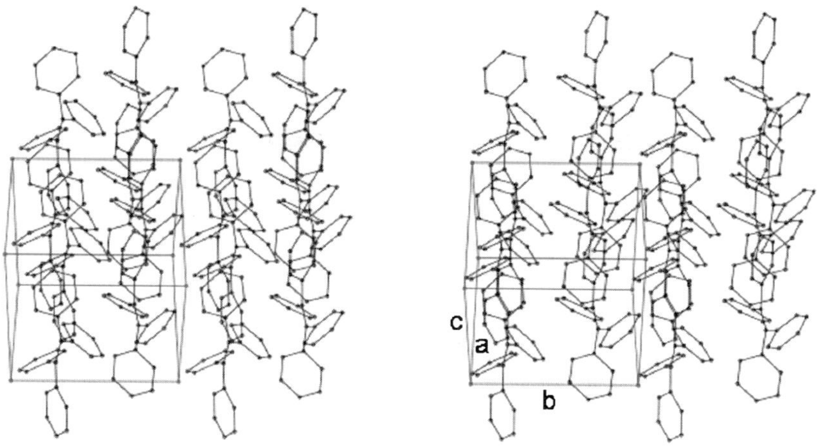

Fig. 4.44. Stereoscopic representation of the crystal packing of the hydrocarbon tetraphenylethene (**3**) on the natural (10-1) face showing poor cleavage planes between monolayers; these extend on (010) despite the bulky molecular form (hydrogens omitted for clarity [42])

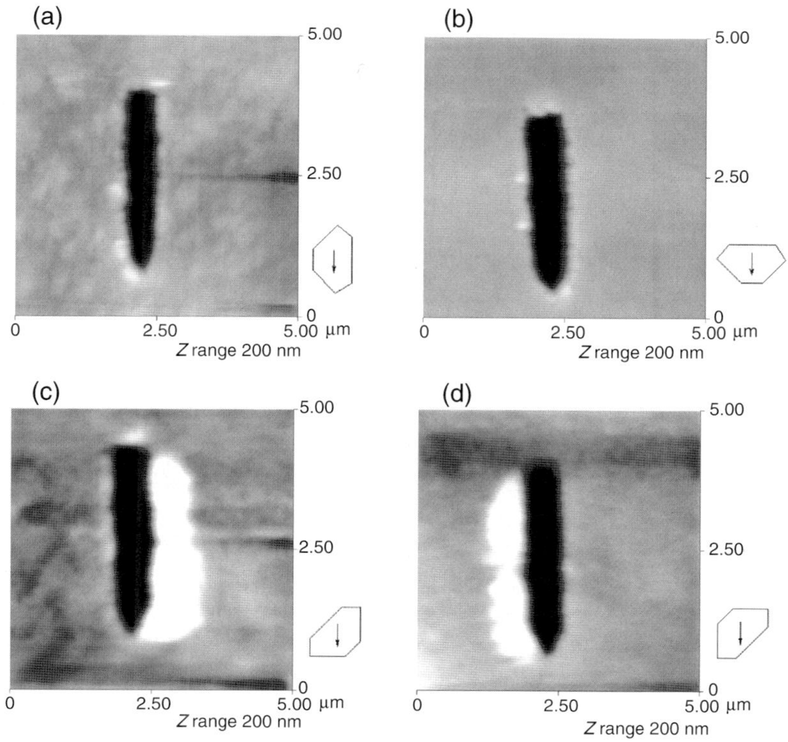

Fig. 4.45. Nanoscratches on (10–1) of tetraphenylethene (**3**) ($P2_1$) in various directions showing complete abrasion in (**a**) and (**b**) and molecular migration to the right in (**c**) or to the left in (**d**), due to vertical cleavage plane and bulkiness of the molecules [42]

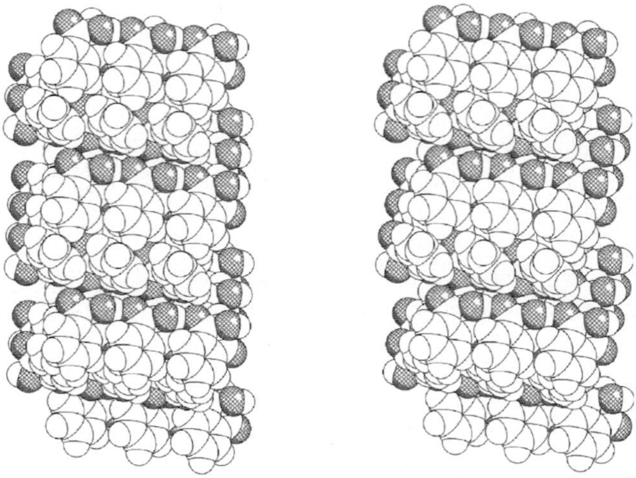

Fig. 4.46. Stereoscopic space filling crystal face model of ninhydrin (**2**) with (110) in front, (1–10) on top and (001) to the left showing the polar character of the (110) face in the vertical direction; oxygen atoms are highlighted

100°–150° (left), 210°–260° (right) and 280°–330° (left). The asymmetry in the ranges means that the push to the molecule from the side (out of the layer) is more important than the push for guiding along the cleavage plane direction and that there are wide ranges without migration [42].

Still another situation is found with polar-axes crystals. For example, the occurrence of a polar axis in the crystals of ninhydrin (**2**) ($P2_1$) must lead to different nanoscratching responses in forward and back direction [42]. Figure 4.46 clearly shows the differences in the up and down directions. The marked differences in Figs. 4.47a, c and b, d derive from the polarity. Tip shape effects have been excluded by exploring both edge and side in front nanoscratching with the cube corner for both directions. It appears that the scratching from top to bottom (in the orientation of Fig. 4.46) makes the flatter-lying molecules penetrate between the steeper ones that are thus pushed aside (Figs. 4.47b,d) in competition with abrasion. On the other hand, scratching in upward direction finds the flatter-lying molecules connected by hydrogen bonding with the steeper ones in the next row and they have to go with the OH groups in front. This process would appear very unfavorable and indeed complete abrasion is observed (Figs. 4.47a,c). Thus very detailed analyses are possible with the powerful nanoscratching tool.

4.6 Scratching of Organic Polymers

Nanoscratching on organic polymers is of high practical importance as these are exposed to various abrasive wear situations in constructions (optical

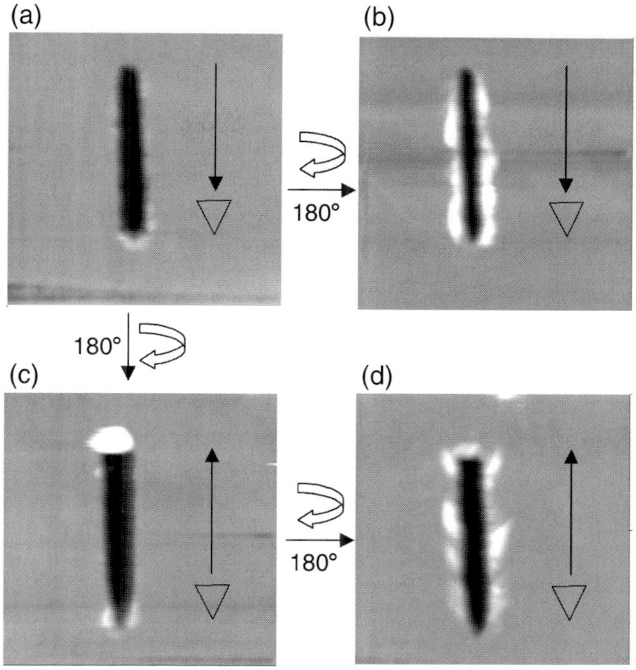

Fig. 4.47. Cube corner nanoscratching on (110) of ninhydrin (**2**) along the polar axis: **(a)**, **(b)** edge in front forward and reverse after a turn of the crystal; **(c)**, **(d)** side in front forward and reverse after the proper turns of the crystal as indicated [42]

glasses, pipes, lacquers, varnishes, automotive clearcoat marring, coatings, roofs, facades, etc., to name a few). Therefore macro- and microscratching have been performed on them. Macroscratching on PMMA provided transitions from viscoplastic scratching to viscoelastic grooving and at higher speed quasi-elastic sliding. The tip speed varied from 0.001 to 0.71 mm s^{-1} [46]. In Microscratching the scratch hardness is judged from the load under specified conditions of an indenter tip at which plastic yield is observed. Correspondingly, the fracture toughness is related to the load at which fracture starts to occur. The best techniques for these determinations are ramp experiments with linearly increasing load at constant rate lateral progression and determination of the start of the permanent scratch by microscopic inspection. But scratch hardness has also been defined in analogy to the indentation hardness as $H_S = 8F_N/\pi d^2$ where d is the scratch width [2]. The mechanic responses cover also surface deformation, fracture toughness, ductility, brittleness, and various transitions of the polymer. Unfortunately, the origin of ductile/brittle transitions is still unknown, and there is a lack of proper investigation of crystallinity. Conclusions as to the bulk mechanical properties are established. It has, however, been reported that biaxially drawn PET exhibits less wear in multipass scratches (sphere with $R = 25\,\mu$m) than the uniaxially drawn PET sample [47]. Polymers exhibit elastic, plastic, viscoelastic, viscoplastic, brittle properties depending on their composition and intrinsic mechanical properties. If there is a time dependence of scratching (strain rate = scratch velocity/scratch width) it relates to the viscous features. These qualities have been studied at various cone angles of the tip and normal loads up to 20 N and documented in deformation maps for polymethylmethacrylate (PMMA) and polycarbonate (PC) [2]. However, an understanding at the molecular level has not been advanced. Scratch resistance at various temperatures on PMMA has been measured by microscratching with a conosphere [48]; however, the scratch width by profilometry with the scratching tip was used without scrutiny with regard to its precision. No quantitative evaluation of the fate of the material was tried, but pileup heights were correlated. But there may also be self-protective mechanisms upon plastic plowing (10 mN load) with "strain-hardened" pileup and high elasto-viscoplastic recovery [20]. Furthermore, attack angles are important in abrasion via chip formation by machining [20].

Viscosity effects are also widespread in polymer nanoscratching. It should be stated here that mechanical interactions of various types (abrasive wear, mechanical destruction, tension, disrupting, cutting, milling, shock waves) to long chain or cross-linked polymers by necessity lead to bond cleavages into radicals that disproportionate to give shorter end alkane and end alkene chains or decrease of cross-linking. This feature has usually not been acknowledged in polymer scratching. Upon continued mechanical interaction the chains become too short or the crosslinked regions too small for enabling further mechanically induced bond cleavages. This may be easily observed in repeated unloading reloading indentations (Fig. 3.3 in Chap. 3) that finally (after several cycles) lead to "stable" viscoelastic behavior. It is only with polymers, which contain C=C double bonds (e.g., rubber), that initial radical formation by mechanically induced chain ruptures leads to an initial increase of cross-linking by addition of the radicals to the double bonds but only until further mechanical degradation reaches the limiting molecular weight of the polymer that does not allow further mechanically induced bond breaking. Basic proof for the radical formations has been amply provided by electron spin resonance spectroscopy [49–52] and chemical analyses of the degradation products [53–56].

In view of these complicated mechanochemical reactions that are responsible for the plastic–viscoelastic transition in the studied F_{N}-range (Sect. 3.11) it is of highest interest that the $F_{\mathrm{L}} \sim F_{\mathrm{N}}^{3/2}$ relationship holds for linear (PC and PMMA) and cross-linked amorphous polymers [CR39 = polymer of diethyleneglycol bis(allylcarbonate)]. Interestingly, the best compromise for obtaining good scratch images with PC, PMMA, and CR39 (here lower cracking) is achieved with a blunt cube corner ("curvature" radius: 200 nm), which has a favorable effective cone angle and does not penetrate too deeply [22].

PC (injection-molded optical lens) gives smooth indents (Fig. 4.48) without materials transport to the surface up to at least 500 µN load at a depth of the scratch of 45 nm. Sidewise pileup starts below 750 µN at 75 nm depth

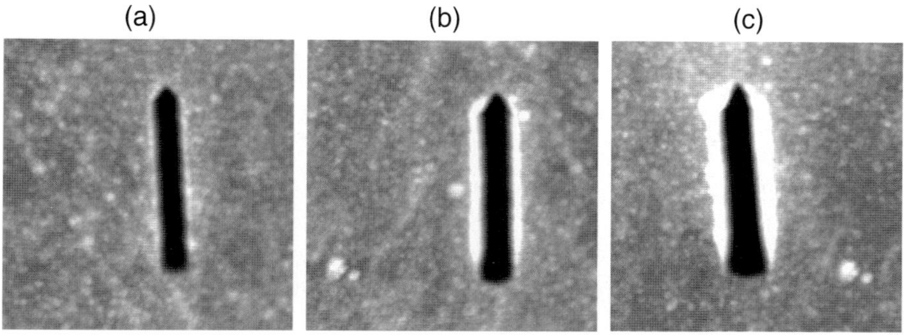

Fig. 4.48. AFM topologies of PC upon nanoscratching with a blunt cube corner (5 µm, z-scale: 200 nm): (**a**) depth 45 nm at 500 µN, (**b**) depth 75 nm at 750 µN, (**c**) depth 100 nm at 1,000 µN normal force [22]

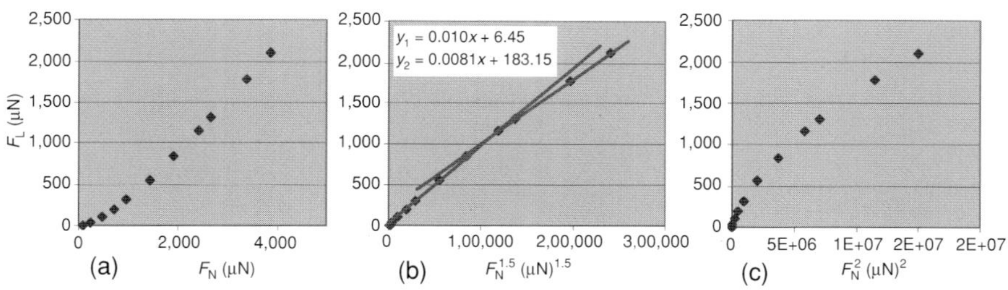

Fig. 4.49. Edge-in-front nanoscratching with a blunt cube corner on PC: (**a**) trial F_L - F_N plot that is invalid as it is strongly curved; (**b**) valid F_L - $F_N^{1.5}$ plot, the kink is the result of the phase transition under pressure, the scratch coefficients K (slopes) are indicated; (**c**) invalid F_L - F_N^2 trial plot that is strongly curved

and is prominent at 1,000 µN and 100 nm depth. This seems to be related to the possibility of density increase by phase transformation or to possible availability of free space in the amorphous polymer. The images in Fig. 4.48 do not indicate abrasion, as the end of the scratches reflects the tip geometry.

The quantitative treatment of the nanoscratching with PC secures again the $F_L \sim F_N^{1.5}$ relation. The trial plots with exponent 1 or 2 are strongly curved (Fig. 4.49). The kink in Fig. 4.49b at 2.3 mN occurs very late and is not very pronounced, but it indicates a phase transformation probably including breakage of polymer chains.

PMMA behaves totally different from PC [22]. Pileup occurs at low load already and is much more prominent at higher load (Fig. 4.50). This observation points to less free space in the amorphous polymer and thus chain breakages with phase transformation at lower load. Consequently, the kink in the linear $F_L \sim F_N^{1.5}$ plot occurs at the lower value of 560 µN (Fig. 4.51). Again, the trial plots with exponent 1 or 2 are curved and therefore invalid.

AFM images of nanoscratched CR39 are not smooth with sharp 60° pyramidal or cube corner tip. A blunt cube corner improves the situation but not

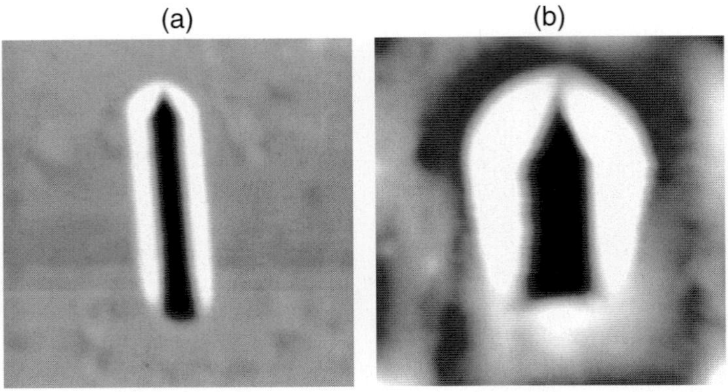

Fig. 4.50. AFM topologies of PMMA upon nanoscratching with a blunt cube corner: (**a**) 100 μN load, 5 μm, z-scale: 300 nm, depth 55 nm; (**b**) 500 μN load, 6 μm, z-scale: 700 nm, depth 260–280 nm [22]

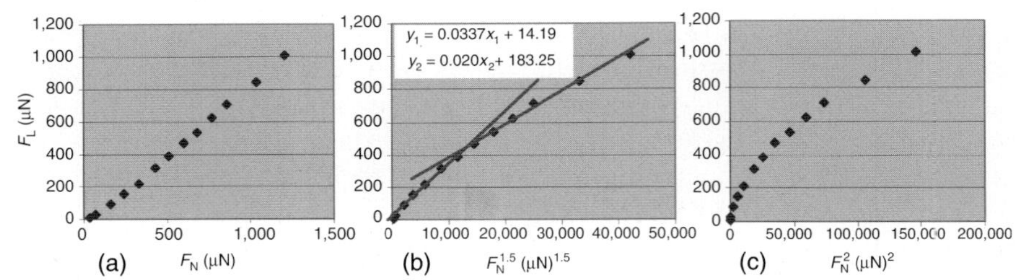

Fig. 4.51. Edge-in-front nanoscratching with a blunt cube corner on PMMA: (**a**) trial $F_{\rm L}$ - $F_{\rm N}$ plot that is invalid as it is strongly curved; (**b**) valid $F_{\rm L}$ - $F_{\rm N}^{1.5}$ plot, the kink is the result of the phase transition under pressure, the scratch coefficients K (slopes) are indicated; (**c**) invalid $F_{\rm L}$ - $F_{\rm N}^2$ trial plot that is strongly curved

completely. A Berkovich tip nanoscratching starts with sliding, followed by shallow grove and deep end with some pileup. Surface roughness plays an important role. The quantitative scratching with a blunt cube corner gives a much more pronounced kink in the nanoscratching of the cross-linked CR39 as compared to PMMA and PC (Fig. 4.52) [22]. It occurs earlier than in the linear polymers at 310 nm load, because the chain breaking is a necessary result if the tip goes down beyond the elastic response. Not unexpectedly, there is chipping in nanoscratching with some pileup in front and irregular cracking along the ditch, which becomes deeper the longer it takes (cf. Fig. 4.2a). The initial scratch coefficients K of PMMA and CR39 are not very different (0.0337 and 0.0315 $\mu N^{-0.5}$, respectively) but they cannot be safely compared, as the former polymer does not experience chipping and cracking under these nanoscratch conditions.

The nanoscratching coefficients K are useful parameters for the judgments of wear behavior of the injection-molded and molded polymers that

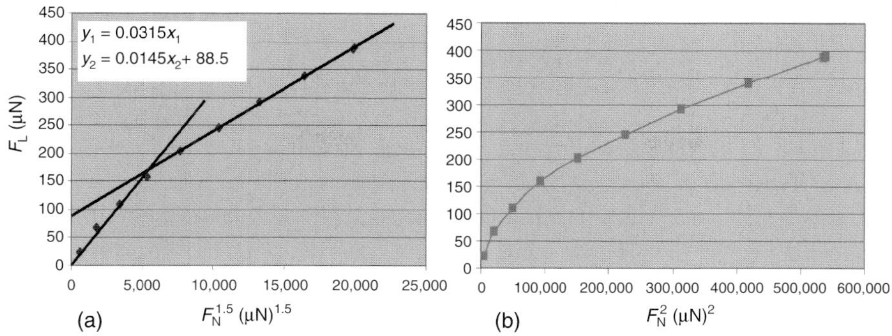

Fig. 4.52. Edge-in-front nanoscratching with a blunt cube corner on CR39: (**a**) valid F_L–$F_N^{1.5}$ plot, the kink is the result of the phase transition under pressure, the scratch coefficients K (slopes) are indicated; (**b**) invalid F_L–F_N^2 trial plot that is strongly curved

are in widest practical use (for example, CDs, contact lenses, optical lenses, glass plates/forms, pipes, bowls, etc.) or extruded PC (transparent roofs for constructions and greenhouses, etc) due to their high transparency and favorable properties. These K-values (4.1) and the nanoindentation coefficients (k-values, (3.18)) are valuable extrapolation tools and they are therefore more characteristic and useful than other mechanical parameters such as hardness and elasticity modulus (see also Table 3.6 in Sect. 3.11).

4.7 Conclusions

Nanoscratching experiments are quantitatively treated by the new general law of (lateral force) ~ (normal force)$^{3/2}$ that applies for virtually all kinds of solid materials, such as amorphous glasses, inorganic oxides, semiconductors, metals, salts, molecular crystals, and polymers. Polarities and special structural features or the bonding of the material or the shape of the (pseudo) conical indenter tips do not alter the validity. The nanoscratching coefficient K with dimension μN$^{-0.5}$ (4.1) is a useful constant that describes the response to the mechanical destruction or change under specified conditions. While the new law still lacks deduction from first principles, it is apparently inherent in the present formalism used for description of microindentation on the basis of previous theories for micromechanics. However, nanoindentation deviates from microindentation (Sect. 3.5/6) with the new quantitative relation (normal force) ~ (normal displacement)$^{3/2}$ so that uncertainty remains about the meaning of the constant parameters. As nanoscratching experiments are always connected to a nanoindentation part both new laws can be combined and lead to the relation (lateral force) ~ (normal displacement)$^{9/4}$, which is also a useful correlation. New unambiguous definitions of specific nanoscratch work (spec W_{sc}) and full scratch resistance ($R_{sc\,full}$) lead to useful relations with lateral force ($F_L \sim R^3_{sc\,full}$) and normal force ($F_N \sim R^2_{sc\,full}$) in the nanoscratching process. They are easily and precisely determined and are particularly useful in the characterization of anisotropy effects in nanoscratching. This includes the description of various types of anisotropic molecular migrations within molecular crystals upon applying pressure by nanoscratching.These relate strictly and sensibly to the crystal packing that vary considerably. Another application of linear F_L–$F_N^{3/2}$ plots is the most reliable detection of phase transformations under the pressure of the translating tip.

The specific nanoscratch work and the nanoscratch coefficient K [$\mu N^{-0.5}$] should be used for mechanical characterizations instead of the previously defined dimensionless "friction coefficient" (ratio of lateral force over normal force), which is no constant as the lateral force and normal force are definitely not linearly related in nanoscratching. The applications of the quantitative treatments are immediately obvious. They allow for useful extrapolations on the basis of the new parameters for all quantities that relate to the lateral force. The nanoscratching results have to be distinguished from the friction force microscopic experiences in the literature. There exist marked differences. Dry wearless elastic friction is described by (friction force) $\sim$ (normal force)$^{2/3}$ and (friction force) $\sim$ (contact area) in most microscopic cases but not by the macroscopic proportionality of (friction force) $\sim$ (normal force). These microscopic relationships are secured by molecular dynamics simulations [57] and deviate enormously from the nanoscratching law with its 3/2 exponent. Clearly, wearless friction force and lateral force in nanoscratching with their different power laws are unrelated. While the wearless friction experiments are very sensitive to impurities (which change the 2/3 exponent), nanoscratching does not meet with similar difficulties, the 3/2 exponent being stable in ambient environment. As nanoscratching is quite versatile, the new relationships may stimulate further approaches toward a unified description of friction, plowing, and wear.

References

1. H. Berns: Microstructural properties of wear-resistant alloys. Wear **181–183**, 271–279 (1995)
2. B.J. Briscoe, E. Pelillo, S.K. Sinha: Characterization of the surface deformation, yield and fracture of polymers during scratching. Deformation, Yield and Fracture of Polymers, International Conference, 10th, Cambridge, UK, Apr. 7–10, 1997, 474–477, Chem. Abstr. 774980 (1997)
3. J. Kleff: Warmritzen metallischer Werkstoffe. *Fortschr.-Ber. VDI*, Reihe 5 Nr. 340 VDI-Verlag GmbH, Düsseldorf 222p (1994), ISBN 3–18–334005–4
4. K.H. Zum Gahr: *Microstructure and Wear of Materials.* Tribology Series **10**, 1987, Elsevier, Amsterdam
5. B. Bhushan (ed.): *Microscratch and Microwear Studies, Springer Handbook of Nanotechnology*, Chapter 26.3.2, pp. 813–822, (1994) Springer, Berlin (1994)

6. A. Rodrigo, H. Ichimura: Analytical correlation of hardness and scratch adhesion for hard films. Argent. Sur. Coat. Technol. **148**, 8–17 (2001)
7. G. Kaupp, J. Schmeyers, U. Pogodda, M. Haak, T. Marquardt, M. Plagmann: AFM for the imaging of large and steep submicroscopic features, artifacts and scraping with asymmetric cantilever tips. Thin Solid Films **264**, 205–211 (1995)
8. G. Kaupp, M. Plagmann: Atomic force microscopy and solid state photolyses: phase rebuilding. J. Photochem. Photobiol. *A* **80**, 399–407 (1994)
9. P. Grau, G. Berg, W. Fränzel, H. Meinhard: Recording hardness testing. Problems of measurement at small indentation depths. Phys. Stat. Sol. A **146**, 537–548 (1994)
10. E. Amitay-Sadovsky, S. Zheng, J. Smith, H. D. Wagner: Directional indentation of transcrystalline polypropylene. Acta Polym. **49**, 588–593 (1998)
11. G. Kaupp, J. Schmeyers, U.D. Hangen: Anisotropic molecular movements in organic crystals by mechanical stress. J. Phys. Org. Chem. **15**, 307–313 (2002)
12. U.D. Hangen: A Comparison of Nano-Hardness and Scratch-Resistance on Mohs Minerals. Z. Metallkd. **92**, 1074–1077 (2001)
13. G. Kaupp, M.R. Naimi-Jamal: Nanoscratching on surfaces: the relationships between lateral force, normal force and normal displacement. Z. Metallkd. **95**, 297–305 (2004)
14. S. Mosch, P. Grau, G. Berg: Nanotribological measurements on sol–gel derived silica films. Mater. Res. Soc. Symp. Proc. **522**, 415–420 (1998)
15. S. Enders, P. Grau, G. Berg: Mechanical characterization of surfaces by nanotribological measurements of sliding and abrasive terms. Mater. Res. Soc. Symp. Proc. **594**, 531–536 (2000)
16. A. C. Fischer-Cripps: *Nanoindentation, Mechanical Engineering Series*, Springer, New York (2002)
17. M.R. Naimi-Jamal, G. Kaupp: Quantitative evaluation of nanoindents: Do we need more reliable mechanical parameters for the characterization of materials? Z. Metallkd, **96**, 1226–1236 (2005)
18. V. D. Jardret, W. C. Oliver: Viscoelastic behavior of polymer films during scratch test: A quantitative analysis. Mater. Res. Soc. Symp. Proc. **594**, 251–256 (2000)
19. J.L. Bucaille, E. Felder: Mechanics of the scratch test on polymers and metals: Modeling and experimental approaches. Mater. Tech. **3–4**, 29–44 (2001)
20. M.J. Adams, A. Allan, B.J. Briscoe, P.J. Doyle, D.M. Gorman, S.A. Johnson: An experimental study of the nano-scratch behaviour of poly(methyl methacrylate). Wear **251**, 1579–1583 (2001)
21. S.V. Prasad, T.H. Kosel: A comparison of carbide fracture during fixed depth and fixed load scratch tests. Proceedings Conference Wear of Materials, April 14–18 1985, Vancouver, (Ed) K.C. Ludema, ASME, New York (1985)
22. G. Kaupp, M.R. Naimi-Jamal: Unpublished results
23. G. Kaupp, M.R. Naimi-Jamal, J. Schmeyers: Quantitative reaction cascades of ninhydrin in the solid state. Chem. Eur. J. **8**, 594–600 (2002)
24. N. A. Stilwell, D. Tabor: Elastic recovery of conical indentations. Proc. Phys. Soc. London, **78**, 169–179 (1961)
25. K. Trachenko, M. Dove, T. Martin: Intermediate state in pressurized silica glass: Reversibility window analogue. Phys. Rev. B **67**, 212203/1–212203/3 (2003)

26. A. Okazaki, N. Ohama, B.T.M. Willis, Y. Iwata, H.J. Scheel, K.A. Mueller: Time-dependent strains near the 105 K transition in strontium titanate ($SrTiO_3$). Phase Transit. **3**, 339–345 (1983)
27. H. Fujishita, Y. Shiozaki, E. Sawaguchi: X-ray crystal structure analysis of low temperature phase of strontium titanate $SrTiO_3$. J. Phys. Soc. Jpn. **46**, 581–586 (1979)
28. A.L. Ruoff: High Pressure Phase Transformations. In *Materials Science and Technology a Comprehensive Treatment.* Vol. **5**, Chapter 8 (eds. R.W. Cahn, P. Haasen, E.J. Kramer), VCH, Weinheim (1991)
29. R.J. Needs, A. Mujica: First-principles pseudopotential study of the structural phases of silicon. Phys. Rev. B **51**, 9652–9660 (1995)
30. A. Kailer, K.G. Nickel, Y.G. Gogotsi: Raman microspectroscopy of nanocrystalline and amorphous phases in hardness indentations. J. Raman Spectr. **30**, 939–946 (1999)
31. Y. Gogotsi, T. Miletich, M. Gardner, M. Rosenberg: Microindentation device for *in situ* study of pressure-induced phase transformations. Rev. Sci. Instrum. **70**, 4612–4617 (1999)
32. C. Riekel, M.C.G. Gutierrez, A. Gourrier, S. Roth: Recent synchrotron radiation microdiffraction experiments on polymer and biopolymer fibers. Anal. Bioanal. Chem. **376**, 594–601 (2003)
33. A. Gourrier, M.C.G. Gutierrez, C. Riekel: Combined microindentation and synchrotron radiation microdiffraction applied to polymers. Macromolecules **35**, 8072–8077 (2002)
34. G. Kaupp, M.R. Naimi-Jamal, A. Gourrier, C. Riekel: Possibilities of studying phase transformation upon microindentation and nanoscratching by synchrotron radiation microdiffraction. Lecture at the 6th European Symposium on Nanomechanical Testing, Aug. 30–Sept. 1 (2005)
35. R. Ahuja, S. Rekhi, B. Johansson: Theoretical prediction of a phase transition in gold. Phys. Rev. B **63**, 212101/1–212101/3 (2001)
36. M.I. Novgorodova, A.I. Gorskov, A.V. Mokhov: Native silver and its new structural modifications. Zap. Vses. Mineralogich. Obshch. **108**, 552–563 (1979)
37. J.A. Xu, J.Z. Hu: X-ray diffraction under high pressure. Wuli Xxuebao **26**, 521–525 (1977)
38. B. Baranowski, M. Tkacz, R. Majchrzak (eds. R. Pucci, G. Piccito). In *Molecular systems under high pressure*, p. 139, Elsevier Science Publishers B.V., North-Holland (1991)
39. B. Matthew Scott: Phase transformation through shearing under pressure with and without oil lubrication (copper, silver, alloying, amorphous alloys, tribology). Diss. Univ. of Virginia, Charlottesville, USA, 141 pp, Avail. UMI Order No. DA9724740 (1997); Diss. Abstr. Int., B **58**, 1470 (1997)
40. H.Iwasaki, T. Fujimura, M. Ichikawa, S. Endo, M. Wakatsuki: Pressure-induced phase transformation in silver–zinc alloy. J. Phys. and Chem. of Solids **46**, 463–468 (1985)
41. C.R. Tellier, G. Benmessaouda: Scratching experiments on quartz crystals: Orientation effects in chipping. J. Mater. Sci. **29**, 3281–3294 (1994)
42. G. Kaupp, M.R. Naimi-Jamal: Mechanically induced molecular migrations in molecular crystals. Cryst. Eng. Commun. **7**, 402–410 (2005)
43. G. Kaupp: Supermicroscopy: AFM, SNOM, and SXM. In *Comprehensive Supramolecular Chemistry*, vol. **8**, 381–423 (eds. J.E.D. Davies, J.A.Ripmeester), Elsevier: Oxford (1996)

44. G. Kaupp: Solid-state molecular syntheses: complete reactions without auxiliaries based on the new solid-state mechanism. CrystEngComm **5**, 117–133 (2003)
45. A.Herrmann, G. Kaupp, T. Geue, U. Pietsch: atomic force microscopy and grazing incidence diffraction investigation of the gas–solid diazotization of sulfanilic acid. Mol. Cryst. Liq. Cryst. **293**, 261–275 (1997)
46. C. Gauthier, R. Schirrer: Viscoelastic-viscoplastic analyses of a scratching and sliding single contact on a polymer surface. Adv. Mech. Behav., Plast. Damage, Proc. Euromat **1**, 213–218 (2000)
47. B.D. Beake, G.J. Leggett: Nanoindentation and nanoscratch testing of uniaxially and biaxially drawn poly(ethylene terephthalate) film. Polymer **43**, 319–327 (2002)
48. P. Morel, V. Jardret: Viscoelastic effects on the scratch resistance of polymers: Relationship between mechanical properties and scratch properties at various temperatures. Mater. Res. Soc. Proc. **710**, 93–98 (2002)
49. J. Becht, H. Fischer: Strain-induced structural changes in highly oriented, partially crystalline polymers observed by ESR studies. Angew. Makromol. Chem. **18**, 81–91 (1971)
50. R.A. Graham, P. M. Richards, R.D. Shrouf: Direct evidence for formation of radicals in a shock-loaded polymer. J. Chem. Phys. **72**, 3421–3422 (1980)
51. S.E. Bresler, S.N. Zhurkov, E.N. Kazbekov, E.M. Saminskii, E.E. Tomashevskii; Investigation of macroradicals arising during mechanical destruction of polymers, Z. Sakh. Promysh. **29**, 358–364 (1959), Chem. Abstr. 108586 (1959)
52. T.N. Kleinert: Stable free radicals in various lignin preparations. Tappi **50**, 120–122 (1967)
53. M.O.W. Richardson, M.W. Pascoe: Mechano-chemistry and poly(vinyl chloride) wear. Polymer Science and Technology, Adv. Polym. Frict. Wear **5B**, 585–597 (1974)
54. G.A. Gorokhovskii, P.A. Chernenko, V.A. Vonsyatskii, I.A. Popov: Mechanical–chemical processes in the contact zone of a metal in a polymer-abrasive environment. Primen. Polim. Kach. Antifricts. Mater. Respub. Nauch.-Tekh. Konf. 189–195 (1971), Chem. Abstr. 48641 (1974)
55. D. Saidov: Physics of elastomer failure. Dokl. Akad. Nauk. Tadzh. SSR **31**, 654–657 (1988)
56. N.K.Baramboim, I.I. Nikitina: Mechanical degradation of unsaturated three-dimensional polyesters. Izv. Akad. Nauk Kazakh. SSR, Ser. Khim. **23**, 77–79 (1973), Chem. Abstr. 479583 (1973)
57. L. Wenning, M. H. Müser: Friction laws for elastic nanoscale contacts. Europhys. Lett. **54**, 693–699 (2001)

Index